Drug Delivery

TOPICS IN CHEMICAL ENGINEERING
A Series of Textbooks and Monographs

Series Editor
Keith E. Gubbins, North Carolina State University

Associate Editors
Mark A. Barteau, University of Delaware
Douglas A. Lauffenburger, MIT
Manfred Morari, ETH
W. Harmon Ray, University of Wisconsin
William B. Russel, Princeton University
Matthew V. Tirrell, University of California, Santa Barbara

DRUG DELIVERY

Engineering Principles for Drug Therapy

W. Mark Saltzman

OXFORD
UNIVERSITY PRESS
2001

OXFORD

UNIVERSITY PRESS

Oxford   New York
Athens   Auckland   Bangkok   Bogotá   Buenos Aires   Calcutta
Cape Town   Chennai   Dar es Salaam   Delhi   Florence   Hong Kong   Istanbul
Karachi   Kuala Lumpur   Madrid   Melbourne   Mexico City   Mumbai
Nairobi   Paris   São Paulo   Shanghai   Singapore   Taipei   Tokyo   Toronto   Warsaw
and associated companies in
Berlin   Ibadan

Copyright © 2001 by Oxford University Press, Inc.

Published by Oxford University Press, Inc.,
198 Madison Avenue, New York, New York 10016
http://www.oup-usa.org

Oxford is a registered trademark of Oxford University Press.

Library of Congress Cataloging-in-Publication Data
Saltzman, W. Mark.
Drug delivery : engineering principles for drug delivery / W. Mark Saltzman.
p. ; cm. — (Topics in chemical engineering)
Includes bibliographical references and index.
ISBN-13 978-0-19-508589-1
ISBN 0-19-508589-2
1. Drug delivery systems   2. Biomedical materials. I. Title. II. Topics in chemical
engineering (Oxford University Press)
[DNLM: 1. Drug Delivery Systems. 2. Biocompatible Materials. 3. Drug Implants. 4.
Pharmacokinetics.   5. Polymers. WB 354 S179d 2001]
RS199.5. S25   2001
615'.7–dc21   00-050140

3 5 7 9 8 6 4

Printed in the United States of America
on acid-free paper

To Emily,
who reads my mind,
every opaque footnote,
each rambling appendix.

Preface

I spend my professional life exploring an intersection of three worlds: chemistry and biology and engineering. This is not always an easy neighborhood, but it is exciting without fail. My own experience suggests that chemical and biomedical engineers adapt easily to life in this niche. But I also believe that individuals working or studying in departments of materials science, pharmaceutics, pharmacology, biology, chemistry, and others may also find this subject matter—if not necessarily this particular treatment—of interest. Therefore, I have tried not to assume too much previous experience; many of the subjects include essential background directly in the text. In a few cases I have omitted detail and provided references, particularly in those cases where good and accessible references exist.

Synthetic materials present a tremendous potential resource in treating human disease. For the rational design of biomaterials, an understanding of polymer chemistry and polymer physics is necessary, but not sufficient. Equally important is a quantitative understanding of the principles that govern rates of drug transport, reaction, and disappearance in physiological and pathological situations. This is the point of the present book: to provide a working foundation in these principles, which I hope you will carry forward into the development of useful methods for drug delivery.

I am grateful to many individuals for encouraging me and facilitating my effort to produce this text. I thank Bob Langer for his unsinkable enthusiasm, and for introducing me to the concepts that underpin this work. I thank Doug Lauffenburger for inciting my interest in authorship, for lending advice and encouragement at critical junctures, and for providing (in his own text) an exceptionally strong example to follow. I thank Mark McHugh and Michael

Shuler for being general sources of inspiration. I thank William Deen, instructor for the first class I took in the field of biomedical engineering, for introducing me to many of the problems surveyed here. I thank Nadya Belcheva, Alison Fleming, Melissa Mahoney, and Jian Tan for reviewing early versions of several chapters and Rupa Patel and Christopher Anker for their careful and thorough reading of the penultimate draft. I thank Claude Cohen for his insightful review of Chapter 4 and Rebecca Willits for assisting with the preparation of Chapter 10. I thank Alexander for his good-natured assistance with the index. And I thank the many students of my courses at Johns Hopkins and Cornell who suffered through early and late (but always incomplete) drafts of this text.

Ithaca, New York   W.M.S.
2000

Contents

Appendix A   Overview of polymeric biomaterials

Appendix B   Useful data and nomenclature

I

INTRODUCTORY MATERIAL

1

Introduction

Who has put wisdom in the inward parts, or given understanding to the
mind?...Or who can tilt the waterskins of the heavens? When the dust runs
into a mass, and the clods cling together?

Job 38:36–38 (NRSV)

1.1 HISTORICAL PERSPECTIVE

Humans have always attempted to improve their health by ingesting or admin-
istering drugs. Examples appear throughout written history, from every con-
tinent and culture. Noah produced alcohol [1] and Christ was offered a
sedative to ease the pain of crucifixion [2]. The use of opium was described
by Theophrastus in the third century B.C., the stimulating power of methyl-
xanthines was exploited by ancient Arabian shepherds and priors, and the
paralyzing properties of curare were recognized by native South Americans
centuries before the arrival of Sir Walter Raleigh [3]. The chemotherapy of
cancer, which many consider a modern development, has existed in some form
for over 500 years [4]. Vaccination, the intentional exposure to pathogens, was
used in China and India to prevent smallpox and other infections [5] centuries
before the birth of either Jenner [6] or Pasteur [7].

Even during the 20th century, drug discovery frequently resulted from
empiricism and happenstance. The anticancer effects of nitrogen mustard
were realized during the development of chemical-warfare agents, and penicillin was discovered after the inadvertent contamination of a bacteriological
plate. As technology advanced, particularly after 1970, methods of drug and
vaccine production became more sophisticated and rational. In parallel with
the rise of modern pharmaceutical technology and the explosive ascent of
biotechnology, the cellular and molecular basis for the action of many drugs
has been uncovered. Today, drug designers benefit from an accumulated base
of scientific knowledge concerning, for example, the interactions between
neurotransmitters and their receptors, the regulation of hormone secretion,
and the sensitivity of tumor cells to specific kinds of chemicals (see [3] for
details).

Table 1.1   Examples of protein drugs produced by modern pharmaceutical technology and biotechnology

Agent	Trade name	Indication
Hepatitis B subunit	Recombivax HB®, Merck & Co., Inc.	Vaccination against hepatitis B
Human insulin	Humilin, Eli Lilly	Treatment of diabetes
Monoclonal antibody against CD3	Orthoclone OKT®, Ortho Biotech	Prevention of organ rejection
Human growth hormone	Protopin®, Genentech, Inc.	Growth failure in children
Human interleukin-2	Proleukin®, Chiron Therapeutics	Metastatic renal cell carcinoma
Erythropoietin	Epogen®, Amgen Procrit®, Ortho Biotech	Anemia
Interferon beta-1a	Avonex®, Biogen Inc.	Multiple sclerosis
Interferon alpha-n3	Alferon®, Immunex Corp.	Intralesional treatment of refractory external condylomata acuminata
Tissue plasminogen activator	Activase®, Genetech, Inc.	Acute myocardial infarction [11]
β-Cerebrosidase	Ceredase®, Genzyme Corp.	Enzyme deficiency
Deoxyribonuclease (DNase)	Pulmozyme®, Genentech, Inc.	Cystic fibrosis [12]

New technology and clearer biological insight have led to new classes of therapeutic and prophylactic agents. Consider some of the new products made available to patients in the United States over the last few years (Table 1.1). A revolution in drug development is clearly upon us. Even more complex agents, such as chimeric antibodies, gene-based drugs, antisense oligonucleotides, and virus-like particles, are emerging as clinically viable entities. New clinical approaches involve cells as well as molecules; the introduction of genetically modified cells into humans has blurred the distinction between conventional pharmacology and transplantation.

This wealth of new technology and the resulting new armaments in the war against disease will incite new strategies for drug and vaccine administration. Most current methods for drug administration are direct descendants of ancient practices [8], and have changed little over the last few centuries. Egyptian physicians employed pills, ointments, salves, and other forms of treatment over 4,000 years ago. Hippocrates—foreshadowing modern sterile procedures—warned against the introduction of unclean agents into wounds in the 4th century B.C. Intravenous injections were first performed in humans in 1665, only a few years following Wren's infusion of opium into dogs and a few decades after Harvey's description of the circulatory system. Subcutaneous injections were introduced by Wood in 1853 and the modern hypodermic syringe was developed by Luer in 1884.

While pills and injections have enabled significant medical advances to be made, these methods are inadequate for the delivery of recombinant proteins, which frequently have short half-lives, poor permeability in membranes, and serious toxicity when delivered systemically in large doses. Similarly, general methods for the delivery of gene-therapy vectors are still unknown since the uptake, biodistribution, expression, and toxicity of oligonucleotides have yet to be systematically studied. Like the new rational approaches to drug and vaccine design, new delivery technologies must exploit findings from basic science.

Biocompatible[1] polymers have become critical components in the design of delivery systems. Polymers have long been used in experimental drug delivery systems to provide controlled release of small organic molecules [9] or macromolecules [10]. Now, many decades after their first description, a handful of polymer-based, controlled drug delivery systems are approved for use in humans (see Table 1.2). The promise of polymeric controlled-release systems is still largely unrealized, because marriages between the new technology and appropriate clinical problems have been difficult to arrange. To achieve this union, an understanding of polymer chemistry and polymer physics is necessary, but not sufficient. A quantitative understanding of the physiological processes related to drug transport, drug action, and drug metabolism is equally important. One important goal of this book is to explore the linkage between (1) the design of drugs and drug delivery systems, and (2) the biology of drug metabolism and distribution in tissues.

1.2 THE AIM AND STRUCTURE OF THIS BOOK

Drug delivery is an area of study in which people from almost every scientific discipline can make a significant contribution. Because of the highly interdisciplinary nature of the field, engineers have a unique role in the development of new strategies for drug delivery. Engineers emerge from training programs that integrate biological science (e.g., quantitative physiology and pathophysiology, cell and molecular biology), engineering technology (e.g., polymer engineering or microfabrication technology), and mathematical analysis (e.g., pharmacokinetics and transport phenomena). Understanding and manipulating the fate of drugs in humans is a classical engineering endeavor, where basic science and mathematical analysis can be used to achieve an important practical end.

---

1. Biocompatibility in materials is not precisely defined, and so all claims of biocompatibility should be read with caution. A useful definition, which is accepted by most experts, is: "Biocompatibility is the ability of a material to perform with an appropriate host response in a specific application" (Williams, D.F., Definitions in Biomaterials. Proceedings of a Consensus Conference of the European Society for Biomaterials, Chester, England, 3–5 March 1986, Vol. 4, Elsevier, New York, 1987). For more information see: *Biomaterials Science: an introduction to materials in medicine.* Ratner, B.D., Hoffman, A.S., Schoen, F.J., and Lemons, J.E. (eds.), Academic Press, 1996.

Table 1.2   Examples of controlled drug delivery systems based on polymers

Drug/polymer	Trade name	Indication
Estradiol/poly[ethylene-*co*-(vinyl acetate)]	Estraderm®, Ciba Pharmaceutical Co.	Estrogen replacement therapy
BCNU/poly[carboxyphenoxy-propane-*co*-(sebacic acid)]	Gliadel®, Rhone-Poulenc Rorer	Recurrent glioblastoma multi-forme
Leuprolide acetate/poly(DL-lactide-*co*-glycolide)	Lupron Depot®, Takeda Chemical Industries	Endometriosis
Levonorgestrel/silicone elastomer	Norplant®, Wyeth-Ayerst Laboratories	Implantable contraceptive
Nitroglycerin/poly[ethylene-*co*-(vinyl acetate)]	Transderm-Nitro®, Summit Pharmaceuticals	Prevention of angina
Pilocarpine/poly[ethylene-*co*-(vinyl acetate)]	Ocusert®, Alza Corp.	Glaucoma therapy
Progesterone/poly[ethylene-*co*-(vinyl acetate)]	Progestasert®, Alza Corp.	Contraceptive intra-uterine device

This text is concerned with the engineering of novel therapies for treating human disease, particularly therapies that involve the use of novel materials. I will attempt to relate the effectiveness of drug therapy to the biophysics and physiology of drug movement through tissues. As an example of the approach, consider the localized delivery of a drug to a tissue by controlled release from an implanted, drug-loaded polymer. In the period following implantation, drug molecules are slowly released from the polymer into the extracellular space of the tissue. The availability of drug to the tissue—i.e., the rate of release into the extracellular space—depends on the characteristics of the device, and can be engineered into the implant by varying the chemical or physical properties of the polymer. To design drug delivery systems like this, one must understand the relationship between polymer chemistry, implant structure, and the rate of drug release.

Once the drug is released into the extracellular space, however, rates of drug migration through the tissue, uptake into the circulatory system, and elimination from the body depend on characteristics of the drug and physiology of the delivery site (Figure 1.1). To understand the barriers to drug delivery, it is helpful to consider anatomical structure at a variety of length scales: the length scale of the organism (1–2 m); an organ (1–100 mm); a tissue segment (1–100 μm); or within the extracellular or intracellular matrix (1–100 nm). Design of an implantable drug delivery system requires more than an understanding of the characteristics of the implant. Appropriate design requires an understanding of the molecular mechanisms for drug transport and elimination, particularly at the site of delivery.

Chemical and physical characteristics of the polymeric materials will be critical in determining the performance of the overall system. Therefore, Appendix A includes an overview of the important aspects of the classes of polymers that are most often used in biomedical applications.

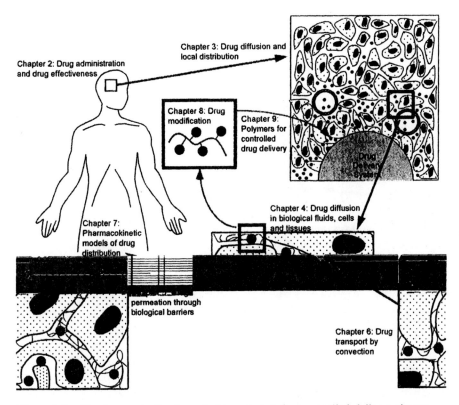

Figure 1.1 Development of polymeric biomaterials for controlled delivery tissues. This book will consider several aspects of the development of delivery systems including the characterization of biocompatible polymers, methods for incorporating bioactive agents into polymer matrices and microspheres, and methods for quantitative analysis of kinetics of drug release. Since these devices are developed for the delivery of agents to tissues, a critical component in this analysis will be the consideration of drug transport through the local tissue surrounding the implant.

REFERENCES

1. Genesis 9:20–21, *The Bible*. New Revised Standard Version ed. Grand Rapids, MI: Zondervan Publishing House.
2. Matthew 27:33–34, *The Bible*. New Revised Standard Version ed. Grand Rapids, MI: Zondervan Publishing House.
3. Goodman Gilman, A., *et al.*, eds, *The Pharmacological Basis of Therapeutics*, 8th ed. New York: Pergamon Press, 1990, 1811 pp.
4. Pratt, W.B., *et al.*, *The Anticancer Drugs*, 2nd ed. New York: Oxford University Press, 1994.
5. Plotkin, S.L. and S.A. Plotkin, A short history of vaccination, in S.A. Plotkin and E.A. Mortimer, *Vaccines*. Philadelphia: W.B. Saunders, 1988, pp. 1–7.
6. Jenner, E., An inquiry into the causes and effects of the variole vaccinae, a disease discovered in some of the western counties of England, particularly Gloucestershire, and known by the name of Cow Pox, reprinted from Low: London, 1798, in T.D.

Brock, *Milestones in Microbiology*, London: American Society for Microbiology, 1975.

7. Pasteur, L., C.-E. Chamberland, and E. Roux, Sur la vaccination charbonneuse. *C.R. Acad. Sci. Paris*, 1881. **92**, 1378–1383.

8. Banker, G.S. and C.T. Rhodes, eds., Modern pharmaceutics, *Drugs and the Pharmaceutical Science*, Vol. 7, New York: Marcel Dekker, 1979.

9. Folkman, J. and D. Long, The use of silicone rubber as a carrier for prolonged drug therapy. *Journal of Surgical Research*, 1964, **4**, 139–142.

10. Langer, R. and J. Folkman, Polymers for the sustained release of proteins and other macromolecules. *Nature*, 1976, **263**, 797–800.

11. The GUSTO Investigators, An international randomized trial comparing four thrombolytic strategies for acute myocardial infarction. *New England Journal of Medicine*, 1993, **329**(10), 673–682.

12. Fuchs, H.J., *et al.*, Effect of aerosolized recombinant human DNase on exacerbations of respiratory symptoms and on pulmonary function in patients with cystic fibrosis. *New England Journal of Medicine*, 1994, **331**(10), 637–642.

2

Drug Administration and Drug Effectiveness

For every atom belonging to me as good belongs to you.
Walt Whitman, *Song of Myself* (1855)

There is some comfort to thinking of the human body as an elaborate bag of chemicals. Chemicals can be produced; chemicals can be added or replaced; chemicals are (sometimes) inexpensive. In fact, several decades ago, it was widely reported that the chemicals in the human body were worth about $1. Other investigators estimate the value at closer to $6 million (particularly if the chemicals are purchased from scientific suppliers). True dollar value aside, we often imagine that our bodies can be supplemented, mended, and improved through the addition of "missing" chemicals.

Perhaps for this reason, the modern practice of healing usually involves medicine, an agent or elixir given as treatment.[1] The new millennium finds us rich in the knowledge of agents; advanced in the art of harvesting or synthesizing remedies; steadfast in the belief that cancer, heart disease, and neurodegeneration will eventually yield to these potions. Our skill in making medicine is far-reaching. Today, it would be difficult to find the person who has not personally experienced the healing force of antibiotics, vaccines, or modern chemotherapy. Unfortunately, it would be equally difficult to find the person who has not endured the premature death of a friend or relative due to untreatable infection or cancer. So we continue to search for better therapeutics.

Better medical treatments do not always require stronger medicine. The effectiveness of chemical agents depends on the method of administration, so

1. For most people, taking medicine into their body is strongly associated with healing. Consequently, clinical tests of new medicine almost always include a placebo group to correct for the positive change in a patient's well-being that occurs from the awareness of taking medicine.

treatments can often be improved by finding optimal drug formulations or delivery systems. For example, Banting and Best demonstrated control of diabetes by insulin injection in 1922. But early use of insulin was difficult: multiple daily injections were required and the effects were difficult to control. Insulin preparations have changed dramatically since that time; recombinant human insulin is now available in addition to highly purified insulin from animals. Today, the formulation of insulin is advanced: various formulations provide rapid or delayed action with long or short duration, so insulin therapy can be tailored to the needs of an individual. Intensive, individualized therapy decreases the long-term consequences of diabetes [1]. Still, in general, diabetics experience no benefit from insulin unless it is directly injected into their tissues.

In recent years, research into new delivery methods for insulin has intensified. Some developments harness new technology to improve existing methods: programmable, implantable pumps provide long-term, continuous, controlled injection of insulin into the peritoneum, which better mimics physiological insulin delivery. Other developments require engineering of the insulin molecule, identification of physiological targets with lower barriers to insulin penetration, or the use of novel materials. Insulin nasal sprays and mouthwashes may be the first alternatives to injection that become available, but they are unlikely to be the last. Similar enhancements to therapy are possible for almost all drugs.

A major objective of this book is to provide a rational foundation that can be used to understand the influence of drug administration on effectiveness and, ultimately, to predict how drug concentrations in the body will be influenced by changes in the way the agent is delivered.

2.1 THE STATE OF THE ART

Modern pharmaceutical science provides many options for administration of a new compound, but most of these choices are limited in important ways.

2.1.1 Drugs are Currently Administered by a Limited Number of Methods

Most of us have experienced multiple forms of drug administration: pills, injections, lotions, and suppositories have been in common use for centuries (Figure 2.1). People eat to survive and usually enjoy (or at least tolerate) the act of ingestion. For this reason, oral dosage forms are usually preferred: they are painless, uncomplicated, and self-administered. Unfortunately, many drugs are degraded within the gastrointestinal tract or not absorbed in sufficient quantity to be effective. Therefore, these drugs must be administered by intravenous, intramuscular, or subcutaneous injection. A variety of other modes of administration, less common than oral or injection, have evolved because of specific advantages for particular agents or certain diseases (see Table 2.1). As introduced in Chapter 1, the sophistication of drug administration has generally followed discoveries in anatomy and physiology.

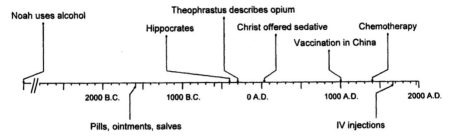

Figure 2.1 Timeline. Humans have always attempted to improve their health by administration of medicines.

2.1.2 Differences in the Mode of Drug Administration Produce Substantial Changes in Drug Concentration

The route and method of drug administration will influence the kinetics of biodistribution and elimination and, therefore, the effectiveness of the therapy. Delivery of identical amounts of drug to tissue sites differing in local anatomy and physiology (e.g., injection into a muscle versus injection into a subcutaneous space) can result in measurable differences in the pattern of delivery. For example, when identical preparations of human immunoglobulin G (IgG) were administered orally, subcutaneously, intramuscularly, or intravenously to human patients, different patterns of IgG concentration in the plasma were observed over time (Figure 2.2) [2]. Since the preparations were identical, the observed differences in plasma concentration must be attributable to differences in the rate of IgG movement and uptake after administration. This observation has important (and probably obvious) clinical consequences; the concentrations of IgG observed in the blood 3 days after subcutaneous or intramuscular injection are substantially different. These differences in concentration probably reflect differences in the stability or rate of migration of IgG molecules in muscle and subcutaneous tissue. A primary goal for this text is to relate these observed differences in drug uptake and distribution to physiological features of the local tissue site.

2.1.3 Delivery Systems can be Engineered to Change the Duration of Therapy or the Localization of Drugs

For all of the methods illustrated in Figure 2.2, the IgG was administered in an aqueous solution. But it is now possible to engineer delivery systems that provide controlled and sustained release of drugs (see the insets in Figure 2.3; delivery systems composed of polymeric materials are described in Chapters 8 and 9). Several delivery systems can be used to release IgG continuously over a long period. The introduction of IgG in different delivery systems changes the temporal pattern of IgG concentration in the body; in this case, the controlled delivery system greatly extends the life time of the

Table 2.1 Common routes of drug administration

Route of administration	Example	Advantages	Disadvantages
Intravenous injection (i.v.)	Antibiotics for sepsis	100% bioavailability	Discomfort to patient Requires health care-provider Risk of overdose or toxicity Risk of infection
Intravenous infusion	Heparin for anti-coagulation	100% bioavailability Continuous control over plasma levels	Requires hospitalization Risk of infection
Subcutaneous injection (s.c.)		Usually high bio-availability	Discomfort to patient
Intramuscular injection (i.m.)	Insulin for diabetes	Usually high bio-availability	Discomfort to patient
Oral (p.o.)	Many	Convenient Self-administered	Drug degradation before absorption Limited absorption of many drugs
Sublingual or buccal	Nitroglycerin for angina	Avoids first-pass metabolism[a] in liver Self-administered	Limited to lipophilic, highly potent agents
Ophthalmic	Pilocarpine for glaucoma	Local delivery Self-administered	Discomfort to some patients Frequent administration
Topical	Antibiotic ointments	Local delivery Self-administered	Limited to agents that are locally active
Intra-arterial injection		Control of vascular delivery to specific regions	High risk
Intrathecal injection		Direct delivery to brain	Limited drug penetration into brain tissue High risk
Rectal		Avoids first-pass metabolism in liver Self-administered	Discomfort leads to poor compliance in some patients
Transdermal	Nitroglycerin patches for angina	Continuous, constant delivery Self-administered	Skin irritation Limited to lipophilic, highly potent agents
Vaginal	Spermicides	Self-administered	Discomfort leads to poor compliance in some patients
Controlled-release implants	Norplant for contraception	Long-term release	Requires surgical procedure

[a] First-pass metabolism occurs when drug molecules enter the circulation through a mucosal surface, and then circulate through the liver, where they can be metabolized, before distributing throughout the rest of the body.

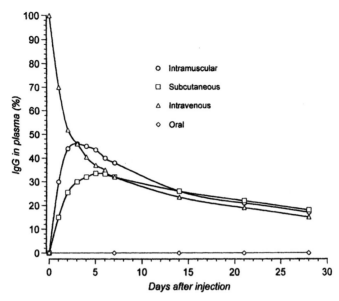

Figure 2.2 Plasma IgG levels following i.m. or s.c. injection in humans. This graph, which is redrawn from Figure 1 in [2], shows the difference in IgG uptake following oral administration and subcutaneous, intramuscular, or intravenous injection. In this particular situation, anti-Rh immunoglobulin G (IgG) was injected into the deltoid or subcutaneously in the buttock. Uptake rates were higher for i.m. injection ($0.43 \pm 0.11\,\mathrm{day}^{-1}$) than for s.c. injection ($0.22 \pm 0.025\,\mathrm{day}^{-1}$). Peak concentrations, which occurred after ~ 3 days, were higher for i.m. injection ($\sim 40\%$) than s.c. injection ($\sim 30\%$). Pharmacokinetic analysis was used to predict average concentrations and rates of equilibration between vascular and extravascular compartments.

administered antibodies in plasma (panel a). However, prediction of IgG concentrations in the blood, after implantation of a device that slowly releases IgG, is surprisingly difficult, even for cases (such as these) in which the delivery system has been characterized.

2.1.4 The Effectiveness and Safety of Drug Delivery Systems Depend on Placement

New drug delivery systems can often be placed in different sites. In panel a of Figure 2.3, a polymer-based delivery system was implanted subcutaneously. In panel b, a similar antibody-delivery system was implanted intracranially, but plasma concentrations were not substantially different than those obtained with an injection of aqueous IgG solution at the same intracranial location. The design of a delivery system that will produce a preselected pattern of IgG concentration in the tissue is difficult, requiring an understanding of the characteristics of the delivery system, the physiology of drug migration in tissue, and the opportunity for interactions between the delivery system and the tissue.

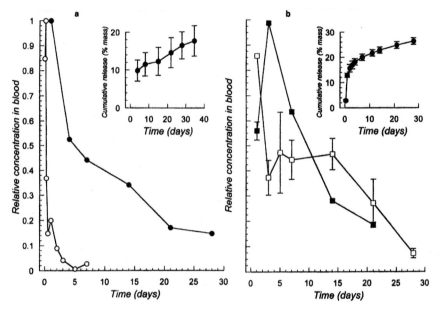

Figure 2.3   IgG levels after administration of drug delivery systems in rats.
Controlled-delivery systems for antibody class IgG. The insert figures show the
release of antibody from the delivery system during incubation in buffered saline. The
panel (a) inset shows release from poly(lactic acid) microspheres; these spherical particles were ~ 10–$100\,\mu$m in diameter. The panel (b) inset shows release from a
poly[ethylene-*co*-(vinyl acetate)] matrix; these disk-shaped matrices were ~ 1 cm in diameter and ~ 1 mm thick. In both cases, molecules of IgG were dispersed throughout
the solid polymer phase. Although the amount of IgG released during the initial 1–2
days is greater for the matrix, the delivery systems have released comparable amounts
after day 5. (a) Comparison of plasma IgG levels after direct injection of IgG (open
circles) or subcutaneous injection of the IgG-releasing polymeric microspheres characterized in the inset (filled circles). The delivery system produces sustained IgG concentrations in the blood [3]. (b) Comparison of plasma IgG levels after direct intracranial
injection of IgG (open squares) or implantation of an IgG-releasing matrix (filled
squares) [4]. The influence of the delivery is less dramatic in this situation, probably
because the rate of IgG movement from the brain into the plasma controls the kinetics
of the overall process.

Experimental results, such as the ones shown in Figure 2.3, can usually be
explained retrospectively; in this example, the influence of the delivery system
is not important because the rate of IgG movement from the brain into the
plasma controls the kinetics of the overall process. In other words, slow release
at the intracranial delivery site does not change the concentration in the
plasma, because the rate of IgG penetration from the brain tissue to the
blood is slower.

   In some situations, the spatial requirements for physical placement of a
delivery system are severely restricted. An obvious example occurs during intravenous injection; a small spatial displacement of the injection needle (> 1 cm)

—so that the needle tip is outside the vessel rather than inside—converts a life-saving, systemic dose of antibiotics into a highly toxic and ineffective pool of fluid in the perivascular tissue. The structure and function of the heart and circulatory system are well known, hence the spatial restrictions and potential hazards of intravenous injection are easily overcome.

Spatial localization of drug is important in other situations, as well. Many of the new drug molecules produced by biotechnology are analogs or recombinant copies of naturally occurring proteins. The protein nerve growth factor (NGF), for example, is found in a variety of tissues. NGF is present at high concentrations in certain regions of the nervous system, where it plays an essential role in development by promoting the survival and differentiation of specific neuronal populations. NGF is highly potent; the concentration required for activity in the brain is not known, but it is active on neurons at ~ 1 ng/mL in cell culture. NGF is also found in non-nervous tissues and in blood, at lower concentration than commonly found in the nervous system, where it can have a variety of activities. For example, NGF influences the acute inflammatory response in rats [5] and intravenous administration of recombinant human NGF in humans causes myalgias whose duration and severity vary in a dose-dependent manner [6].

Like many recombinant proteins that are being considered for therapeutic use, NGF is a troublesome drug candidate. It is high in molecular weight and water soluble; therefore, it permeates through biological barriers (such as the brain capillary endothelial barrier) very poorly. It has activities at a variety of sites; therefore, administration of large quantities to treat disease in one organ can cause toxic effects in another. Finally, NGF has a short half-life after administration; therefore, use for chronic disease requires frequent readministration.

These difficulties can be overcome with the use of locally implanted controlled-release systems (Figure 2.4). Small implants that slowly release NGF can be implanted at the site of action, producing long-lasting concentrations of biologically active protein in the tissue near the implant (see Figure 2.4, panel b). But these implants have a surprisingly short range of action (panel c), probably due to the short half-life of NGF in the tissue. As a consequence, NGF-releasing implants can be used to deliver high doses of protein at local sites, but these implants must be placed within ≈ 1 mm of the site of action.

In summary, the effectiveness of new medicines is limited by the mode of administration, the duration of action, and the requirements for physical placement. Modern engineering has greatly extended the number of available delivery technologies. But so far—in most cases—the influence of these new delivery systems cannot be predicted in advance of the experiment.

2.2 TWO VIEWS OF THE FUTURE

2.2.1 Pharmacokinetics, Drug Design, and Drug Delivery

A drug is not effective unless it is present at its site of action for an adequate period of time. In the past, most drug doses were determined empirically,

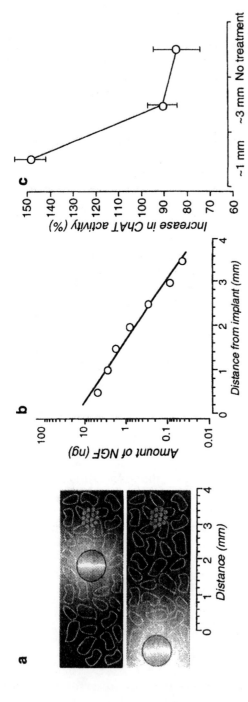

Figure 2.4 Millimeter scale positioning of NGF-releasing implants. (a) Positioning of NGF-releasing implant. The same anatomical treatment site was stereotactically defined in all animals: 1 mm rostral to the bregma and 2–3 mm to the right of the
bregma). The site of treatment was injected with a population of donor cells (gray circles). An NGF-containing polymer was
implanted either adjacent to (~ 1–2 mm) or separate (~ 3 mm) from the site of treatment. (b) Concentration of NGF as a
function of distance from the implant site (data replotted from [7]). (c) Effect of NGF positioning on a treatment site that
contained donor cells. Levels of choline acetyltransferase (ChAT) activity were significantly elevated above that of controls
when NGF-containing polymers were implanted local to the site of treatment. Control animals received blank polymers.
Adapted from [8].

through clinical experiments in which drug effectiveness and toxicity were monitored as the dose was gradually increased. This approach remains an important step in the approval of compounds for clinical use. But the development of pharmacokinetic principles has enabled us to rationalize the changes in drug concentration that occur in various tissues in the body after administration. Pharmacokinetics has been extremely influential; the ability to design dosage schedules that maximize drug effectiveness (and minimize drug toxicity) allowed the use of novel molecular medicines, such as those for cancer chemotherapy. Today, because more is known about the relationship between chemical properties and movement throughout the body, drug designers can consider the pharmacokinetic properties of agents much earlier in the development process.

This practice is sensible; drug development is expensive and, therefore, agents that appear unlikely to work, because they are difficult to administer, should be avoided in favor of alternative agents. On the other hand, our reliance on a few simple methods for drug administration complicates the use of categories of agents that have undesirable pharmacokinetic properties. As mentioned in the previous section, nucleic acids and proteins are (in general) poor drug candidates, because they are large, unstable, and active at non-target sites (i.e., potentially toxic). However, nucleic acids and proteins have exceptional promise as therapeutics; therefore, the historical barriers to consideration of unconventional drug delivery methods are dropping rapidly. With new delivery technologies, one can imagine alternate routes to the design of effective molecular medicines: *selecting* among drug candidates to find the one with the most "acceptable" pharmacokinetic properties, *modifying* drug molecules to create analogs that have "improved" properties, or *packaging* agents into delivery systems that are designed for particular agents and tissue sites. In this last category, the design of delivery system and drug are intimately related.

2.2.2 A Proliferation of Drug Targets

Hundreds of drugs are approved for use in the United States. The *Physician's Desk Reference*, which lists the characteristics of all prescription drugs, currently extends over 3,000 pages. But these agents act on surprisingly few different biological targets (Table 2.2). Most drugs act by inhibition of an enzyme or interaction with a receptor; a few interact directly with DNA.

Within a few years, the human genome will be completely mapped; the raw information used by cells to operate and replicate will be available. This information is likely to be highly useful in drug design; as we learn new mechanisms for control of cell function, new targets for drug action will be uncovered. It is reasonable to expect that many of these targets will not be readily accessible, perhaps because they are within a cell compartment (such as the nucleus) or expressed in only a few cells within a tiny volume of a tissue. As in the past, the discovery of these new targets will require invention of new drug-delivery technology.

Table 2.2   Classic drug targets

	Channel	DNA	Enzyme	Factors and hormones	Nuclear receptor	Receptor	Unknown
Synaptic and neuroeffector junctional sites and CNS	8		12			115	1
Inflammation	1		19	1		26	2
Renal and cardiovascular functions	11		17		1	15	3
Gastrointestional functions	1		1	1		10	3
Uterine motility	1		2	1		4	
Bacterial infections			23				9
Fungal infections			4				1
Viral infections			10				
Parasitic infectious diseases	1	2	8			1	7
Neoplastic diseases		5	20	6	3	4	1
Immunomodulation		2	2	6	1	4	1
Blood and blood-forming organs			2	29		7	
Hormones and hormone antagonists	2		14	11	3	25	
Vitamins			1		2	4	3
Total	25	9	135	55	10	215	31

Adapted from a table that appeared in *Nature Biotechnology* (Drews, J. and Ryser, S.), which was based on information in [9].

SUMMARY

- This text provides a foundation for the rational design of drug delivery systems. The factors that influence rates and patterns of drug movement throughout tissues will be considered; drug modifications and delivery systems will be presented in the context of normal patterns of drug movement and clearance.

- Since diffusion is an important mechanism of drug transport in tissues, Chapter 3 presents an analysis of rates and patterns of diffusive transport.

- In Chapter 4, diffusion models are used as the basis for relating local rates of drug movement with local tissue architecture.

- Chapters 5 and 6 focus on the special cases of transport through membranes and transport due to fluid flows.

- In Chapter 7, pharmacokinetic models are used to relate local rates of drug transport to the distribution of the drug throughout an organism.

- In the final chapters, Chapters 8 through 10, this information will be used as the basis for the design of new drug delivery systems.

REFERENCES

1. Shamoon, E.A., The effect of intensive treatment of diabetes on the development and progression of long-term complications in insulin-dependent diabetes-mellitus. *New England Journal of Medicine*, 1993, **14**, 977–986.
2. Smith, G.N., *et al.*, Uptake of IgG after intramuscular and subcutaneous injection. *Lancet*, 1972, **i**, 1208–1212.
3. Kuntz, R.M. and W.M. Saltzman, Polymeric controlled delivey for immunization. *Trends in Biotechnology*, 1997, **15**, 364–369.
4. Salehi-Had, S. and W.M. Saltzman, Controlled intracranial delivery of antibodies in the rat, in J. Cleland and R. Langer, *Protein Formulations and Delivery*, Washington, DC: ACS Symposium Series, 1994, pp. 278–291.
5. Otten, U. and R.A. Gadient, Neurotrophins and cytokines—intermediaries between the immune and nervous systems. *International Journal of Developmental Neuroscience*, 1995, **13**(3/4), 147–151.
6. Petty, B.G., *et al.*, The effect of systemically administered recombinant human nerve growth factor in healthy human subjects. *Annals of Neurology*, 1994, **36**(2), 244–246.
7. Saltzman, W.M., *et al.*, Intracranial delivery of recombinant nerve growth factor. *Pharmaceutical Research*, 1999. **16**, 232–240.
8. Mahoney, M.J. and W.M. Saltzman, Millimeter-scale positioning of a nerve-growth-factor source and biological activity in the brain. *Proceedings of the National Academy of Sciences*, 1999, **96**, 4536–4539.
9. Goodman Gilman, A., *et al.*, eds, *The Pharmacological Basis of Therapeutics*, 8th ed. New York: Pergamon Press, 1990, 1811 pp.

II

FUNDAMENTALS

3

Diffusion and Drug Dispersion

You air that serves me with breath to speak!
You objects that call from diffusion my meanings, and give them shape!
Walt Whitman, *Song of the Open Road* (1856)

Most biological processes occur in an environment that is predominantly water: a typical cell contains 70–85% water and the extracellular space of most tissues is 99%. Even the brain, with its complex arrangement of cells and myelinated processes, is ≈ 80% water. Drug molecules can be introduced into the body in a variety of ways (recall Table 2.1); the effectiveness of drug therapy depends on the rate and extent to which drug molecules can move through tissue structures to reach their site of action. Since water serves as the primary milieu for life processes, it is essential to understand the factors that determine rates of molecular movement in aqueous environments. As we will see, rates of diffusive transport of molecules vary among biological tissues within an organism, even though the bulk composition of the tissues (i.e., their water content) may be similar.

The section begins with the random walk, a useful model from statistical physics that provides insight into the kinetics of molecular diffusion. From this starting point, the fundamental relationship between diffusive flux and solute concentration, Fick's law, is described and used to develop general mass-conservation equations. These conservation equations are essential for analysis of rates of solute transport in tissues.

3.1 RANDOM WALKS

Molecules that are initially localized within an unstirred vessel will spread throughout the vessel, eventually becoming uniformly dispersed. This process, called diffusion, occurs by the random movement of individual molecules; molecular motion is generated by thermal energy. Einstein demonstrated that the average kinetic energy of particles in a system depends on the absolute temperature T [1]:

$$\left\langle \frac{mv_x^2}{2} \right\rangle = \frac{k_B T}{2} \tag{3-1}$$

where v_x is the velocity of the particle on one axis, m is the mass of a particle, and k_B is Boltzmann's constant ($k_B T = 4.4 \times 10^{-14}$ g $\cdot$ cm^2/s^2 at 300 K). The root-mean-square (r.m.s.) velocity can be found from this relationship:

$$v_{rms} = \sqrt{\langle v_x^2 \rangle} = \sqrt{\frac{k_B T}{m}} \tag{3-2}$$

Albumin, a protein of 68,000 M_w, has a predicted r.m.s. velocity of 600 cm/s at 300 K. A molecule moving at this speed should travel 2 m (the height of a very tall person) in about 0.33 s, if it moved in a straight line. Experience teaches us that molecules do not move this rapidly. In fact, rates of diffusive transport are much slower than v_{rms} would suggest. Diffusion is slow because molecules do not travel in a straight path; individual molecules collide with other molecules and change directions frequently, producing a pattern of migration known as a random walk (Figure 3.1). Since air and water have different molecular densities, the number of collisions per second—and hence the rate of overall dispersion—differs for these fluids.

These patterns of migration can be simulated by examining particles that follow simple rules for movement: random walkers. Many of the important characteristics of diffusive processes can be understood by considering the dynamics of particles executing simple random walks. The excellent book by Berg [2] provides a useful introduction to the random walk and its relevence in biological systems, which is followed here. Whitney provides a complete, tutorial introduction to a variety of random processes, including the random walk [3].

Figure 3.1
Typical pattern of
migration for a particle
executing a random
walk.

Consider, for example, a particle constrained to move on a one-dimensional axis (Figure 3.2). During a small time interval, τ, the particle can move a distance δ, which depends on the particle's velocity v_x: $\delta = v_x\tau$. Further assume that at the end of each period τ the particle changes its direction of movement, randomly moving to the right or the left with equal probability. At each step in the walk, the decision to move left or right is completely random and does not depend on the particle's previous history of movement. If N identical particles, each moving according to these simple rules, begin at the origin of the coordinate system at $t = 0$, one-half of the particles will be at position $-\delta$ and one-half of the particles will be at $+\delta$ at the end of the first interval τ (Figure 3.2). If the location of one particular particle, indicated by the index i, after $n - 1$ such intervals is designated $x_i(n - 1)$, then the position of that same particle after n intervals is easily determined:

$$x_i(n) = x_i(n - 1) \pm \delta \qquad (3\text{-}3)$$

The mean particle position can be predicted by averaging Equation 4-3 over the ensemble of particles, which becomes:

$$\langle x(n)\rangle = \frac{1}{N}\sum_{i=1}^{N} x_i(n) = \frac{1}{N}\sum_{i=1}^{N}(x_i(n - 1) \pm \delta) = \frac{1}{N}\sum_{i=1}^{N} x_i(n - 1) + \frac{1}{N}\sum_{i=1}^{N}(\pm\delta)$$

$$(3\text{-}4)$$

upon substitution of Equation 3-3. The averaging operation, designated $\langle \bullet \rangle$, is a conventional arithmetical average:

$$\frac{1}{N}\left(\sum_{i=1}^{N} \bullet_i\right).$$

The second term on the right-hand side of Equation 3-4 is zero, since an equal number of particles will move to the right $(+\delta)$ and to the left $(-\delta)$, which produces:

$$\langle x(n)\rangle = \langle x(n - 1)\rangle \qquad (3\text{-}5)$$

Therefore, the average position of the randomly migrating particles in this ensemble does not change with time. For a group of particles that begin at the origin, $x = 0$ as in Figure 3.2, the average position will always be at the origin.

Although this random migration does not change the average position of the particle ensemble, it does tend to spread the particles over the axis. The extent of spread can be determined by examining the mean square displacement of the particle ensemble, $\langle x^2(n)\rangle$:

$$\langle x^2(n)\rangle = \frac{1}{N}\sum_{i=1}^{N}(x_i(n - 1) \pm \delta)^2 = \frac{1}{N}\sum_{i=1}^{N}(x_i^2(n - 1) \pm 2\delta x_i(n - 1) + \delta^2) \quad (3\text{-}6)$$

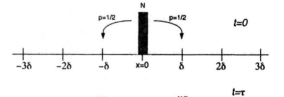

Figure 3.2
Coordinate system for a
one-dimensional ran-
dom walk. Consider a
group of N particles
that are all located at
the original ($x = 0$) at
some initial time ($t = 0$).
In each increment of
time, τ, each particle
randomly moves a fixed
distance either to the
left ($-\delta$) or the right
($+\delta$).

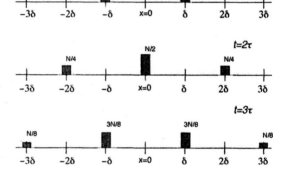

When the summation is expanded, the sum resulting from the middle term on
the right-hand side is zero, because an equal number of particles move to the
left and right. The first and last terms can be reduced, to yield:

$$\langle x^2(n)\rangle = \langle x^2(n-1)\rangle + \delta^2 \tag{3-7}$$

Since the mean square displacement is initially zero, Equation 3-7 indicates
that the mean square displacement increases linearly with the number of steps
in the random walk:

$$\langle x^2(0)\rangle = 0$$
$$\langle x^2(1)\rangle = \delta^2$$
$$\langle x(2)\rangle = 2\delta^2 \tag{3-8}$$
$$\vdots$$
$$\langle x^2(n)\rangle = n\delta^2$$

Since the total elapsed time t is equal to $n\tau$, the mean square displacement also
increases linearly with time:

$$\langle x^2(t)\rangle = \left(\frac{\delta^2}{\tau}\right)t \tag{3-9}$$

and the r.m.s. displacement, x_{rms}, increases with the square root of time:

$$x_{rms} \equiv \langle x^2(t)\rangle^{1/2} = \sqrt{2Dt} \tag{3-10}$$

where the diffusion coefficient D is defined:

$$D = \delta^2/2\tau \qquad (3\text{-}11)$$

Experimental measurements, obtained by measuring the rate of spreading of protein molecules in solution (or by a variety of other methods described in Chapter 4), suggest that D for albumin at 300 K is $\sim 8 \times 10^{-7}\,\text{cm}^2/\text{s}$. Since v_x is the particle velocity between changes of direction, which is equal to δ/τ, the distance between direction changes, δ, can be estimated as $2D/v_x$. For albumin, with $v_x = v_{\text{rms}} = 600\,\text{cm/s}$, this yields a value for δ of $3 \times 10^{-9}\,\text{cm}$ (or 0.3 Å, small compared to the diameter of albumin, which is $\approx 60\,\text{Å}$) and a time between direction changes τ of $4 \times 10^{-12}\,\text{s}$. This model suggests a different view of albumin movement than obtained by only considering the r.m.s. velocity of individual particles. The random walk of albumin will require $2.5 \times 10^{10}\,\text{s}$ or 800 years to travel 2 m (from Equation 3-10), a significantly longer time than the r.m.s. velocity would suggest. This simple calculation illustrates the slow progress of diffusing molecules over meter-scale distances (and explains why humans have a circulatory system).

Note that these calculations suggest several important characteristics for diffusion processes:

- since the r.m.s. displacement increases with the square root of time, a diffusing particle that takes T min to diffuse L mm will take $4T$ to diffuse $2L$;
- the diffusion velocity, which might be defined as x_{rms}/t, is inversely proportional to the square root of time and therefore not a useful measure of molecular speed.

This analysis can be extended to two- and three-dimensional random walks by assuming that particle motion in each dimension is independent. The mean square displacement and r.m.s. displacement for higher dimension random walks become:

$$\begin{aligned}
\langle r^2 \rangle_{2D} &= \langle x^2 \rangle + \langle y^2 \rangle = 4Dt &&\text{or} && \langle r^2 \rangle_{2D}^{1/2} = \sqrt{4Dt} \\
\langle r^2 \rangle_{3D} &= \langle x^2 \rangle + \langle y^2 \rangle + \langle z^2 \rangle = 6Dt &&\text{or} && \langle r^2 \rangle_{3D}^{1/2} = \sqrt{6Dt}
\end{aligned} \qquad (3\text{-}12)$$

For particles executing a random walk, like albumin molecules in buffered water, the calculations above suggest that individual steps in the random walk occur very quickly, over a short time interval. As a consequence, during a typical observation time, each particle takes many steps on the axis of Figure 3.2. The probability that a random walking particle took a total of k steps to the right after a sequence of n steps in the random walk is provided by the binomial distribution:

$$P\left\{\begin{array}{l} \text{particle took } k \\ \text{steps to the right} \\ \text{after } n \text{ steps} \end{array}\right\} = P(k; n, q) = \frac{n!}{k!(n-k)!}(q)^k(1-q)^{n-k} \qquad (3\text{-}13)$$

where q is the probability of moving to the right at each step; $q = 1/2$ for an unbiased random walk. For example, after four steps in an unbiased random walk, the binomial distribution can be used to determine the probabilities of finding particles at any location on the axis, as shown in Table 3.1. The probabilities calculated by Equation 3-13 agree with those indicated in Figure 3.2.

By using the binomial distribution, it is possible to determine $p(x)$, the probability of finding a particle at a position between x and $x + dx$, as a function of time after a large number of individual steps in the random walk have occurred [2]:

$$p(x) = \frac{1}{\sqrt{4\pi Dt}} e^{-x^2/4Dt} dx \tag{3-14}$$

Equation 3-14 describes a Gaussian distribution; the random walk model predicts that a group of diffusing particles, initially placed at the origin, will spread over time so that the mean position is unchanged, but the variance in the distribution increases as $2Dt$. In the section below, we will see this same distribution emerge from the solution to the macroscopic equations for molecular diffusion (as we will see in Figure 3.5).

The rate of movement of particles at any particular location can be estimated by considering the net rate of particle movement at any specific position during the random walk. During a single step in the random walk, which occurs over the time interval τ, the net rate of particle movement between two adjacent positions x and $x + \delta$ is $N(x)/2 - N(x + \delta)/2$, where $N(\bullet)$ gives the number of particles at a given location, because half of the particles at position x will move to the right (ending at $x + \delta$) and half the particles at $x + \delta$ will move to the left (ending at x). The particle flux j_x is defined as the net rate of particle movement per unit area, so that:

$$j_x = -\frac{1}{2A\tau}[N(x + \delta) - N(x)] \tag{3-15}$$

Rearranging slightly gives:

$$j_x = -\frac{\delta^2}{2\tau}\left[\frac{C(x + \delta) - C(x)}{\delta}\right] \tag{3-16}$$

Table 3.1 Characterization of random walks with binomial distribution

k	0	1	2	3	4
$n - k$	4	3	2	1	0
x	-4δ	-2δ	0	$+2\delta$	$+4\delta$
$P(k; n, 1/2)$	0.0625	0.25	0.375	0.25	0.0625

The binomial distribution, Equation 3-13, was used to calculate the probability of finding a random walker at position x after 4 steps in an unbiased random walk. The position on a coordinate axis, x, was determined from the number of steps to the right, k, and number of steps to the left, $n - k$: $x = \delta(k - (n - k))$.

where $C(x, N(x)/A\delta)$, is the concentration of particles at a given location, x. Taking the limit of Equation 3-16 as the distance between the adjacent points becomes small yields:

$$j_x = -D\frac{\partial C}{\partial x} \qquad (3\text{-}17)$$

where the diffusion coefficient D is defined as in Equation 3-11. Thus, this random walk model also provides a relationship between the spatial concentration distribution and the particle flux. Particles move towards locations with lower concentration; the flux of particles is directly proportional to the concentration gradient and the diffusion coefficient is the constant of proportionality. This expression, often called Fick's law [4], will be of considerable practical value in predicting rates of molecular transport through cells and tissues.

3.2 EQUATIONS FOR THE DIFFUSIVE FLUX (FICK'S LAW)

This expression for the diffusive flux can be obtained more rigorously by considering the velocities of individual species in a multi-component system. The mass average flux of component A in a mixture, $\overline{j_A}$, is defined based on the velocity of the species of interest, $\overline{v_A}$, relative to the mass average velocity of the system, $\overline{v}$:

$$\overline{j_A} = \rho_A(\overline{v_A} - \overline{v}) \qquad (3\text{-}18)$$

where $\overline{j_A}$, $\overline{v_A}$, and $\overline{v}$ are now vector quantities, and the mass average velocity is defined by weighting the velocities of each component by the mass density of that component in the system.[1] The flux defined in this manner is due to the presence of concentration gradients in the system, and depends on the diffusion coefficient D_A:

$$\overline{j_A} = -\rho D_A \nabla \omega_A \qquad (3\text{-}19)$$

where ρ is the total mass density of the system and ω_A is the mass fraction of component A. There are many equivalent ways to express Fick's law for molar or mass fluxes with respect to molar or mass average velocities (see a text on transport phenomena, such as [5] or [6], for examples). In all cases, however, the diffusion coefficient, D_A, is identical. Strictly speaking, this definition of D_A applies to binary mixtures where D_A is the diffusion coefficient of A in a mixture of A and B. In this book, which is primarily concerned with the diffusion of dilute species in an aqueous system containing many additional dilute components, we will use D_A to denote the diffusion coefficient of the species of interest in a multi-component complex system. When used in this

---

1. For more complete definitions, the derivation of conservation of mass equations, and application of these equations in conventional chemical engineering analysis, the reader is referred to the classical textbook on transport phenomena [5]. The notation used in the present text follows the notation by Bird *et al.*

manner, D_A is an effective binary diffusion coefficient for A in a multi-component system [5].

Equation 3-19 defines the diffusion coefficient D_A, but is inconvenient for many calculations since the flux $\bar{j}_A$ is defined with respect to a moving coordinate system. For binary systems, $\bar{j}_A$ is simply related to the mass flux of A with respect to a stationary coordinate system, $\bar{n}_A$:

$$\bar{n}_A = \omega_A(\bar{n}_A + \bar{n}_B) + \bar{j}_A = \omega_A(\bar{n}_A + \bar{n}_B) - \rho D_A \nabla \omega_A \qquad (3\text{-}20)$$

The mass flux $\bar{n}_A$ is the sum of the diffusive flux $\bar{j}_A$, which accounts for the movement of A due to concentration gradients, and the quantity $\omega_A(\bar{n}_A + \bar{n}_B)$, which accounts for the movement of A due to bulk motion of the fluid. Equation 3-20 is the appropriate form of the diffusive flux equation to use for most engineering problems, since quantities are usually measured with respect to some fixed frame of reference. For many of the problems considered here, A is a dilute species in water, in which case $\omega_A \approx 0$ and the bulk motion term can be neglected.

3.3 EQUATIONS OF MASS CONSERVATION
(FICK'S SECOND LAW)

Consider the migration of species A through a homogeneous region of space, like a small volume of tissue (Figure 3.3). General differential equations can be developed to describe the variation of concentration of compound A, C_A, with time and position in the tissue by first considering the conservation of mass within this small volume.

Molecules of A can enter through any face of the volume element; they can be generated or consumed within the volume by chemical reaction; they can accumulate within the volume. A general balance equation (in − out + generation = accumulation) yields:

Figure 3.3
Differential mass balance on a characteristic element. This diagram shows a simple mass balance in a rectangular coordinate system. Not shown are the possible sources of consumption or generation of mass within the volume element, which may be important if the element is a small section of tissue.

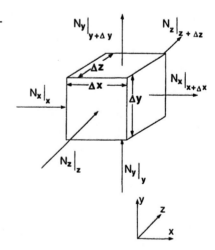

$$\left[n_x\big|_x \Delta y \Delta z - n_x\big|_{x+\Delta x} \Delta y \Delta z \right] + \left[n_y\big|_y \Delta x \Delta z - n_y\big|_{y+\Delta y} \Delta x \Delta z \right]$$

$$+ \left[n_z\big|_z \Delta x \Delta y - n_z\big|_{z+\Delta z} \Delta x \Delta y \right] \qquad (3\text{-}21)$$

$$+ \psi_A \Delta x \Delta y \Delta z = \frac{\partial \rho_A}{\partial t} \Delta x \Delta y \Delta z$$

where the first three bracketed terms account for the movement of A through the faces of the volume element, ψ_A is the rate of generation (or negative the rate of consumption) of A per unit volume, and ρ_A is the mass density of A in the volume ($\rho_A = \omega_A \rho$). Dividing each term by the volume ($\Delta x \Delta y \Delta z$) and taking the limit of the resulting expression as the differential volume becomes very small yields:

$$-\frac{\partial n_x}{\partial x} - \frac{\partial n_y}{\partial y} - \frac{\partial n_z}{\partial z} + \psi_A = \frac{\partial \rho_A}{\partial t} \qquad (3\text{-}22)$$

which can be written more simply as:

$$-\nabla \cdot \overline{n_A} + \psi_A = \frac{\partial \rho_A}{\partial t} \qquad (3\text{-}23)$$

which was derived by considering a volume element in rectangular coordinates; identical expressions could be obtained in cylindrical or spherical coordinates.

The mass balance procedure produced a partial differential equation with two dependent variables, solute flux and concentration. To complete this analysis, an appropriate form of Fick's law must be substituted into Equation 3-23. Substituting Equation 3-20 yields:

$$-\nabla \cdot (\omega_A (\overline{n_A} + \overline{n_B}) - \rho D_A \omega_A) + \psi_A = \frac{\partial \rho_A}{\partial t} \qquad (3\text{-}24)$$

By recognizing that the sum of the fluxes $\overline{n_A} + \overline{n_B}$ is equal to the density times the mass average velocity, $\rho \bar{v}$, this equation reduces to:

$$-\nabla \cdot (\rho_A \bar{v}) + \nabla \cdot (\rho D_A \nabla \omega_A) + \psi_A = \frac{\partial \rho_A}{\partial t} \qquad (3\text{-}25)$$

which is a general expression of the conservation of mass of solute A. Expressions equivalent to Equation 3-23 through Equation 3-25 could also be written for component B of a binary system; the sum of these individual mass balance equations produces the total mass conservation, or continuity, equation:

$$-\nabla \cdot \rho \bar{v} = \frac{\partial \rho}{\partial t} \qquad (3\text{-}26)$$

The aqueous systems considered in this text are incompressible (i.e., the total density ρ is constant). For incompressible fluids, Equations 3-25 and 3-26 can be written as:

$$\cdot D_A \nabla \rho_A + \psi_A = \frac{D\rho_A}{Dt} \tag{3-27}$$

$$\nabla \cdot \bar{v} = 0 \tag{3-28}$$

where Equation 3-28 has been used to eliminate the term involving velocity gradient in Equation 3-27 and the substantial derivative is defined:

$$\frac{D}{Dt}(\bullet) = \frac{\partial}{\partial t}(\bullet) + \bar{v} \cdot \nabla(\bullet) \tag{3-29}$$

Equations 3-27 and 3-28 provide the most compact and general form of the equation of conservation of mass. These expressions are limited only by the assumption that the fluid is incompressible; this assumption is reasonable for all of the problems discussed in this text.

Equations 3-20 to 3-28 can be written in terms of molar concentrations and molar fluxes, as well. A summary of the most important equations, expressed in both molar and mass units, is provided in Table 3.2.

The equations provided in Table 3.2 assume that diffusion is isotropic: i.e., that the flux of diffusing molecules through some particular plane within a system is proportional to the concentration gradient measured orthogonal to the plane. This assumption is not true for anisotropic media, in which the diffusion characteristics depend on the direction of diffusion. Anisotropic

Table 3.2   The basic equations for mass transfer, derived on a mass and molar basis for isotropic diffusion

	Mass	Molar
Average velocity	$\bar{v} = \frac{1}{\rho}\sum_{i=1}^{u}\rho_i\bar{v}_i$	$\bar{V} = \frac{1}{c}\sum_{i=1}^{u}c_i\bar{v}_i$
Flux with respect to coordinate system moving at average velocity	$\bar{j}_A = \rho_A(\bar{v}_A - \bar{v}) = -\rho D_A \nabla \omega_A$	$\bar{J}_A = c_A(\bar{v}_A - \bar{V}) = -cD_A \nabla x_A$
Flux with respect to stationary coordinate system (binary system)	$\bar{n}_A = -\rho D_A \nabla \omega_A + \omega_A(\bar{n}_A + \bar{n}_B)$	$\bar{N}_A = -cD_A \nabla x_A + x_A(\bar{N}_A + \bar{N}_B)$
General form of conservation of mass for component A (binary system)	$-\nabla \cdot (\rho_A \bar{v} - \rho D_A \nabla \omega_A) + \psi_A = \frac{\partial \rho_A}{\partial t}$	$-\nabla \cdot (c_A \bar{V} - cD_A \nabla x_A) + \psi_A^m = \frac{\partial c_A}{\partial t}$
Conservation of mass for component A (binary system, incompressible fluid, constant D_A)	$D_A \nabla^2 \rho_A + \psi_A = \frac{D\rho_A}{Dt}$	$D_A \nabla^2 c_A + \psi_A^m = \frac{Dc_A}{Dt}$

The symbols are defined in the text: ψ_A^m indicates molar rate of generation of A (mol/cm$^3 \cdot$ s)

media usually have oriented physical structures, such as crystals or polymers, that provide varying resistances of diffusion in different directions. Anisotropic diffusion may be important in diffusion in biological systems, since the cytoplasm of cells or the extracellular space of tissues can contain highly organized and oriented structures. Consider, for example, the diffusion of compounds in the extracellular space of the brain, which may be anisotropic in regions where myelinated fibers run in a particular direction.

In the case of anisotropic diffusion, Fick's law (Equation 3-19), must be written:

$$j_{A_x} = -\rho \left[D_{11} \frac{\partial \omega_A}{\partial x} + D_{12} \frac{\partial \omega_A}{\partial y} + D_{13} \frac{\partial \omega_A}{\partial z} \right]$$

$$j_{A_y} = -\rho \left[D_{21} \frac{\partial \omega_A}{\partial x} + D_{22} \frac{\partial \omega_A}{\partial y} + D_{23} \frac{\partial \omega_A}{\partial z} \right] \qquad (3\text{-}30)$$

$$j_{A_z} = -\rho \left[D_{31} \frac{\partial \omega_A}{\partial x} + D_{32} \frac{\partial \omega_A}{\partial y} + D_{33} \frac{\partial \omega_A}{\partial z} \right]$$

where fluxes depend not only on the concentration gradient in the orthogonal direction, but may also depend on gradients in the other directions as well. The diffusion coefficients in Equation 3-30, D_{nm}, indicate the significance of diffusion in the n-direction due to gradients in the m-direction. Equation 3-30 reduces to Fick's law for isotropic diffusion when the off-diagonal terms are 0—i.e., $D_{nm} = 0$ for $n \neq m$—and the diagonal terms are all equal—i.e., $D_{nm} = D_A$ for $n = m$. In some cases, appropriate coordinate transformations can be used to reduce the equations describing diffusion in anisotropic media to analogous equations in isotropic media; these techniques are discussed by Crank [7].

The conservation of mass equations listed in the final row of Table 3.2 are frequently used to describe the movement of solutes through tissues and cells. These equations were developed by assuming that the tissue is homogeneous throughout the region of interest. Diffusing solute molecules must have equal access to every possible position within the volume of interest and D_A must be constant with respect to both space and time. This assumption is not valid for the diffusion of certain molecules in tissues; for example, consider a molecule that diffuses through the extracellular space of the tissue and does not readily enter cells. The limitations of this assumption are discussed in Chapter 4.

3.4 SOLUTIONS TO THE DIFFUSION EQUATION WITH NO SOLUTE ELIMINATION OR GENERATION

When appropriately applied, the equations derived in the section above can be used to predict variations in drug concentration within a tissue following administration. For the description of molecular transport in cells or tissues, the mass conservation equations must be simplified by making appropriate assumptions that are specialized for the geometry of interest. For example, the mass conservation equation in terms of molar fluxes assuming no bulk flow ($\bar{v} = 0$) and no chemical reaction ($\psi_A = 0$) is:

$$D_A \nabla^2 c_A = \frac{\partial c_A}{\partial t} \tag{3-31}$$

a commonly used form that is frequently referred to as Fick's second law. When solved subject to the appropriate boundary and initial conditions, this differential equation yields the concentration c_A as a function of time and position within the volume of interest. Since equations equivalent to Equation 3-31 arise in a variety of physical settings, detailed solutions for many situations already exist (see the excellent books by Crank [7] and Carslaw and Jaeger [8]).

3.4.1 Rectangular Coordinates

Consider a solute A that is neither consumed nor generated, but diffuses with a constant diffusion coefficient, D_A. Assume that a bolus of N molecules is injected into a long cylinder (Figure 3.4a), so that all of the molecules are present within an infinitesimal volume at the origin of the coordinate system at $t = 0$. After injection, the molecules will diffuse along the axis of the cylinder, where bulk flow is negligible. This situation will not occur frequently in physiological systems, although it may be a reasonable model for the microinjection of inert tracers into the axon or dendrite of a neuron [9]. On the other hand, *in vitro* experimental systems are often intentionally arranged into this geometry, which greatly simplifies the subsequent analysis [10, 11].

The concentration of A as a function of time and distance from the site of initial injection can be predicted by solving Equation 3-31, expressed in a one-dimensional rectangular coordinate system:

$$D_A \frac{\partial^2 c_A}{\partial x^2} = \frac{\partial c_A}{\partial t} \tag{3-32}$$

subject to the following initial and boundary conditions:

$$
\begin{aligned}
c(x, t) &= N\delta(x) & -\infty < x < \infty \quad & t = 0 \\
c(x, t) &= 0 & x \to \infty \quad & t > 0 \\
c(x, t) &= 0 & x \to -\infty \quad & t > 0
\end{aligned} \tag{3-33}
$$

where $\delta(x)$ is the unit impulse or Dirac delta function centered at $x = 0$. This equation can be solved by Laplace transform techniques to produce the solution:

$$c_A(x, t) = \frac{A}{\sqrt{t}} e^{-x^2/4D_A t} \tag{3-34}$$

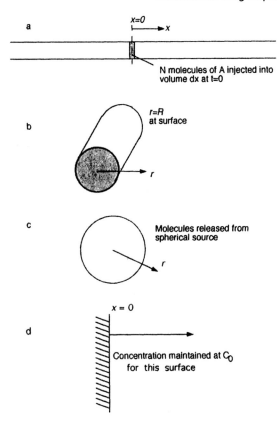

Figure 3.4
Typical problems of diffusive transport. Many real examples of diffusion in organs and tissues can be analyzed in terms of simple solutions to the diffusion equation in rectangular, cylindrical, or spherical coordinates: (a) a bolus of molecules is injected into a cylindrical volume of infinite extent; (b) a cylindrical source of molecules in an infinite volume; (c) a spherical source of molecules in an infinite volume; or (d) drug concentration is maintained at a constant value at the surface of a semi-infinite medium.

where A is a constant of integration. The constant A can be determined by integrating Equation 3-34 with respect to x, to obtain the total number of diffusing molecules:

$$\int_{-\infty}^{\infty} c_A(x, t) \cdot \pi R^2 dx = \pi R^2 \frac{A}{\sqrt{t}} \int_{-\infty}^{\infty} e^{-x^2/4D_A t} dx = 2\pi R^2 A \sqrt{\pi D_A} = \frac{N}{N_{Av}} \quad (3-35)$$

where R is the radius of the cylinder cross-section and N_{Av} is Avogadro's number. This definite integral reduces to $\sqrt{4D_A t \pi}$. Since the value of the constant A can be determined from Equation 3-35, the complete solution is:

$$c_A(x, t) = \frac{N}{N_{Av}} \frac{1}{2\sqrt{\pi D_A t}} \frac{1}{2\pi R^2} e^{-x^2/4D_A t} \quad (3-36)$$

This expression is plotted in Figure 3.5. The similarity of Equation 3-36, which is a solution to the conservation of mass equations, to the Gaussian distribution obtained from random walk calculations (see Equation 3-14) is obvious.

A similar analysis can be used to predict concentration profiles in the region on one side of a planar boundary after suddenly raising the concentration on the opposite side (Figure 3.4d). In this case, the differential equation is still Equation 3-32, but the boundary conditions must be modified slightly:

Figure 3.5
Concentration profiles
for diffusion from a
point source. When a
concentrated bolus of
solute is deposited
within a small region of
an infinitely long cylin-
der, as shown in Figure
3.4a, the molecules
slowly disperse along
the axis of the cylinder.
The curves shown here
are realization of
Equation 3-34 for a
solute with
$D_A = 10^{-7}\,cm^2/s$,
$R = 0.1\,cm$, $N = N_{Av}$,
and $t = 6$, 24, and 72 h.

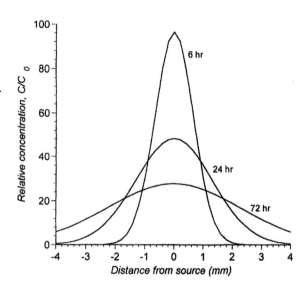

$$c(x, t) = 0; \quad \text{for } 0 < x < \infty; \quad t = 0$$
$$c(x, t) = c_0; \quad \text{for } x = 0; \quad\quad t > 0 \tag{3-33'}$$
$$c(x, t) = 0; \quad \text{for } x \to \infty; \quad\quad t > 0$$

The solution of Equation 3-32 in this situation yields:

$$\frac{c_A}{c_0} = \text{erfc}\left(\frac{x}{2\sqrt{D_A t}}\right) \tag{3-37}$$

where erfc(•) is the error function complement, which is simply related to the
error function, erf(•):

$$\text{erf}(z) = 1 - \text{erfc}(z) = \frac{2}{\sqrt{\pi}} \int_0^z \exp(-\eta^2)d\eta \tag{3-38}$$

The error function occurs frequently in solutions to the diffusion equation:
extensive tables of error functions are available as well as series expansions
for approximation [12]. Some values are tabulated in Appendix B. Equation 3-
37 can be used to examine the penetration of drug molecules into a tissue when
suddenly presented at a surface (Figure 3.6).

Similarly, if a drug is initially confined to a linear band of width $2w$ and
concentration c_0, the drug molecules will spread in both directions with time:

$$\frac{c_A}{c_0} = \frac{1}{2}\left[\text{erf}\left(\frac{w+x}{2\sqrt{D_A t}}\right) + \text{erf}\left(\frac{w-x}{2\sqrt{D_A t}}\right)\right] \tag{3-39}$$

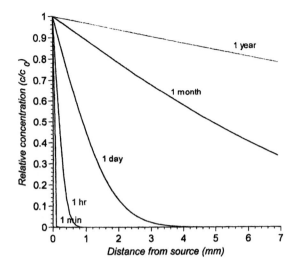

Figure 3.6
Concentration profiles for diffusion in a semi-infinite medium. When the concentration of solute is increased within one region of an unbounded space, as shown in Figure 3.4d, the molecules slowly penetrate into the adjacent region. The curves shown here are realizations of Equation 3-37 for a solute with $D_A = 10^{-7}$ cm^2/s, $R = 0.1$ cm, $N = N_{Av}$.

3.4.2 Cylindrical Coordinates

Consider the diffusion of solute A from the surface of a cylinder of radius R into a homogeneous tissue (Figure 3.4b). For example, the cylinder might represent the external surface of a capillary that contains a high concentration of a drug. The concentration within the tissue, in the region $r > R$, can be determined by solving Equation 3-31 in cylindrical coordinates:

$$D_A \frac{1}{r} \frac{\partial}{\partial r} \left[r \frac{\partial c_A}{\partial r} \right] = \frac{\partial c_A}{\partial t} \tag{3-40}$$

If the concentration of A in the tissue outside the cylinder is initially zero, and the concentration at the outer surface of the cylinder is maintained at c_0, the initial and boundary conditions can be written as:

$$\begin{aligned}
c(r, t) &= 0; &&\text{for } R \le r; &&t = 0 \\
c(r, t) &= c_0; &&\text{for } r = R; &&t > 0 \\
c(r, t) &= 0; &&\text{for } r \to \infty; &&t > 0
\end{aligned} \tag{3-41}$$

Solving Equation 3-40 subject to Equation 3-41 yields:

$$\frac{c_A}{c_0} = 1 + \frac{2}{\pi} \int_0^\infty e^{-D_A u^2 t} \frac{J_0(ur) Y_0(ua) - J_0(ua) Y_0(ur)}{J_0^2(ua) - Y_0^2(ua)} \frac{au}{u} u \tag{3-42}$$

where J_0 and Y_0 are Bessel functions of the first and second kind of order 0. Equation 3-42 is shown graphically in Figure 3.7. When the concentration of the solute is maintained at c_0, solute will continue to penetrate into the adjacent medium. For a solute with $D_A = 10^{-7}$ cm^2/s and a cylinder with $R = 0.1$ cm, the solute will penetrate ≈ 0.2 mm in the first 6 hr, ≈ 0.4 mm in the first 24 h, and ≈ 0.6 mm in the first 100 h.

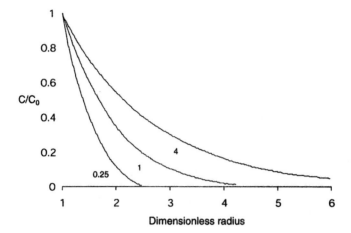

Figure 3.7 Concentration of solute diffusing from the surface of a cylinder. Redrawn from [7], p. 87. Each curve represents the concentration profile at a different dimensionless time $(D_A t/a^2)$. Dimensionless concentration is shown as a function of dimensionless distance from the center of the cylinder (r/R). For $D_A = 1 \times 10^{-7}$ cm^2/s and $R = 0.1$ cm, the dimensionless times correspond to 7, 28, and 110 h.

3.4.3 Spherical Coordinates

Biological systems often exhibit spherical symmetry, or something near it. For example, certain cells are assumed to be spherical as are vesicles within cells and synthetic vesicles. Therefore, it is frequently convenient to examine the transport of solutes in a spherical coordinate system. Equation 3-31 can be expressed in spherical coordinates:

$$D_A \frac{1}{r^2} \frac{\partial}{\partial r} \left[r^2 \frac{\partial c_A}{\partial r} \right] = \frac{\partial c_A}{\partial t} \qquad (3\text{-}43)$$

where gradients in the θ and ϕ directions are neglected due to symmetry, so that diffusion occurs only in the radial direction. This differential equation can be simplified by making the substitution $u = rc_A$, which produces:

$$D_A \frac{\partial^2 u}{\partial r^2} = \frac{\partial u}{\partial t} \qquad (3\text{-}44)$$

an equation identical to the one obtained for one-dimensional diffusion in a rectangular coordinate system, Equation 3-32.

Consider the flux of solute A towards a spherical cell in suspension (Figure 3.4c). If solute is consumed at the surface of the cell, and the rate of consumption is rapid, solute concentrations in the vicinity of the cell can be predicted by solving Equation 3-43 subject to the following conditions:

$$c_A(r, t) = c_0; \quad \text{for } r \geq R; \quad t = 0$$
$$c_A(r, t) = 0; \quad \text{for } r = R; \quad t > 0 \quad \quad (3\text{-}45)$$
$$c_A(r, t) = c_0; \quad \text{for } r \to \infty; \quad t > 0$$

where the initial concentration of solute in the medium surrounding the cell is c_0. The solution to these equations is closely related to the solution of the equations for one-dimensional diffusion in a semi-infinite medium (see Chapter 3 of reference [7]):

$$\frac{c_A}{c_0} = 1 - \left(\frac{R}{r}\right) \text{erfc}\left[\frac{r - R}{2\sqrt{D_A t}}\right] \quad \quad (3\text{-}46)$$

For long observation times, Equation 3-46 reduces to the steady-state solution:

$$\frac{c_A}{c_0} = 1 - \frac{R}{r} \quad \quad (3\text{-}47)$$

Solute concentration in the vicinity of a sphere is shown in Figure 3.8. Again, D_A is assumed to be $10^{-7} \text{cm}^2/\text{s}$ and R to be 1 mm. The change in concentration penetrates ~ 0.5 mm in the first 1 h of absorption by the sphere; as time increases, the decrease in concentration moves progressively deeper into the medium. For the conditions specified in Figure 3.8, the approach to steady

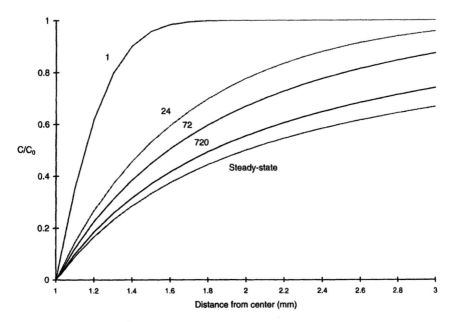

Figure 3.8   Concentration profiles in the vicinity of absorbing sphere. When a perfectly absorbing sphere is placed within an infinite medium, the concentration varies near the surface of the sphere. Concentration profiles are indicated at 1, 24, 72 and 720 h for a diffusing solute with $D_A = 10^{-7} \text{cm}^2/\text{s}$ for a sphere of radius 1 mm.

state is slow; even after 720 h (30 days) the concentration profile is still significantly different from the steady-state profile.

The rate of solute diffusion to the surface of the cell can be calculated from the flux:

$$\text{Rate of disappearance} = -D_A\left(4\pi R^2\right)\frac{dc_A}{dr}\bigg|_{r=R} = 4\pi D_A R c_0 \qquad (3\text{-}48)$$

Equation 3-48 permits the definition of a rate constant for diffusion-limited reaction at a cell surface, k_+. The rate constant, defined as (rate of disappearance) $= k_+ c_0$, is

$$k_+ = 4\pi D_A R \qquad (3\text{-}49)$$

This expression for the rate constant, first described by Smoluchowski [13], will be useful in the analysis of rates of diffusion and reaction during ligand–receptor binding (see Chapter 4).

3.5 SOLUTIONS TO THE DIFFUSION EQUATION WITH SOLUTE BINDING AND ELIMINATION

In many cases, the concentration of solute or drug within a region of tissue will depend not only on diffusion, but also on other physiological processes. For example, a solute may be generated or consumed within a cell or tissue region by a chemical reaction, usually one mediated by enzymes. Alternatively, the solute could be eliminated from the diffusion process by immobilization to some fixed element, binding to the cytoskeleton or an organelle, or by partitioning from the extracellular space into a capillary, where it enters the circulatory system. Similarly, when considering diffusion through the extracellular space of a tissue, the solute could be eliminated by enzymatic conversion to another form, immobilized by some fixed element in the extracellular space, eliminated by internalization into cells or capillaries, or generated by secretion from a cell. When a diffusing solute is generated or consumed homogeneously within some region of interest in a tissue, the differential mass balance in molar concentrations (see Table 3.2) becomes:

$$D_A \nabla^2 c_A + \psi_A^m = \frac{\partial c_A}{\partial t} \qquad (3\text{-}50)$$

where the diffusion coefficient is assumed constant, bulk flow is neglected ($v = 0$), and ψ_A^m is the molar rate of generation (or consumption) of solute per volume within the differential volume element.

3.5.1 Diffusion with Reversible Binding to Immobilized Elements

Consider a solute that diffuses within the extracellular space of the tissue, but also interacts with the tissue by reversibly binding to some fixed component of the tissue. For example, the diffusing solute might bind to a protein on the cell surface or to a protein in the extracellular matrix. When bound the solute may be considered to be completely immobilized; when released it is free to diffuse again within the extracellular space. If the conversion of the diffusible solute, A, to the bound form, B, is rapid and characterized by an equilibrium constant $K_b = c_B/c_A$, the local rate of elimination of solute A, $-\phi_A^m$, is then equal to the rate of formation of the bound form B, which is $\partial c_B/\partial t$. Therefore, this situation can be represented by a form of Equation 3-50:

$$D_A \nabla^2 c_A - \frac{\partial c_B}{\partial t} = \frac{\partial c_A}{\partial t} \tag{3-51}$$

Upon substitution of the equilibrium relationship and the assumption that K_b is constant with time, this equation becomes

$$\frac{D_A}{K_b + 1} \nabla^2 c_A = \frac{\partial c_A}{\partial t} \tag{3-52}$$

Therefore, the net effect of a homogeneous, rapid, reversible reaction is to retard the rate of diffusion of solute through the tissue. Solutions to this equation are identical to solutions of the pure diffusion equation (compare Equation 3-31 with Equation 3-52), except that the diffusion coefficient is reduced by a factor equal to the binding constant plus unity. These same equations can be used to evaluate penetration into tissues when more complicated equilibrium expressions are appropriate, by substituting the non-linear equilibrium expression into Equation 3-50 and solving the resulting equation (see [7]).

Numerical values for the term K_b, defined above, can be obtained from information on the equilibrium association (K_a) or dissociation (K_d) constants for receptor–ligand pairs. These constants are usually defined:

$$L + R \underset{k_r}{\overset{k_f}{\rightleftarrows}} L - R \tag{3-53}$$

$$K_d = \frac{1}{K_a} = \frac{[L][R]}{[L-R]} = \frac{k_r}{k_f}$$

where L is the ligand, R is the receptor and L–R is the ligand–receptor complex. Consider the case of a drug molecule, which can be assigned the role of binding ligand $[L] = c_A$, interacting with a binding site R, which is present on the surface of cells or stationary molecules in the extracellular space, to form bound complexes $[L-R] = c_B$. Assuming that the receptor concentration is constant in the tissue, the binding constant K_b, is easily related to the dissociation constant:

$$K_b = \frac{c_B}{c_A} = \frac{[L-R]}{[L]} = \frac{[R]}{K_d} \tag{3-54}$$

Some values for the dissociation constant are presented in Table 3.3.

Table 3.3   Values for receptor dissociation constants

Ligand	Receptor	$K_d(M)$	R_T	Reference
NGF	trkA	10^{-11}		[18]
NGF	p75^{NTR}	10^{-9}		[19]
Insulin	Insulin receptor	1.2×10^{-8}	100,000 to 250,000/liver cell	[20]
Biotin	Avidin	10^{-15}		
Transferrin	Transferrin receptor	3.3×10^{-8}	50,000/HepG2 cell	Reported in [21]
FNLLP	Chemotactic peptide receptor	2×10^{-8}	50,000/neutrophil	Reported in [21]
TNF	TNF receptor	1.5×10^{-10}	6,600/A549 cell	Reported in [21]
Hydroxybenzylpindolol	β-Adrenergic	1×10^{-10}	?/turkey RBC	Reported in [21]
Insulin	Insulin receptor	2.1×10^{-8}	10^5/rat fat-cell	Reported in [21]
EGF	EGF receptor	6.7×10^{-7}	25,000/fetal rat lung	Reported in [21]
Fibronectin		8.6×10^{-7}	5×10^5/fibroblast	Reported in [21]
IgE	Fcsub(epsilon)	4.8×10^{-10}	?	Reported in [21]
	β-Adrenergic	10^{-9}	1,300–1,800/frog RBC	[20]
	Cholinergic	10^{-7}	10^{11}/cell	
	Glucagon	1.5×10^{-9}	110,000/liver cell	
	TSH	1.9×10^{-9}	500/thyroid cell	

R_T is the number of receptors on the cell population. Collected from a variety of sources.

3.5.2 Diffusion with Elimination

Upon substitution of an appropriate kinetic expression for the rate of generation or consumption of solute within the tissue space, Equation 3-50 can be solved to determine concentration as a function of time and position. Full analytical solutions are generally difficult to obtain, unless both the kinetic expression and the geometry of the system are simple. For example, consider the linear diffusion of solute from an interface where the concentration is maintained constant (as in Figure 3.4d). If the diffusing solute is also eliminated from the tissue, such that the volumetric rate of elimination is first order with a characteristic rate constant k, Equation 3-51 can be reduced to:

$$D_A \frac{\partial^2 c_A}{\partial x^2} - k c_A = \frac{\partial c_A}{\partial t} \tag{3-55}$$

This equation can be solved subject to the initial and boundary conditions:

$$
\begin{aligned}
c_A(x, t) &= 0: &&\text{for } x \geq 0; &&t = 0 \\
c_a(x, t) &= c_0; &&\text{for } x = 0; &&t > 0 \\
c_A(x, t) &= 0; &&\text{for } x \to \infty; &&t > 0
\end{aligned}
\tag{3-56}
$$

to yield:

$$\frac{c}{c_0} = \frac{1}{2}\exp\left\{-x\sqrt{k/D_A}\right\}\text{erfc}\left\{\frac{x}{2\sqrt{D_A t}} - \sqrt{kt}\right\} + \frac{1}{2}\exp\left\{x\sqrt{k/D_A}\right\}$$
$$\text{erfc}\left\{\frac{x}{2\sqrt{D_A t}} + \sqrt{kt}\right\}$$

(3-57)

The corresponding steady-state solution to Equation 3-55 can be found:

$$\frac{c_A}{c_0} = \exp\left(-x\sqrt{\frac{k}{D_A}}\right)$$

(3-58)

Concentration profiles predicted by Equations 3-57 and 3-58 are shown in Figure 3.9. In panel (a), the approach to steady state is shown for the situation where $D = 1 \times 10^{-7}\,\text{cm}^2/\text{s}$ and $k = 1 \times 10^{-6}/\text{s}$. After approximately sufficient time, the transient equations are identical to the steady-state solution. In panel (b), steady-state solutions are shown for the same D, with k varying between 1×10^{-8} and $1 \times 10^{-4}/\text{s}$. Clearly, the rate of elimination of a solute from the tissue space can have a profound influence on the ability of the solute to penetrate from a localized source.

3.6 A FEW APPLICATIONS

Solutions to the diffusion equation are helpful in interpreting a variety of biological phenomena. This idea is illustrated with examples from developmental biology, drug design, and neuroscience.

Application 1: Developmental Biology. Maternal effect genes are segregated to defined regions in the developing embryo; one of these genes, which encodes a protein called bicoid, is concentrated at the anterior end of *Drosophila* embryos [14]. Bicoid produced at the anterior end diffuses toward

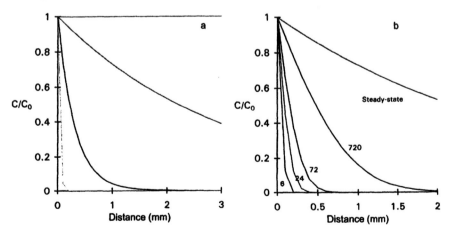

Figure 3.9 Concentration of solute diffusing with first-order elimination. (a) Steady-state profiles for $D_A = 10^{-7}\,\text{cm}^2/\text{s}$ and k varying from 0, 10^{-8}, 10^{-6}, $10^{-4}\,\text{s}^{-1}$; (b) the approach to steady state for a solute with $D_A = 10^{-7}\,\text{cm}^2/\text{s}$ and $k = 10^{-8}\,\text{s}^{-1}$.

the posterior pole; as the protein diffuses, it can be metabolized. Simultaneous diffusion and elimination produce a stable protein gradient (Figure 3.10); the cells of the embryo respond to the local concentration of bicoid by expressing certain genes. In this way, the bicoid gradient provides positional information to cells throughout the embryo. These events are critical for the formation of structures throughout the organism; they are achieved by a mechanism for gradient formation—diffusion with homogeneous elimination—which can be explained by the simple steady-state model of Equation 3-58.

Application 2: Drug Penetration in Tissue. The diffusion equation can be used to develop a simple, quantitative method for predicting the extent of drug penetration into a tissue following the introduction of a local source. Consider the simple geometry shown in Figure 3.4d, where drug is maintained at a constant value, c_0, at the interface of a semi-infinite medium. From the steady-state solution, Equation 3-58, it is possible to

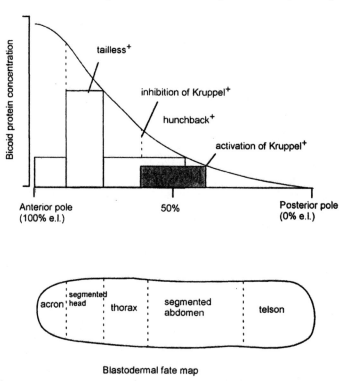

Blastodermal fate map

Figure 3.10   Concentration gradient of bicoid in the developing fruitfly. Concentration of bicoid protein has been measured in developing embryos as a function of location on the anterior–posterior axis. When present at high levels, biocoid protein activates the gene tailless[+] at intermediate levels, inhibits the gene Kruppel[+], and activates hunchback[+]. Spatial localization of gene expression is thereby achieved with a simple mechanism for protein-gradient formation.

quantify drug penetration in the local tissue. When a gradient of drug concentration is present, the region of tissue nearest the implant will be exposed to high, possibly toxic, drug levels; the region farthest from the implant will be untreated. We define the effectiveness of drug delivery, η, as the ratio of average drug concentration within a given tissue volume to drug concentration at the implant/tissue interface, c_0 (i.e., the maximum concentration of drug in the tissue):

$$\eta = \frac{\bar{c}_A}{c_0} \quad \text{where} \quad \bar{c}_A = \frac{\int_0^L c_A(x)\,dx}{\int_0^L dx} \tag{3-59}$$

where L is the distance into the tissue that requires treatment with the drug. For values of η near unity, the overall concentration profile is nearly flat, and solute delivery to this region of tissue is effective. Steep concentration profiles yield values of η near zero, or ineffective delivery.

Substitution of Equation 3-58 into Equation 3-59 gives:

$$\eta = \frac{1}{\phi}\left(1 - e^{-\phi}\right) \tag{3-60}$$

where the dimensionless parameter ϕ is defined:

$$\phi^2 = \frac{L^2 k}{D} \tag{3-61}$$

Equation 3-61 is shown graphically in Figure 3.11.[2] As ϕ increases, the rate of drug diffusion decreases with respect to the rate of elimination. This results in steep concentration gradients, or ineffective solute delivery to the tissue. Similar calculations can be performed in other geometries; these calculations are useful for evaluating the influence of physiochemical properties of the drug, which determine ϕ, on the extent of drug penetration in tissue [16].

Application 3: Neurotransmitter Diffusion Across the Synaptic Cleft. These methods can be used to predict the movement of molecules in more complex geometrical arrangements. Consider a portion of the synaptic cleft, which is shown schematically in Figure 3.12a. In this region of space between two neurons—the presynaptic neuron sending a signal and the postsynaptic neuron receiving a signal—several characteristic geometries are present: the rectangular distance across the cleft, the spherical vesicle that releases the neurotransmitter, and the cylindrical patch of receptors that sense the signal on the postsynaptic neuron. The concentration of neurotransmitter within the synaptic cleft, $n(x, y, z, t)$, can be found from the diffusion equation, obtained from Table 3.2 and written on a molar basis in three dimensions:

$$D_n\left(\frac{\partial^2}{\partial x^2} + \frac{\partial^2}{\partial y^2} + \frac{\partial^2}{\partial z^2}\right)n(x, y, z, t) + f(x, y, z, t) = \frac{\partial}{\partial t}n(x, y, z, t) \tag{3-62}$$

2. Note the similarity of this analysis to the work of Thiele on diffusion and reaction in heterogeneous catalysis [15].

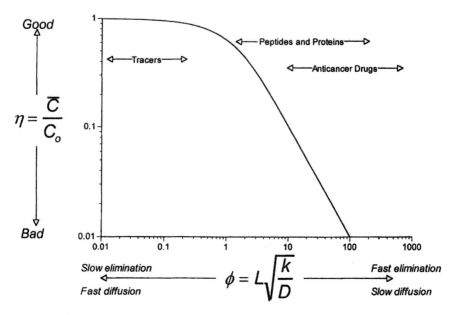

Figure 3.11   The extent of drug penetration in tissue as a function of properties of the drug. Figure redrawn from [16] showing the one-dimensional rectangular coordinate system result.

where D_n, the diffusion coefficient of the neurotransmitter, is assumed constant. The function $f(x, y, z, t)$ is a source term accounting for neurotransmitter release from a single vesicle (Figure 3.12a). A continuous source function that is consistent with existing experimental data is:

$$f(x, y, z, t) = q \cdot \exp\left\{-\frac{x^2 + y^2}{b} - \frac{z^2}{c}\right\} \cdot t^\alpha \exp[-\beta t] \qquad (3\text{-}63)$$

where q is related to the number of neurotransmitter molecules initially in the vesicle, b and c are Gaussian variances in the lateral and z-directions (limited by the size of the vesicle opening and cleft width d_{syn}), and α and β are parameters (Figure 3.12b). Neurotransmitter molecules diffuse in the cleft and do not cross the membrane boundaries:

$$\frac{\partial}{\partial z} n(x, y, z, t) = 0 \quad \text{at } z = 0 \text{ and } z = d_{syn} \qquad (3\text{-}64)$$

These equations were solved [17] to obtain:

$$n(x, y, z, t) = \frac{qb\sqrt{c}}{4D_n} \sum_{j=-\infty}^{\infty} \int_0^{4D_n t} \frac{(t - s/4D_n)^\alpha \exp\{-\beta(t - s/4D_n)\}}{\sqrt{c + s}(b + s)}$$

$$\exp\left\{-\frac{r^2}{b + s} - \frac{(z - 2jd_{syn})^2}{c + s}\right\} ds \qquad (3\text{-}65)$$

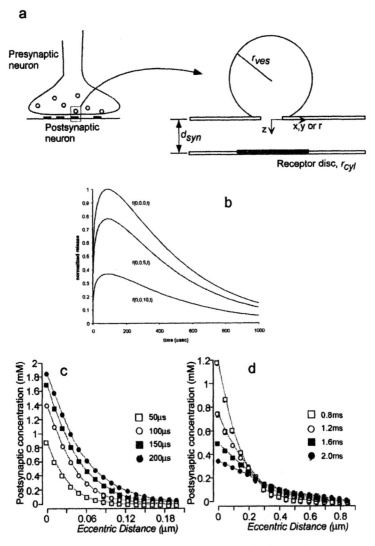

Figure 3.12 Model of diffusion in synaptic cleft. Model from 17. (a) Neurotransmitter molecules are released from vesicles in the presynaptic neuron into the synaptic cleft. The vesicle diameter r_{ves} is 20 nm and the cleft width d_{syn} is 20 nm. The origin of the coordinate system is centered at the opening of the vesicle. (b) Release of molecules from the vesicle is simulated by a source $f(x, y, z, t)$. (c) and (d) Concentration of neurotransmitter at the receptor disk, as a function of lateral displacement of the disk center from the vesicle opening. The model was evaluated using parameters that were reasonable for glutamate synapses: $D = 3 \times 10^{-7}\,\mathrm{cm^2/s}$; $r_{cyl} = 50\,\mathrm{nm}$; $Q = 2{,}000$ glutamate molecules/vesicle (concentration = 100 mM); $\beta = 1/360\,\mu s^{-1}$; $\alpha = 0.25$.

where $r = \sqrt{x^2 + y^2}$ and the coefficient q is determined by mass balance on the total number of molecules initially contained in the vesicle, Q:

$$Q = q\pi^{3/2}b\sqrt{c}\int t^\alpha \exp\{-\beta t\}\mathrm{d}t \qquad (3\text{-}66)$$

Equation 3-65 can be integrated numerically to calculate neurotransmitter concentrations within the receptor disk as a function of relative position of the disk with respect to the vesicle opening (Figure 3.12d). This model demonstrates that receptor fields on the postsynaptic membrane must be near the vesicle to receive sufficient amounts of the released transmitter chemical to become activated. Receptor disks that are laterally displaced > 200 nm from the vesicle opening will not be affected by neurotransmitter.

SUMMARY

- Molecules disperse in quiescent aqueous media by random walk migration, which is driven by thermal fluctuations in the solvent. As a result, the diffusive flux is proportional to the concentration gradient.

- Equations for describing changes in concentration with time and location can be obtained using Fick's law in combination with a differential mass balance.

- Solutions to the diffusion equation can be used to describe the movement of endogenous chemicals or drugs in physiological situations.

REFERENCES

1. Einstein, A., A new determination of molecular dimensions. *Annalen der Physik*, 1906, **19**, 289–306.
2. Berg, H.C., *Random Walks in Biology*. Princeton, NJ: Princeton University Press, 1983.
3. Whitney, C.A., *Random Process in Physical Systems*. New York: John Wiley, 1990.
4. Fick, A., Ueber Diffusion. *Ann. der Physick*, 1855, **94**, 59–86.
5. Bird, R.B., W.E. Stewart, and E.N. Lightfoot, *Transport Phenomena*. New York: John Wiley, 1960, 780 pp.
6. Welty, J.R., C.E. Wicks, and R.E. Wilson, *Fundamentals of Momentum, Heat, and Mass Transfer*. 3rd ed. New York: John Wiley, 1984.
7. Crank, J., *The Mathematics of Diffusion*. 2nd ed. Oxford: Oxford University Press, 1975, 414 pp.
8. Carslaw, H.S. and J.C. Jaeger, *Conduction of Heat in Solids*. 2nd ed. Oxford: Oxford University Press, 1959, 510 pp.
9. Popov, S. and M.M. Poo, Diffusional transport of macromolecules in developing nerve processes. *Journal of Neuroscience*, 1992, **12**(1), 77–85.
10. Radomsky, M.L., et al., Macromolecules released from polymers: diffusion into unstirred fluids. *Biomaterials*, 1990, **11**, 619–624.
11. Lauffenburger, D., C. Rothman, and S. Zigmond, Measurement of leukocyte motility and chemotaxis parameters with a linear under-agarose migration assay. *Journal of Immunology*, 1983, **131**, 940–947.

12. Abramowitz, M. and I.A. Stegun, *Handbook of Mathematical Functions with Formulas, Graphs, and Mathematical Tables*. Washington, DC: National Bureau of Standards, 1964, 1046 pp.

13. Smoluchowski, M., Versuch einer mathematischen Theorie der Koagulationskinetik kolloider Losungen. *Z. Physik. Chem.*, 1917, **92**(9), 129–168.

14. Driever, W. and C. Nusslein-Volhard, A gradient of bicoid protein in Drosophila embryos. *Cell*, 1988, **54**, 83–93.

15. Thiele, E.W., Relation between catalytic activity and size of particle. *Industrial and Engineering Chemistry*, 1939, **31**(7), 916–920.

16. Saltzman, W.M. and M.L. Radomsky, Drugs released from polymers: diffusion and elimination in brain tissue. *Chemical Engineering Science*, 1991, **46**, 2429–2444.

17. Kleinle, J., *et al.*, Transmitter concentration profiles in the synaptic cleft: an analytical model of release and diffusion. *Biophysical Journal*, 1996, **71**, 2413–2426.

18. Kaplan, D., *et al.*, The *trk* proto-oncogene product: a signal transducing receptor for nerve growth factor. *Science*, 1991, **252**, 554–557.

19. Dobrowsky, R.T., *et al.*, Activation of the sphingomyelin cycle through the low-affinity neurotrophin receptor. *Science*, 1994, 1596–1599.

20. Cooper, J.R., F.E. Bloom, and R.H. Roth, *The Biochemical Basis of Neuropharmacology*, 7th ed. New York: Oxford University Press, 1996.

21. Lauffenburger, D.A. and J.J. Linderman, *Receptors: Models for Binding, Trafficking, and Signaling*. New York: Oxford University Press, 1993, 365 pp.

4

Diffusion in Biological Systems

"Faith" is a fine invention
When Gentlemen can see—
But *Microscopes* are prudent
In an Emergency.

Emily Dickinson, *Number 185* (ca. 1860)

Drug diffusion is an essential mechanism for drug dispersion throughout biological systems. Diffusion is fundamental to the migration of agents in the body and, as we will see in Chapter 9, diffusion can be used as a reliable mechanism for drug delivery. The rate of diffusion (i.e., the diffusion coefficient) depends on the architecture of the diffusing molecule. In the previous chapter a hypothetical solute with a diffusion coefficient of 10^{-7} cm^2/s was used to describe the kinetics of diffusional spread throughout a region. Therapeutic agents have a multitude of sizes and shapes and, hence, diffusion coefficients vary in ways that are not easily predictable.

Variability in the properties of agents is not the only difficulty in predicting rates of diffusion. Biological tissues present diverse resistances to molecular diffusion. Resistance to diffusion also depends on architecture: tissue composition, structure, and homogeneity are important variables. This chapter explores the variation in diffusion coefficient for molecules of different size and structure in physiological environments. The first section reviews some of the most important methods used to measure diffusion coefficients, while subsequent sections describe experimental measurements in media of increasing complexity: water, membranes, cells, and tissues.

4.1 MEASUREMENT OF DIFFUSION COEFFICIENTS

Diffusion coefficients are usually measured by observing changes in solute concentration with time and/or position. In most situations, concentration changes are monitored in laboratory systems of simple geometry; equally simple models (such as the ones developed in Chapter 3) can then be used to determine the diffusion coefficient. However, in biological systems, diffusion almost always occurs in concert with other phenomena that also influence solute concentration, such as bulk motion of fluid or chemical reaction. Therefore, experimental conditions that isolate diffusion—by eliminating or reducing fluid flows, chemical reactions, or metabolism—are often employed.

Certain agents are eliminated from a tissue so slowly that the rate of elimination is negligible compared to the rate of dispersion. These molecules can be used as "tracers" to probe mechanisms of dispersion in the tissue, provided that elimination is negligible during the period of measurement. Frequently used tracers include sucrose [1, 2], iodoantipyrene [3], inulin [1], and size-fractionated dextran [3, 4]. Since the tracer is not eliminated, the diffusion coefficient is obtained by measuring time-dependent spatial concentration patterns after controlled introduction of the tracer. Quantitative autoradiography and fluorescence microscopy of labeled drug analogs are convenient methods for accomplishing this measurement, although care must be exercised to ensure that elimination is truly negligible and to verify that the fluorescent or radioactive label remains associated with the drug.

Consider an application of this approach, in which fluorescence microscopy was used to measure D for proteins in water (Figure 4.1). Polymer matrices containing fluorescently labeled protein were "implanted" into capillary tubes filled with buffered water, which were periodically scanned by fluorescence microscopy to obtain concentration profiles [5, 6]. Since proteins were constrained within the volume of the tube, which was kept absolutely still during the measurement, diffusion was the only mechanism for protein migration in the tube. Proteins were also stable under the conditions of the experi-

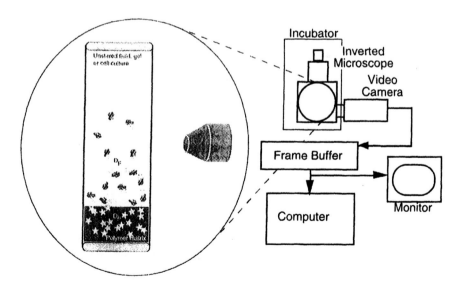

Figure 4.1   Diffusion coefficients estimated by fluorescence microscopy. Adapted from [6].

ment; therefore, diffusion coefficients for these "tracers" were obtained by fitting an appropriate solution to the diffusion equation to measured concentration profiles (the solution is similar to Equation 3-37, shown in Figure 3.6). This technique yields D for the probes in water. Autoradiographic and fluorescence techniques have been applied in a similar fashion to obtain D for tracers in cultured cell aggregates [7], cytoplasm of cultured neurons [8–11], brain slices [12], and whole animals [13, 14].

The diffusion of radiolabeled compounds within tissues is frequently measured by quantitative autoradiography [2, 3, 15–17]. In this technique, tissue sections containing radiolabeled compounds are placed next to photographic emulsions; the intensity of exposure at different locations in the emulsion is related to the local concentration of radiolabel at the corresponding position within the section. When this technique is used with a radiolabeled protein, estimation of the diffusion coefficient is complicated by the fact that the protein is also metabolized during the time it takes for molecules to move sufficient distances by diffusion. As a result, diffusion coefficients must be estimated by comparing measured profiles to solutions of the diffusion equation that account for simultaneous elimination, such as Equation 3-58, which is plotted in Figure 3.9. This technique is illustrated in Figure 4.2.

Autoradiography has several advantages for examination of diffusion in tissues. Low tracer concentrations can be measured; in addition, inhomogeneities in diffusion in different anatomical regions of the tissue can be observed visually. But the technique also has disadvantages. A large number of animals must be examined, because autoradiography requires the use of a single animal

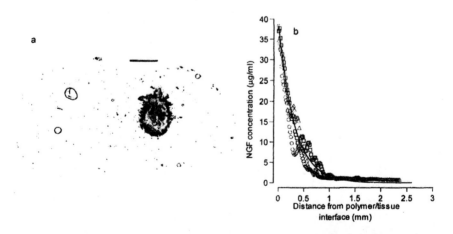

Figure 4.2   Estimation of diffusion coefficients by autoradiography of tissue sections. (a) Digitized autoradiographic image of radiolabeled NGF released from a polymeric implant in a rat brain, reproduced from [18]. (b) The resulting concentration profile, obtained by scanning intensities from the autoradiographic image, and using calibrated standards to determine the local concentration; the solid line indicates Equation 3-58.

for each measurement. These methods are slow, since concentration differences must be measured at spatial separations of $\approx 100\,\mu m$; the characteristic time for diffusion ($t = L^2/D$) in these circumstances is several hours. In addition, autoradiographic images can take weeks to develop, although rapid phosphor-imaging methods are becoming more widely used [19]. Rapid changes in diffusion rates, such as one might expect to observe in a dynamic environment such as a living tissue, usually cannot be resolved.

Fluorescence microscopy, which has been applied by Jain and co-workers in their studies of interstitial diffusion [20, 21] and lymphatic flow, can be used for measurements within the tissue of a living animal, provided that the tissue can be accessed by light. This access can sometimes be obtained by installing window chambers in the tissue [22]. Multi-photon fluorescence imaging, an important new technique introduced by Webb and colleagues [23–25], promises to broaden the applications of this technique, since quantitative fluorescence imaging can be performed in three-dimensional specimens, even specimens that scatter light.

Elimination in the tissue usually cannot be totally neglected. Elimination occurs over periods ranging from minutes to hours, which is the same time scale as the measurement techniques described above. If rapid measurement techniques are used, then elimination can be neglected. Fluorescence photo-bleaching recovery (FPR), also known as fluorescence recovery after photo-bleaching (FRAP), is an alternative technique for measuring dispersion of fluorescent molecules within a microscopic tissue volume.[1] FPR techniques were developed as a method for probing the mobility of molecules in cell membranes (for review, see [26]), and remain important tools for examining membrane dynamics [27, 28]. FPR has also been used to examine the diffusion of intracellular tracers [29, 30]. Fluorescence techniques, involving either measurement of concentration profiles within certain transparent tissues or FPR, can be used to estimate diffusion coefficients within the tissues of living organisms [20, 21, 31, 32].

FPR measurements are rapid, so dynamic processes are examined in real time. The most important limitation of the method is interference from out-of-focus light and photodamage of the diffusing molecule or the medium due to the high intensity during the photobleach [33]. An improved technique uses two photons—with twice the wavelength and thus one-half the energy—for excitation [24, 34]. This modification reduces photodamage outside of the focal region. Further, only fluorescently labeled molecules inside the focal volume receive enough energy to fluoresce, which reduces scatter from out-of-focus light and improves the signal-to-noise ratio. FPR has been used to measure diffusion coefficients in water [35], agarose gels [36], cells in culture [37], living tissue slices [38], and animals [39].

An alternative method for determining diffusion coefficients, pioneered by Nicholson [40], employs ion-selective microelectrodes (Figure 4.3). A marker

---

1. A more complete description of FPR is provided in Chapter 5.

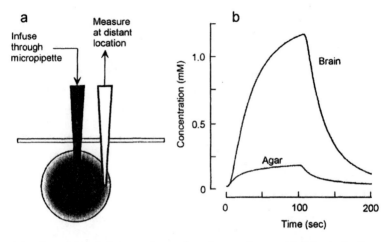

Figure 4.3 Sample trace from an iontophoresis experiment. (a) Schematic diagram of an instrument for making an iontophoretic measurement, reproduced from [42]. (b) The concentration–time profile resulting from iontophoresis in an agarose gel and brain tissue, from [45]. Diffusion coefficients are obtained by comparing the experimental data to solutions of the diffusion equation evaluated at a fixed distance. For example, Equation 3-36 could be evaluated at position $x = x^*$ to obtain a function c_A (t) that is compared to the experimental data. In practice, the solution would be obtained using a coordinate system, boundary conditions, and initial condition that were appropriate for the experimental system under study.

ion is ejected from a micropipette using either an iontophoretic current or pressure injection. The ion migrates away from the source; concentration is measured by an ion-selective microelectrode positioned 50–150 μm away. Concentration–time profiles are plotted and a suitable diffusion equation, which neglects elimination and assumes the tissue is homogeneous and isotropic, is adjusted to represent best the data: the volume of distribution and diffusion coefficient are obtained in this procedure [41]. Iontophoretic measurements are rapid, so sudden changes in the tissue dynamics, such as those that occur during anoxia [42], can be resolved. This technique has been used to measure the diffusion coefficients of ions, most frequently tetramethylammonium (TMA), in agarose gels [42], living brain slices [43], and animals [41, 42]. Iontophoresis is moderately sensitive to ions such as dopamine [44] ($\sim$ 20 ng/mL), but the major drawback of iontophoresis is the limited number of ions that can be detected.

Magnetic resonance (MR) imaging may also be useful for measuring diffusion coefficients of certain compounds within tissues [46]. MR imaging has important advantages, particularly since it is non-invasive; therefore, serial measurements can be performed on the same animal. Unfortunately, the technique is limited by the requirement that all compounds must contain a strong paramagnetic group to be detected.

Table 4.1 Survey of methods used to measure diffusion coefficients in liquids

Method	Reference
Quasielastic light scattering spectroscopy	[65, 135, 136]
Stabilization of an inverse density gradient	[137]
Boundary spreading	[138, 139]
Diffusion between reservoirs separated by macroporous or microporous membranes	[64, 140]
Measurement of transient concentration profiles of deuterated polymers in polymer blends using forward recoil spectroscopy	[141]
Measurement of transient concentration profiles of fluorescently labeled molecules in capillary tubes using epifluorescence microscopy	[5, 6]
Measurement of transient concentration profiles of fluorescently labeled molecules in biological tissues using epifluorescence microscopy	[20]
Fluorescence recovery (or redistribution) after photobleaching (FPR)	[26, 35, 142]
NMR measurement of solvent self-diffusion coefficients in polymer solutions	[143]
NMR measurement of protein diffusion coefficients in solution and in synthetic membranes	[66]
In situ determination of diffusion in gels using radio-tracers	[144]
Holographic relaxation spectroscopy for measuring protein diffusion coefficients in polyacrylamide gels	[145]
Holographic interferometry	[54]
Spreading of DNA within agarose gels	[55]

Some of the methods that have been used to measure diffusion coefficients in liquids are listed, with representative references.

A variety of other techniques have been used to measure the diffusion coefficients in specific situations; some of these methods are collected in Table 4.1. Many of these techniques are now routinely used to measure the diffusion coefficients of biologically important molecules, like proteins, in buffered water or in native biological fluids.

4.2 DIFFUSION IN WATER

For a solute diffusing in water, the diffusion coefficient depends primarily on temperature and solute size. Consider a particle moving through a simple Newtonian fluid, such as water. The velocity of the particle in the x-direction, v_x, is related to the force required to move it through the fluid, F_x:

$$v_x = \frac{F_x}{f} \qquad (4\text{-}1)$$

where f is the frictional drag coefficient. Einstein demonstrated that the diffusion coefficient is related to the absolute temperature, T, and the frictional drag coefficient [47]:

$$D_A = \frac{k_B T}{f} \qquad (4\text{-}2)$$

where k_B is Boltzmann's constant, 1.38054×10^{-23} J/K. Equation 4-2, also known as the Nernst–Einstein equation, can be used to predict the diffusion coefficient for particles with known size and shape. For example, in the case of spherical particles moving through a continuum, the equations of fluid motion can be solved to yield an analytical relationship between the particle radius, a, and the frictional drag coefficient:

$$f = 6\pi\mu a \qquad (4\text{-}3)$$

where μ is the viscosity of the fluid. Combining these last two expressions yields the most commonly used equation for predicting the diffusion coefficients of large particles in a fluid:

$$D_A = \frac{k_B T}{6\pi\mu a} \qquad (4\text{-}4)$$

which is known as the Stokes–Einstein equation. It is widely used for predicting the diffusion coefficients of proteins based on their size, or in the event that the diffusion coefficient is known but the molecular size unknown, for predicting the size of proteins. When a for a solute is calculated from a known D_A, the resulting radius is usually called a hydrodynamic radius or Stokes–Einstein radius, to distinguish it from molecular sizes measured using static techniques (e.g., microscopy).

While the Stokes–Einstein equation is strictly applicable only in cases where the diffusing particle is large when compared to the surrounding solvent molecules (so that the fluid can be considered a continuum), it has proven to be useful for solute–solvent pairs in which the radius, a, is only two to three times the solvent radius. For a solute with radius comparable to the solvent radius, the 6 in Equations 4-3 and 4-4 should be replaced by a 4, since the assumption of "no slip" at the solute surface is no longer valid [48].

4.2.1 Diffusion of Proteins and Nucleic Acids in Water

Figure 4.4 shows diffusion coefficients for a variety of proteins plotted as a function of molecular weight. In all cases, the protein molecules are much larger than the solvent molecules, which are of water, so the assumptions underlying Equation 4-4 should be satisfied. For spherical particles, the molecular radius is proportional to the cube root of molecular volume: $\sqrt[3]{\tilde{V}/N_{Av}}$, where $\tilde{V}$ is the molar volume of the protein. Since the molecular volume of a globular protein increases in direct proportion to M_w, Equation 4-4 can be rewritten as a power law [49]:

$$D_A = A(M_w)^{-\frac{1}{3}} \qquad (4\text{-}5)$$

where A is a constant. While other equations for correlating protein diffusion coefficients have been evaluated (for a review of some typical correlations, see [50]), none are significantly better than Equation 4-5. The dashed line in Figure 4.4 indicates another useful correlation for predicting protein diffusion coefficients [51]:

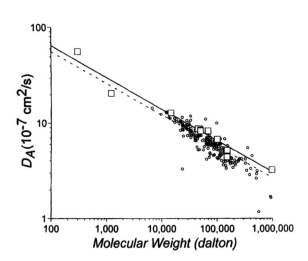

$$D_A = \frac{9.40 \times 10^{-15}T}{\mu M_w^{1/3}} \tag{4-6}$$

where D_A is in cm^2/s if T is in K, μ is in Pa$\cdot$s, and M_w is in daltons.

Protein diffusion coefficients depend on the pH and ionic composition of the medium. For example, in a 1% bovine serum albumin (BSA) solution, the diffusion coefficient of BSA increased by a factor of 4 when KCl concentration in the solution was decreased below 0.01 M[54].

Unlike globular proteins, fibrous proteins and nucleic acids do not behave as spheres. In some cases, the approach described above can be extended to particles of other shapes, provided that the frictional drag coefficient can be determined. For example, the drag coefficient for prolate ellipsoids with a major axis length, a, much larger than the minor axis length, b ($a^2 \gg b^2$) has been determined:

$$f = \frac{6\pi\mu a}{\ln(2a/b)} \tag{4-7}$$

Alternatively, empirical correlations based on the functional form predicted by the Stokes–Einstein equation can be employed (Figure 4.5). An empirical correlation is frequently used for linear DNA molecules with M_w up to 4×10^6 Da diffusing in water [55]:

$$D_A = \frac{0.116RT}{(1 - \tilde{V}\rho)M_w^{0.675}} \tag{4-8}$$

Equation 4-8 was obtained from a correlation of sedimentation coefficients with molecular weight. The sedimentation coefficient is related to the diffusion coefficient by the Svedberg equation:

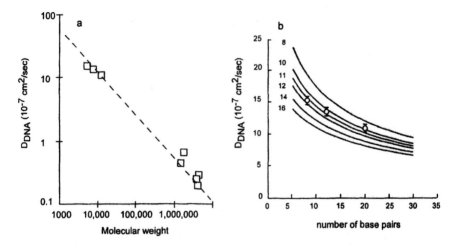

Figure 4.5   Diffusion coefficients for DNA molecules in water. (a) DNA diffusion data are collected in Table 4.2. Dashed line represents the correlation indicated above (Equation 4-8). (b) DNA diffusion coefficients predicted as a function of molecular weight in base pairs. Redrawn from [57].

$$D_A = \frac{s_0 RT}{M_w(1 - \tilde{V}\rho)} \qquad (4\text{-}9)$$

where s_0 is the sedimentation coefficient extrapolated to zero concentration, ρ is the density of water, and $\tilde{V}$ is the specific volume of the molecule of interest, which in this case is DNA. At 20°C, the specific volume of DNA is 0.55 mL/g and the density of water is 0.9982 g/mL; this same expression can be used to estimate protein diffusion coefficients from sedimentation data by using the specific volume of proteins: 0.74 mL/g [56].

Linear DNA molecules in solution appear to behave as random coils with a significant persistence length (p, a measure of the distance along the chain between changes in direction). Therefore, a globular protein of equivalent molecular weight is significantly more compact than a nucleic acid molecule. For example, the radius of gyration (R_G) for DNA can be predicted from:

$$R_G = \left[\frac{p}{3}(L - 3p)\left(\frac{L}{2p}\right)^{\epsilon}\right]^{\frac{1}{2}} \qquad (4\text{-}10)$$

where L is the contour length of the chain and the exponent ϵ is usually assumed to be ~ 0.05 for DNA in ≈ 1 mM salt solutions. Under these conditions, the persistence length of linear DNA is ~ 50 nm, which produces the values for R_G collected in Table 4.2.

The effect of polymer structure on the hydrodynamic radius is illustrated in Figure 4.6. The hydrodynamic radius was calculated from diffusion coefficients in water (using Equation 4-4) for globular proteins, DNA,

Table 4.2 Comparison of the molecular sizes of DNA and protein molecules of equivalent molecular weight

DNA	Molecular weight		Molecular shape and size		$D_A(10^{-7}\,cm^2/s)$
	Length (bp)	$M_w{}^a$	L (nm)	R_G (nm)	
d(CGCGCGCG)	8	5,200	2.9	1.4	15.3
> d(CGCGCGCGCGCG)	12	7,800	4.2	1.7	13.4
d(CGTACTAGTTAAC TAGTACG)	20	13,000	6.9	2.1	10.9
Linear pLH2311	2,311	1.5×10^6		104.4	0.442 [67]
Linear pUC19	2,686	1.75×10^6		119	0.38[b]
Linear pBR322	4,363	2.84×10^6		160	0.27[b]
Linear pBR325	5,996	3.90×10^6		191	0.22[b]
SV40[c]	5,100	3.3×10^6	1700		
Linear ColE1		4.35×10^6			0.216 [146]
Linear ColE1		4.3×10^6			0.198 [147]
Relaxed ColE1		4.15×10^6			0.245
Supercoil ColE1		4.59×10^6			0.289
Single-stranded Fd DNA		1.87×10^6			0.663 [148]
E. coli chromosome[c]	4×10^6	2.6×10^9	14×10^5		

Protein	M_w	R_H (nm)	D at 25 °C $(10^{-7}\,cm^2/s)$
FNNLP	1,200	1.2	21
Lactalbumin	14,500	1.9	13
Ovalbumin	45,000	2.8	8.7
Bovine serum albumin	68,000	3.0	8.3
Human IgG	150,000	5.5	5.2
Human IgA	150,000	4.7	4.4
IgM	$\sim 1 \times 10^6$	7.5	3.2
Glutamate dehydrogenase, bovine liver	$\sim 1 \times 10^6$		2.5
Turnip yellow mosiac virus protein	$\sim 3 \times 10^6$		1.5

DNA sizes, R_G, were determined as described in [55]; hydrodynamic radii for proteins were determined from diffusion coefficients (see Figure 4.4) by using the Stokes–Einstein equation (Equation 4-4). Diffusion coefficients are at 20 °C, unless otherwise indicated.

[a] Assuming 650 Da per base pair.

[b] Estimated as described in [55].

[c] E. coli and SV40 from [58], pp. 82–83. Others from [59].

and molecular weight fractions of dextran. For a polymer with given molecular weight, the random coil dextran and DNA polymers diffuse more slowly and, therefore, have a larger effective radius than globular proteins. On a per mass basis, dextran and DNA are less compact than globular proteins.

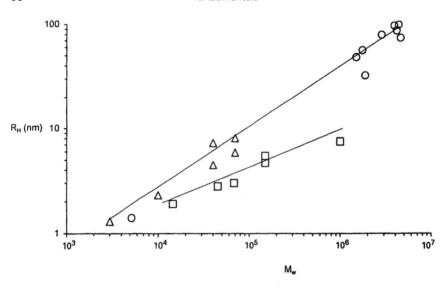

Figure 4.6 Hydrodynamic radius for globular proteins, dextran and DNA polymers. Diffusion coefficients were obtained as described in Table 4.2. The hydrodynamic radius was calculated for proteins (squares), DNA (circles), and dextran (triangles).

4.2.2 Diffusion of Small Molecules in Water

For diffusing solute molecules that are identical to the solvent, it is reasonable to assume that the molecules in the liquid are organized on a cubic lattice, such that every molecule is in contact with six identical nearest neighbors. In this case, the distance between particle centers is simply twice the molecular radius, a, which can be determined from the molar volume, $\tilde{V}$, and Avogadro's number, N_{Av}. Replacing this factor for radius, and using the coefficient 4 (rather than 6) since the molecules are all the same size, Equation 4-4 becomes:

$$D_{AA} = \frac{k_B T}{2\pi\mu} \sqrt[3]{\frac{N_{Av}}{\tilde{V}}} \tag{4-11}$$

where the double subscript indicates that this is the self-diffusion coefficient for solvent molecules,[2] which is accurate to within $\pm 12\%$ for a variety of liquids [48].

Empirical correlations are often used to predict the diffusion coefficient when the molecular radii of the solute and the solvent are similar (Figure 4.7); a good correlation for dilute liquid solutions is:

---

2. "Self"-diffusion strictly applies to the diffusion of solvent molecules within a solvent environment; i.e., all molecules are identical. It can be used, however, to estimate "tracer" diffusion coefficients for infinitely dilute solutions in which the solute and solvent are similar.

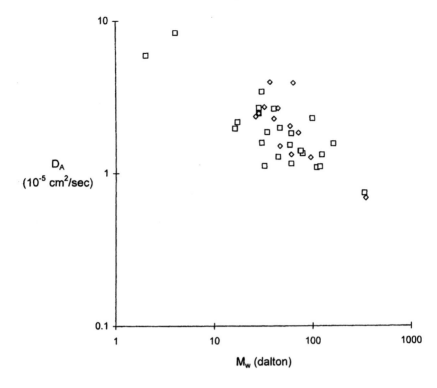

Figure 4.7 Diffusion coefficients for a variety of small molecules in water. See Table B.9 for values. Note the considerable variation in the diffusion coefficient, even for solutes with the same molecular weight.

$$D_A = 7.4 \times 10^{-8} \frac{T\sqrt{\psi_{solvent}M_{w,solvent}}}{\mu \tilde{V}_A} \tag{4-12}$$

where D_A has units of cm²/s if T is in K, μ is in centipoise, and $\tilde{V}$ is in cm³/mol. For water, the association constant $\psi_{solvent}$ is 2.6, and Equation 4-12 provides diffusion coefficients within ±10% [60].

4.2.3 Multi-component Diffusion

In the preceding sections, equations for the diffusive flux are based upon binary diffusion; a single solute is diffusing through a solvent medium. Although biological fluids contain many components, we assumed that the diffusion of a particular solute could be approximated by treating all of the other components as an effective "solvent." This assumption is adequate for most purposes, but in some situations it is useful to consider the equations for diffusion in multi-component systems. Following the discussions provided in other sources [48, 61], and writing the flux equations in molar terms, Fick's law for a multi-component system can be written as an extension of Equation 3-19 (see also Table 3.2 and note that this is the molar flux):

$$J_i = c_i(\overline{v}_i - \overline{V}) = \sum_{j=1}^{n-1} -D_{ij}\nabla c_j \qquad (4\text{-}13)$$

where D_{ij} are the multi-component diffusion coefficients [61]. In general, the relationship between the multi-component diffusion coefficients and the binary diffusion coefficients (D_A in the discussions above) is known only in a few cases, such as ternary systems of ideal gases. For more complicated systems, the main-term coefficients, D_{ii}, are approximately equal to corresponding binary diffusion coefficients, while the cross-term coefficients, D_{ij}, are typically much smaller and not symmetrical ($D_{ij} \neq D_{ji}$). These cross-terms can be important, however, since they describe the flux of one component due to gradients of other components: i.e., D_{ij} measures the influence of gradients of j on the flux of component i. Note that Equation 4-13 implies $(n-1)^2$ diffusion coefficients for an n-component system, since one of the components is defined as the solvent, and therefore its presence is implied in each of the D_{ij}. Using theories from irreversible thermodynamics, supported by appropriate experiments, it can be shown that the diffusion coefficients can be expressed as:

$$D_{ij} = \sum_{k=1}^{n-1}\sum_{l=1}^{n-1} L_{ik}\left(\delta_{kl} + \frac{c_l \tilde{V}_k}{c_n \tilde{V}_n}\right)\left(\frac{\partial \mu_l}{\partial c_j}\right)_{c_{m\neq j,n}} \qquad (4\text{-}14)$$

where the subscript n refers to the solvent, the number of other components is $(n-1)$, μ_l is the chemical potential of component l and δ_{kl} is the Kronecker delta ($\delta_{kl} = 0$ or 1 when $k \neq l$ or $k = l$, respectively). The coefficients L_{ij} are Onsager coefficients that obey the reciprocal relation:

$$L_{ij} = L_{ji} \qquad (4\text{-}15)$$

In principle, Equations 4-14 and 4-15 could be used to determine the relationship between the cross-term diffusion coefficients, although the thermodynamic information necessary to evaluate the partial derivative of chemical potential with concentration is usually not available.

Multi-component diffusion coefficients have been determined in certain liquids of biological interest [61]. For example, the ternary system KCl ($0.06\,\text{g/cm}^3$)/sucrose ($0.03\,\text{g/cm}^3$)/water—components 1, 2, and 3, respectively—has the following coefficient at 25 °C: $D_{11} = 1.78 \times 10^{-5}\,\text{cm}^2/\text{s}$, $D_{12} = 0.02 \times 10^{-5}\,\text{cm}^2/\text{s}$, $D_{21} = 0.07 \times 10^{-5}\,\text{cm}^2/\text{s}$, and $D_{22} = 0.50 \times 10^{-5}\,\text{cm}^2/\text{s}$. In this system, the cross-terms are much smaller than the main terms and the system could be adequately described using binary diffusion coefficients. In other systems, particularly when there are ionic or other interactions between the diffusing species, the cross-terms are significant. In equimolar mixtures of hexadecane/dodecane/hexane, for example, the cross-terms are $\sim 25\%$ of the main terms, and should be included in quantitative descriptions of diffusion in this system.

4.2.4 Tracer Diffusion versus Mutual Diffusion

The diffusion coefficients used to describe multi-component diffusion are *mutual* diffusion coefficients. In the multi-component system, mutual diffusion coefficients are defined by Equation 4-13; the matrix of diffusion coefficients depends on the concentration of individual components. The diffusion coefficients used in the earlier sections of the chapter, however, describe solute molecules diffusing in a medium at infinite dilution. The isolated molecule is called a tracer; these *tracer* diffusion coefficients are defined by the physics of random walk processes, as described in Chapter 3. The self-diffusion coefficient, used in Equation 4-11, is a tracer diffusion coefficient in the situation where all of the molecules in the system are identical. The self-diffusion coefficient, D_{AA}, is defined by (recall Equation 3-12) [62]:

$$D_{AA} = \lim_{t \to \infty} \frac{\langle [r(t - t_0) - r(t_0)]^2 \rangle}{6t} \tag{4-16}$$

A tracer diffusion coefficient is defined in a similar fashion:

$$D_A^* = \lim_{t \to \infty} \frac{\langle [r^*(t - t_0) - r^*(t_0)]^2 \rangle}{6t} \tag{4-17}$$

where the superscript "$*$" is a reminder that the "tracer" is infinitely dilute and surrounded by dissimilar solute molecules.

In real situations, where the concentration of solute is finite, the mutual diffusion coefficient is often the relevant measure of transport rate. Mutual diffusion coefficients provide a quantitative measure of the rate of molecular diffusion when gradients are present: i.e., when solute and solvent molecules are both diffusing in an attempt to eliminate differences in chemical potential. The mutual diffusion coefficient is defined by Fick's law (recall Equation 3-19):

$$D_A = \frac{-\bar{j}_A}{\rho \nabla \omega_A} \tag{4-18}$$

Mutual diffusion coefficients are measured experimentally by observing the change in concentration as the gradient dissipates. Tracer diffusion coefficients, on the other hand, are measured by adding a small quantity of a labeled tracer to a solution where the overall solute concentration is uniform; this tracer diffusion coefficient is associated with the displacement of particles randomly migrating without any macroscopic concentration gradients.

There is no simple relationship that permits calculation of the mutual diffusion coefficient from the tracer diffusion coefficient for a binary mixture; the diffusion coefficients represent molecular mixing in different physical situations. However, the mutual diffusion coefficient should reduce to the tracer diffusion coefficient as the solute concentration approaches zero. For example, a frequently used formula for determining the mutual diffusion coefficient from the tracer diffusion coefficients is:

$$D_A = D_A^* \left(1 + \frac{\partial \ln \gamma_A}{\partial \ln c_A}\right) \tag{4-19}$$

where γ_A is the activity coefficient. This expression can be obtained by thermodynamic arguments, ignoring the effect of intraparticle friction coefficients [63]. The application of Equation 4-19, or similar expressions, is further limited by the availability of the thermodynamic information necessary to determine the derivative of activity coefficient, γ_A, with concentration.

In tracer diffusion, however, where the attention is focused upon individual solute molecules that are labeled in some manner, molecular interactions between solvent and tracer, solvent and unlabeled solute, and tracer and unlabeled solute will all contribute to the observed rate of tracer diffusion. Tracer diffusion can be analyzed by considering a ternary system: solvent, solute (A), and labeled solute (tracer, A^*). The molar fluxes of the individual solute species can be written directly from Equation 4-13:

$$\begin{aligned} J_A &= -D_{AA}\nabla c_A - D_{AA} \cdot \nabla c_{A^*} \\ J_{A^*} &= -D_{A^*A}\nabla c_A - D_{A^*A^*}\nabla c_{A^*} \end{aligned} \tag{4-20}$$

In a typical tracer diffusion experiment, the total concentration of labeled and unlabeled solute is uniform throughout the system, so that the fluxes and the local concentration gradients are equal and opposite, and the tracer diffusion coefficient, $D_{A^*}^*$, is equal to the main-term diffusion coefficients minus a cross-term [64]:

$$D_{A^*}^* = D_{A^*A^*} - D_{A^*A} = D_{AA} - D_{AA^*} \tag{4-21}$$

To identify the differences between mutual diffusion coefficients and tracer diffusion coefficients, consider the mutual diffusion of A or A^* in water. Since these solutes are chemically indistinguishable, and therefore interact with the solvent in the same way, the binary mutual diffusion coefficients, D_A and D_{A^*}, are identical [61]:

$$D_A(\sigma_{Aw}, \epsilon_{Aw}) = D_{A^*}(\sigma_{A^*w}, \epsilon_{A^*w}) \tag{4-22}$$

and equal to the main-term coefficients plus a cross-term [64]:

$$D_A = D_{AA} + D_{A^*A} \tag{4-23}$$

In addition, for chemically identical species A and A^*, the Onsager reciprocal relations, Equations 4-14 and 4-15, yield [64]:

$$\frac{D_{AA^*}}{c_A} = \frac{D_{A^*A}}{c_{A^*}} \tag{4-24}$$

The tracer diffusion coefficient in Equation 4-21 is less than the mutual diffusion coefficient in Equation 4-24 by a factor equal to the sum of the cross-term diffusion coefficients defined in Equation 4-15. If the cross-term diffusion coefficients are small, then the tracer and mutual diffusion coefficients are approximately equal.

4.2.5 Diffusion Coefficients as a Function of Concentration

Diffusion coefficients depend on solute concentration, as described in the previous section. When diffusion coefficients for albumin and hemoglobin were measured using a diffusion-cell technique, the diffusion coefficients were indistinguishable for a range of concentrations corresponding to protein volume fractions from 0 to 20% [64] (Figure 4.8a,b). Using light scattering, however, mutual diffusion coefficients for BSA were found to be substantially lower than tracer diffusion coefficients at high concentrations [65] (Figure 4.8c). Diffusion coefficients for other proteins have been measured as a function of concentration; for example, tracer diffusion coefficients for ovalbumin as a function of concentration were measured by pulsed-field gradient NMR [66].

Diffusion of DNA molecules in solution is also concentration dependent. The diffusion coefficient for short DNA duplexes (8 to 20 base pairs) increases slightly with concentration over the range 0 to 20 mg/mL [59]. The diffusion coefficient for linear DNA fragments also increases with concentration: for example, the diffusion coefficient for a 2,311 base pair restriction fragment increases over the concentration range 0 to 450 μg/mL according to:

$$D_A(c) = D_{A,0}(1 + A_d\phi_v + \cdots) \tag{4-25}$$

where A_d is the virial concentration coefficient and ϕ_v is the volume fraction of DNA in solution. The volume fraction is related to the concentration by:

$$\phi_v = \frac{N_{Av}cV_h}{M_w} \tag{4-26}$$

where V_h is the partial molar volume of the DNA. A number of studies suggest that the virial coefficient for DNA fragments of M_w between 1×10^6 and 5×10^6 is in the range 1.12 to 1.42 [67], near the value predicted for particles with hard-sphere interactions (1.45) [68].

4.3 DIFFUSION IN POLYMER SOLUTIONS AND GELS

Most biological systems are predominantly water, with other components conferring important structural and mechanical properties. The complexity of the fluid can have a substantial impact on rates of diffusional transport. For example, Chapter 5 discusses the consequences of having self-organized phospholipid phases (i.e., membrane bilayers) in systems that are primarily composed of water. Membranes separate the medium into smaller aqueous compartments, which remain distinct because the membrane permits the diffusion of only certain types of molecules between the compartments. Complex fluid phases have diverse roles in biological systems: hyaluronic acid forms a viscoelastic gel within the eye (vitreous humor) that provides both mechanical structure and transparency; actin monomers and polymers within the cytoplasm control cell shape and internal architecture. Drug molecules often must diffuse through these complex fluids in order to reach their site of action.

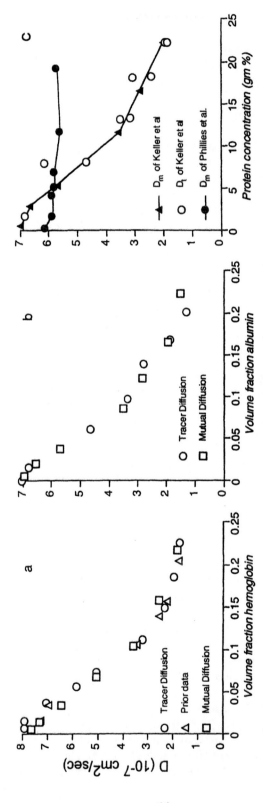

Figure 4.8 The tracer diffusion coefficient as a function of protein concentration. Tracer and mutual diffusion coefficients measured by a variety of techniques illustrate the dependence of protein diffusion on concentration. (a) Diffusion coefficients for hemoglobin [64]; (b) diffusion coefficients for albumin [64]; (c) diffusion coefficients for albumin [64, 65].

Therefore, this section considers the diffusion of molecules through complex fluids: membranes, polymer solutions, and polymer gels.

4.3.1 Membranes

Geometrically, the simplest membranes are thin sheets with cylindrical pores. The diffusion of spherical solutes through cylindrical pores in a membrane can be described quite accurately, even for solutes that are nearly as large as the pore [69]. One commonly used semi-empirical expression for the diffusion of solutes in porous materials was developed by Renkin [70] to describe the diffusion of proteins in cellulose membranes. According to Renkin, the reduced diffusion coefficient is given by:

$$\frac{D_{A,membrane}}{D_{A,\infty}} = (1 - \lambda)^2 [1 - 2.1044\lambda + 2.089\lambda^3 - 0.948\lambda^5] \qquad (4\text{-}27)$$

where $D_{A,membrane}$ is the diffusion coefficient of solute A in the membrane, $D_{A,\infty}$ is the diffusion coefficient of the solute in water, and λ is the ratio of the radius of the solute to the average pore radius of the material (a/r_p). This equation can be derived by calculating the drag on a spherical particle within a cylindrical pore under Stokes flow conditions, by assuming that the spherical particles are confined to the centerline of the pore [71]. While more sophisticated models for hindered diffusion in liquid pores are available (see [69] for a review, which is discussed more completely in Chapter 5), the expression shown in Equation 4-27 captures many of the important features. Although derived for the situation where a spherical probe is diffusing through a perfect, cylindrical pore, this equation has also been used to characterize the diffusion of probes through fibrous gels. In these cases, the pore diameter, $2r_p$, is usually interpreted as the mean fiber-to-fiber distance within the gel network [5].

4.3.2 Polymer Solutions

Ogston *et al.* [72] developed a model for the diffusion of spherical particles through a randomly oriented array of straight, cylindrical fibers, with radius r_f, occupying a volume fraction ϕ:

$$\frac{D_{A,p}}{D_{A,\infty}} = \exp\left(-\phi^{1/2}\frac{a}{r_f}\right) \qquad (4\text{-}28)$$

which is similar in form to the stretched exponential that is frequently used to correlate diffusion coefficients in polymer solutions [73]:

$$\frac{D_{A,p}}{D_{A,\infty}} = \exp(-\alpha c^\nu) \qquad (4\text{-}29)$$

where α and ν are constants and c is the polymer concentration. In most cases, α varies with the radius of the diffusing species, a; the variation is sometimes linear (ν varies between 0.4 and 2 [73, 74]). For probe diffusion through polymer solutions, where the probe has a radius a and the polymer solution has concentration c and molecular weight M_w, the prefactor α and exponent ν have been measured or predicted for a number of systems of biological interest (Table 4.3).

A more general form of the stretched exponential:

$$\frac{D_{A,p}}{D_{A,\infty}} = \exp\left(-ac^{\nu}M^{\gamma}I^{\beta}R^{\delta}\right) \qquad (4\text{-}30)$$

is used for probe diffusion in a solution containing neutral polymer or poly-electrolytes, where I is the ionic strength, and R is the radius of the probe molecule. Experimentally, ν is in the range 0.5–1; β is ~ -0.8 for polyelectro-lytes and 0 for non-electrolytes; δ is ~ 0.3–0.5 for polyelectrolytes and 0 for non-electrolytes; γ is ~ 0.8 [75]. In certain cases, a hydrodynamic scaling model can be used to predict the coefficients for polymer self-diffusion [73].

When diffusion coefficients for BSA were measured in dextran solutions by holographic interferometry, the BSA diffusion coefficient decreased by less than a factor of 2 as dextran concentration was increased from 0 to 0.08 g/mL [54]; the diffusion coefficient was not a function of dextran mole-cular weight (the range tested was 9,300 to 2×10^6). The BSA diffusion coeffi-cient was described very well by using Brinkman's equation to estimate the influence of hydrodynamic screening due to dextran molecules in the solution. The dextran "fibers" were assumed to have a radius r_f of 1 nm; the hydraulic permeability, κ, of the dextran solution was estimated from the semi-empirical relationship:

$$\frac{\kappa}{r_f^2} = 0.31\phi^{-1.17} \qquad (4\text{-}31)$$

Table 4.3 Factors influencing diffusion of probes in polymer solutions

Probe/polymer solution	α	ν	Reference
Proteins/water/dextran or water/ hyaluronic acid	$\sim a^{-1}$	0.5	Laurent *et al.*, as cited in [73]
Polystyrene spheres/dextran/water	Independent of a	1.0	Turner and Hallett as cited in [73]
Probes/PEO/water	—	2/3	Langevin and Rondelez as cited in [73]
Polystyrene spheres/BSA/water	0.0044–0.008	0.96–0.99	[73]
Fluorescein isothiocyanate (FITC)-BSA/DNA solutions (0 to 35 mg/mL)	0.018–0.024 mL/mg	~ 1.0	[149]

Data compiled from [73].

where ϕ is the volume fraction of dextran. Brinkman's equation was used to estimate the diffusion coefficient of BSA in the presence of the dextran:

$$\frac{D_{A,p}}{D_{A,\infty}} = \left[1 + \left(\frac{a^2}{\kappa}\right)^{\gamma_2} + \frac{1}{3}\left(\frac{a^2}{\kappa}\right)\right]^{-1} \tag{4-32}$$

This approach—which uses Brinkman's equation, with an appropriate correlation to permit estimation of the hydraulic permeability from the structural characteristics of the medium—provides a straightforward method for estimating the influence of hydrodynamic screening in polymer solutions; predicted diffusion coefficients for probes of 3.4 and 10 nm in dextran solutions ($r_f = 1\,\text{nm}$) are shown in Figure 4.9. This approach should be valid for cases in which probe diffusion is much more rapid than the movement of fibers in the network, although it appears to work well for BSA diffusion in dextran solutions, even though the dextran molecules diffuse as quickly as the BSA probes [54].

4.3.3 Polymer Gels

When polymer molecules are completely dissolved in solution, the solution behaves as a liquid. Liquids, for example, continuously deform in response to a continuously applied stress. Polymers can also form gels, which do not behave as liquids. The individual polymer chains in a gel form a continuous network within the liquid phase. The network is maintained by chemical or physical interactions between polymer chains; these interactions may involve covalent cross-links, hydrogen bonds, or physical entanglement of the molecules. The network restricts the response of the gel to an applied stress. As a result, the mechanical properties of a gel are qualitatively different from the mechanical properties of a fluid.

The extracellular space of tissues is an aqueous gel of proteins and polysaccharides. This gel potentially provides an additional resistance to the diffusion of molecules in the extracellular space due to volume exclusion and hydrodynamic interactions. Reconstituted gels of extracellular matrix components (e.g., collagen) are often used to evaluate the magnitude of this resistance. In general, the diffusion coefficient for a protein depends on the properties of the gel and the size of the protein (Figure 4.10).

It is difficult to predict the rate of diffusion of a solute through a fibrous gel: fibers in the gel form a porous sieve with a distribution of interfiber spacings. However, in most cases, the fibers are not immobile: motion of fibers due to thermal fluctuations, or due to the forces created by the motion in the fluid, must be considered. Some progress has been made in describing the diffusion of probes in dilute polymer solutions, as described above. Although these theories have been extended to describe the diffusion of small molecules in biological gels [76], they are not as appropriate for large diffusing species like proteins, which are nearly as large as the characteristic distance between fibers in the gel.

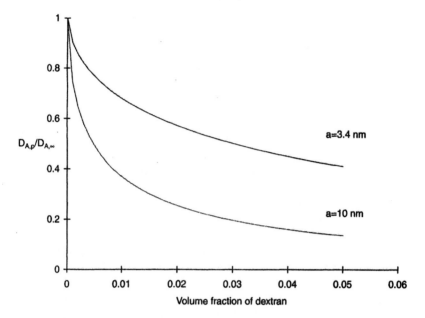

Figure 4.9 Estimation of reduced diffusion coefficient by effective medium approximation. Reduced diffusion coefficient for solutes of radius 3.4 and 10 nm in a solution of dextran ($r_f = 1$ nm).

One of the earliest, and still most frequently used, descriptions of diffusion through gels is due to Ogston (Equation 4-28). Although Ogston developed this expression for the diffusion of particles in polymer solutions, it is equally applicable to particle diffusion through gels. Similarly, the partition coefficient (the ratio of concentration within the gel to concentration within an external aqueous solution) can be estimated as:

$$\Phi = \exp\left[-\phi\left(1 + \frac{a}{r_f}\right)^2\right] \tag{4-33}$$

where, again, ϕ is the volume fraction of fibers, a is the solute radius, and r_f is the radius of the fibers.

Hydrodynamic interactions of the solute with the gel were not considered by Ogston. In general, the diffusion of probe molecules through a gel can be described as depending primarily on steric factors, i.e., on the volume excluded by the fiber matrix of the gel:

$$\frac{D_{A,g}}{D_{A,\infty}} = 1 - \alpha\phi_0 + \cdots \tag{4-34}$$

where ϕ_0 is the volume fraction of the gel matrix, or on hydrodynamic interactions between the diffusing probe and the gel matrix:

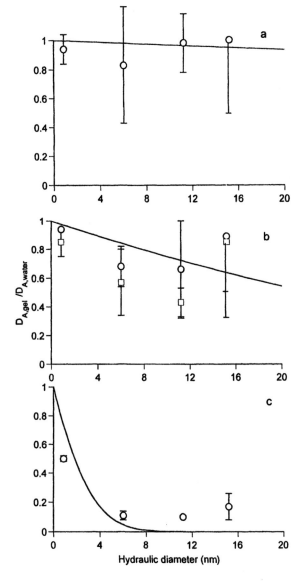

Figure 4.10
Diffusion coefficients for fluorescein, BSA, and human antibodies (IgG and IgM) in biological gels. Data from [5]. The diffusing probes were fluorescein, BSA, human IgG, and human IgM. The gels were formed from (a) collagen (1.0 mg/mL), (b) freshly collected human cervical mucus, and (c) gelatin (100 mg/mL). Lines indicate fits of data to Equation 4-27 with r_p equal to 500, 75, and 6 nm.

$$\frac{D_{A,g}}{D_{A,\infty}} = 1 - \beta\phi_0^{1/2} + \cdots \tag{4-35}$$

where the coefficients, α and β depend on the radius of the diffusing species, a. The diffusion of probes through gels has also been characterized using an exponential function:

$$\frac{D_{A,g}}{D_{A\infty}} = \exp\left\{-a\left(\frac{3\pi\lambda_f}{\ln(L/r_f)}\right)^{1/2}\right\} \tag{4-36}$$

where λ_f is the length density of the fibers ($\mu m/\mu m^3$), r_f is the fiber diameter (μm), and L is the fiber length (μm). A stretched exponential form, similar to Equations 4-29 or 4-30, has also been used. In general, these equations are used empirically, and are most useful for correlation of data obtained for probes diffusing within specific gel systems. It is difficult to extrapolate from these correlations to predict rates of probe diffusion in other types of gels.

An effective medium approach, which uses hydraulic permeabilities to define the resistance of the fiber network to diffusion, has been used to estimate reduced diffusion coefficients in gels [77]. For a particle diffusing within a fiber matrix, the rate of particle diffusion is influenced by steric effects (due to the volume excluded by the fibers in the gel, which is inaccessible to the diffusing particle) and hydrodynamic effects (due to increased hydrodynamic drag on the diffusing particle caused by the presence of fibers). Recently, it was proposed that these two effects are multiplicative [78], so that the diffusion coefficient observed for particles in a fiber mesh can be predicted from:

$$\frac{D_{A,g}}{D_{A,\infty}} = F\left(\frac{a}{\sqrt{\kappa}}\right)S(f) \tag{4-37}$$

where the function $F(\bullet)$ represents the hydrodynamic reduction in the diffusion coefficient and the function $S(\bullet)$ represents the steric reduction. The hydrodynamic effect can be obtained using Brinkman's equation, Equation 4-32, in which κ is the Darcy permeability of the fiber matrix (i.e., the conductance of the medium for flow of water driven by a pressure gradient) and $\kappa^{1/2}$ is a hydrodynamic screening length, approximately equal to the fiber spacing. While hydraulic permeabilities for the gel, κ, are typically not available for samples of interest, they can be estimated in many cases (Equation 4-31 is an example of an equation that is often used for estimation). Alternatively, numerical simulations have been used to predict the hydrodynamic interactions between a diffusing spherical solute and a random array of slender cylindrical fibers. The simulations, which yield short-time diffusion coefficients, compare favorably with the stretched exponential function (Equation 4-29) (Figure 4.11), with appropriate values of the parameters α and ν used to account for the systems of different fiber radius-to-solute radius ratio ($\lambda = r_f/a$) (Table 4.4).

The steric effect depends on the steric factor, f, which depends on the volume fraction of the fiber matrix. The definition of volume fraction is expanded to include the volume of the matrix that is inaccessible to the particle center due to the finite particle size:

$$f = \left(1 + \frac{a}{r_f}\right)^2 \phi \tag{4-38}$$

The steric effect can be calculated for a variety of fiber arrangements, including square fiber arrays, in which diffusion is perpendicular to the axis of the cylindrical fibers [80]:

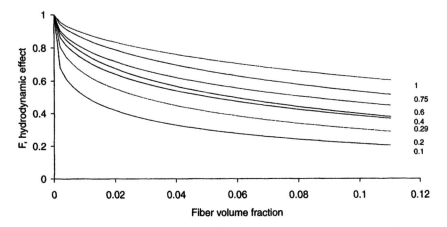

Figure 4.11   Effect of hydrodynamic interactions for spherical solutes in cylindrical fiber media. Adapted from [79]; parameters for Equation 4-29 were obtained by fitting numerical simulation results. Each of the lines represents the variation of F with fiber volume fraction for systems with a particular value of λ (which varies between 0.1 and 1 in this figure).

$$S(f) = \frac{1}{1-f}\left[1 - 2f\left(1 + f - \frac{0.305827f^4}{1 - 1.402958f^8} - 0.013362f^8\right)^{-1}\right] \quad (4\text{-}39)$$

or random arrays [81]:

$$S(f) = \exp\left(-0.84f^{1.09}\right) \quad (4\text{-}40)$$

in which the exponential function is fit to results obtained by dynamic simulations, or

Table 4.4   Parameters used in stretch exponential to estimate the short-time diffusion coefficient

λ	α	v
0.1	3.483	0.354
0.2	3.248	0.434
0.29	2.871	0.477
0.4	3.146	0.532
0.6	2.526	0.518
0.75	2.500	0.600
1.0	1.900	0.600
2.0	2.114	0.1719

From [79].

$$S(f) = \frac{1}{\left[1 - \left(\frac{2}{3}\right)\ln(f)\right]} \tag{4-41}$$

The use of these three equations is illustrated in Figure 4.12.

The effective medium approach, including both hydrodynamic and steric interactions (Figure 4.13), compares favorably with experimental results for the diffusion of proteins and polysaccharides through agarose gels [81, 82].

For certain biological macromolecules, particularly extended polymers such as nucleic acids, migration through a polymer gel can occur by reptation, which was first proposed as a mechanism for polymer diffusion through polymer melts [83]. Gosnell and Zimm measured the diffusion of linear DNA in agarose gels [55]. $D_{A,g}$ varies with the -1.0 power of the molecular length in base pairs (Table 4.5 and Figure 4.14), which is intermediate between the molecular weight dependence expected for free diffusion (exponent of -0.667, see Equation 4-6) and for reptation (exponent of -2). Reptation of DNA polymers has been observed experimentally by using optical tweezers to pull individual DNA molecules through a concentrated DNA solution; by pulling with the optical tweezers, the DNA becomes fully extended in the solution. After release, the fluorescently labeled DNA polymer can be observed to relax along the path described by the tube-like, extended DNA molecule [84].

In summary, models with a variety of assumptions have been developed to explain diffusion in polymer gels. A recent review of the models for solute diffusion in hydrogels [85] compares data from the literature with a variety of models, including models that attempt to account for reduced diffusion by reduction in free volume (see, e.g., [86]), by the increase in hydrodynamic drag experienced by the solute and created by polymer chains (see, e.g., Equation 4-32), and by the presence of physical obstructions to solute diffusion (see, e.g.,

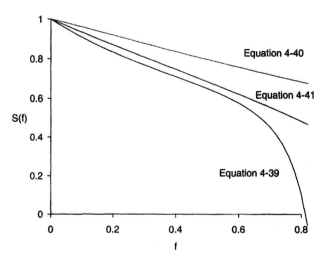

Figure 4.12   Three expressions were used to calculate the steric effect for a porous medium. From Equations 4-39, 4-40, and 4-41.

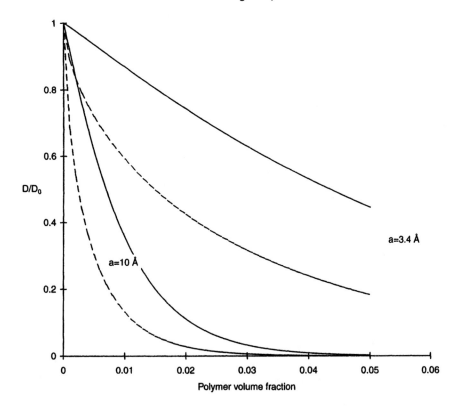

Figure 4.13 Estimation of reduced diffusion coefficient by effective medium approximation. Combination of steric and hydrodynamic effects on reduced diffusion coefficient. The solid lines represent hydrodynamic effect for probe radii of 3.4 and 10 Å calculated using Brinkman's equation (see Figure 4.9). The dashed lines represent the combined steric and hydrodynamic effect using Equation 4-40 for the steric effect.

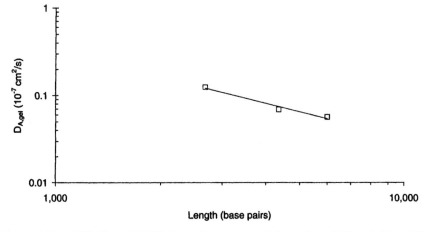

Figure 4.14 Diffusion of DNA through agarose gel. Data from [55], as indicated in Table 4.5

Table 4.5   Diffusion of DNA molecules in water and
agarose gels

DNA	D_A (10^{-7} cm^2/s)	$D_{A,g}$ (10^{-7} cm^2/s)
Linear pUC19	0.38	0.125
Linear pBR322	0.27	0.0692
Linear pBR325	0.22	0.0569

Data from [55].

Equation 4-28). Although no single model is adequate to describe all of the experimental data, hydrodynamic models appear to be more appropriate for gels with flexible polymer chains and obstruction models appear to be more appropriate for gels with rigid chains.

4.4 DIFFUSION IN THE EXTRACELLULAR SPACE

Frequently in biological systems, the medium through which solutes are transported is not homogeneous. Solutes diffusing through the tissues, for example, encounter an environment where a continuous phase—the extracellular space—is filled with discrete obstacles to diffusion—the cells. If the solute does not enter the cell readily, it diffuses primarily within the extracellular space. Diffusion within the tissue, i.e., diffusion over a large length scale as shown in Figure 4.15, is slower than diffusion through the medium filling the extracellular space, i.e., diffusion over a length scale much smaller than the cells in the tissue. A complete description of the physics of diffusion through a chaotically organized structure, such as the extracellular space of a tissue, is extremely cumbersome. Effective diffusion coefficients are frequently used to estimate rates of transport in these situations.

4.4.1 Experimental Measurements

The rate of diffusion of molecules through intact tissues in an animal is difficult to measure, so the amount of information currently available is limited. Diffusion coefficients for size-fractionated dextrans, albumin, and antibodies have been measured in granulation tissue and tumor tissue [20, 21]; similar measurements have been made in slices of brain tissue [87]. In both cases, the diffusion coefficient was estimated by fitting solutions to the diffusion equation, similar to Equation 3-36, to data obtained by direct visualization of fluorescent tracers in the interstitial space. These measurements, as well as others made by a variety of techniques, are compiled in

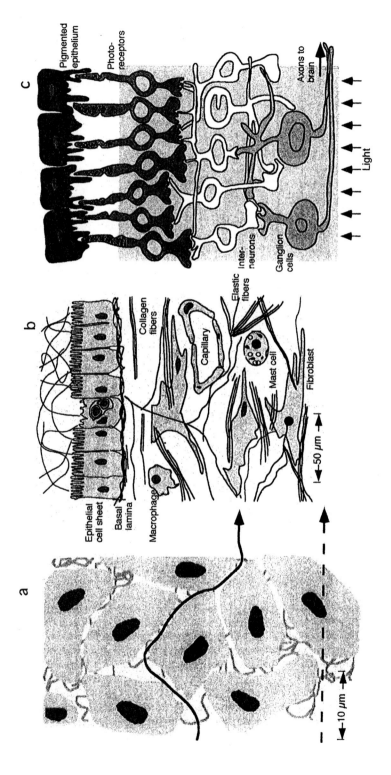

Figure 4.15 Illustrations of tissue structure depicting the complex paths over which diffusion must occur. (a) Undifferentiated tissue, based on structure of tissue in the developing limb. A water-soluble molecule diffusing through this tissue (solid line) would encounter different obstacles than a lipid-soluble molecule (dashed line). (b) Structure of a loose connective tissue underlying an epithelium. (c) Structure of the retina. The light gray area surrounding the interneurons and ganglion cells is populated with supporting cells, which are not shown.

Figure 4.16. In general, the diffusion coefficient decreases with increasing molecular weight.

Compounds with different lipid solubility have different fates in tissue. This is most clearly demonstrated in the brain using ventriculocisternal perfusion [88]. In these experiments, solutes delivered into the cerebrospinal fluid permeate through the ependyma into the extracellular space of the brain. Three classes of compounds, with different patterns of local distribution, have been identified: (a) water-soluble compounds that remain in the extracellular space of the brain, occupying a volume fraction of 15–20% (e.g., sucrose and EDTA), (b) large, lipid-soluble compounds that have slow capillary transport, but quickly enter the cells of the brain, occupying a volume fraction of 50–200% (e.g., mannitol, creatinine, cytosine arabinoside), and (c) small, lipid-soluble compounds that are rapidly removed from the brain by capillary transport (e.g., H_2O, ethanol, 1,3-bis(2-chloroethyl)-1-nitrosourea (BCNU). Similar behavior probably occurs in extracranial tissues as well.

Diffusion coefficients in tissue can change with disease and injury; these changes often correlate with changes in tissue structure. Again, this effect has been observed in the brain. The normal volume fraction of the brain extracellular space is ~ 0.18 [89, 90], but it decreases after ischemic injury to the brain (to ~ 0.07), probably due to osmotic swelling of brain cells upon injury.

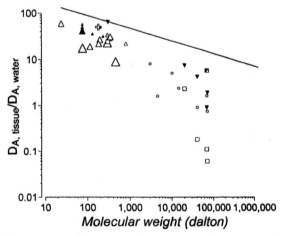

Figure 4.16 Diffusion coefficients for molecules in the tissue interstitial space. The diffusion coefficients were measured by quantitative autoradiography and quantitative fluorescence methods, iontophoresis, and FPR. The measurements were made in brain slices (circles), cell cultures (cross), brain of animals (triangles), tissues in animals (squares), and tumors in animals (inverted triangles). Data for the diffusion of size fractionated dextrans, BSA, and IgG in granulation tissue and tumor tissue within a chamber in the rabbit ear [20, 21]. Symbol size indicates species: in order of increasing size: rats, rabbits, turtle, skate, dog, monkey. A previous versions of this plot has appeared [52]. The solid line represents the best-fit diffusion coefficient vs M_w (see Figure 4.4).

This change is accompanied by a corresponding decrease in the diffusion coefficient for ions. Similarly, the apparent diffusion coefficient for water in normal rat cortex and caudate putamen is 6×10^{-6} and $5 \times 10^{-6}\,cm^2/s$, respectively [91]. The diffusion coefficient increases in rats with experimental brain tumors ($8 \times 10^{-6}\,cm^2/s$) and experimental edema ($9 \times 10^{-6}\,cm^2/s$), consistent with the increase in extracellular volume that occurs during these states [91].

In general, the diffusion coefficient in tissues is significantly slower than the diffusion coefficient for the same solute in water (Figure 4.17). As the size of the diffusing solute increases, this difference becomes more pronounced. However, small molecules can diffuse significantly more slowly in tissues. Glucose diffusion has been measured in connective tissue and in the fibrous connective tissue formed around silicone implants [92]; in all cases, the diffusion of glucose was significantly slower in the connective tissue than in water (Figure 4.17). Similar results were obtained for the diffusion of fluorescein through fibrous capsules of various density that formed around subcutaneously implanted materials [93]; fluorescein diffusion decreased by up to a factor of 4 in the densest fibrous tissue.

4.4.2 Effective Diffusion Coefficients and Tortuosity

The composition of the extracellular space will influence the rate of diffusion of compounds through tissues; this effect is particularly severe for macromolecules, since the diffusion of large molecules is most affected by the concentrated proteins within the extracellular matrix (ECM) gel (see Figure 4.10). However,

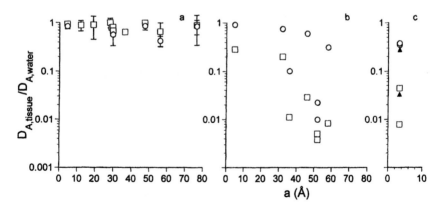

Figure 4.17 Reduced diffusion coefficient as a function of molecular size. Hydrodynamic radii were determined from the diffusion coefficient in water according to Equation 4-4. (a) Diffusion of proteins and peptides in mid-cycle human cervical mucus [5]. Measurements were performed by FPR (squares) or quantitative imaging of fluorescence profiles (circles). (b) Protein and dextran diffusion through granulation (squares) or tumor tissue (circles) in the rabbit ear [20, 21]. (c) Glucose diffusion through capsular tissue (squares) and cartilage (circles and triangles) [92]. Each symbol represents a separate measurement.

the architecture of tissue can also influence overall rates of diffusion through tissues.

Consider the different tissue structures shown schematically in Figure 4.15. For molecules that diffuse through the extracellular space, the rate of movement through each of these tissues could be substantially different. In addition, a molecule that can dissolve in the extracellular fluid and also permeate cell membranes will encounter less geometrical complexity in its diffusional path, but it will diffuse through microscopic regions that differ substantially in composition. Prediction of effective diffusion coefficients in either situation is difficult.

The effective diffusion coefficient for solute diffusion in a complex microstructure is related to the diffusion coefficient in an unbounded fluid:

$$\frac{D_{A,\text{eff}}}{D_{A,\text{pore}}} = h(\mathfrak{R}) \tag{4-42}$$

where $D_{A,\text{eff}}$ is an effective diffusion coefficient, $D_{A,\text{pore}}$ is the diffusion coefficient for A in unbounded pore space fluid, and $\mathfrak{R}$ indicates a complete geometric description of the material. This effective diffusion coefficient could be substituted into Fick's law, and subsequently used in describing changes in concentration throughout macroscopic regions of heterogeneous media, using the same methodology described in Chapter 3. If the exact geometry of the pore space was known, an analytical description of $\mathfrak{R}$ could be developed. If this geometric description was sufficiently simple, it could be used as boundary conditions for the diffusion equation, and the effective diffusion coefficient could be predicted exactly.

Analytical expressions for effective diffusion coefficients in complex media can be obtained only when the geometry is simple. Consider, for example, the diffusion of solute through a periodic array of spherical obstructions, in which the solute diffuses through the continuous interstitial space with a diffusion coefficient $D_{A,\text{pore}}$ and through the spheres with a diffusion coefficient $D_{A,s}$. The effective diffusion coefficient for such a composite material can be determined exactly [94]:

$$\frac{D_{A,\text{eff}}}{D_{A,\text{pore}}} = \frac{\frac{2}{D_{A,s}} + \frac{1}{D_{A,\text{pore}}} - 2\epsilon_s\left(\frac{1}{D_{A,s}} - \frac{1}{D_{A,\text{pore}}}\right)}{\frac{2}{D_{A,s}} + \frac{1}{D_{A,\text{pore}}} + \epsilon_s\left(\frac{1}{D_{A,s}} - \frac{1}{D_{A,\text{pore}}}\right)} \tag{4-43}$$

where ϵ_s is the volume fraction of spheres in the composite. This equation was derived under the assumption that the spheres are sparse, so that they do not interact. This result has been extended to other simple geometries [95]. A similar approach is frequently used to find effective transport coefficients for evaluating the resistance of a porous medium to fluid flow. One of the first geometric models of a porous medium was developed by assuming a simple geometrical model for the porous material—a parallel bundle of capillaries with different diameters but constant length [96]. More information on the construction of models for flow through porous media is available in review papers [97–99] and textbooks [100–102].

For diffusion in a porous material, the effective diffusion coefficient is assumed to depend on two factors, pore shape and diffusion path tortuosity [103]:

$$\frac{D_{A,\text{eff}}}{D_{A,\text{pore}}} = \frac{1}{F\tau} \qquad (4\text{-}44)$$

where F is a shape factor and τ is the tortuosity. The effective diffusivity is less than the molecular diffusion coefficient of the solute for two reasons: (i) the diffusional path length is increased because of windiness in the diffusional path $(\tau > 1)$,[3] and (ii) the pore structure is constricted, creating local sites of decreased permeability $(F > 1)$. While tortuous and constricted diffusion paths physically retard the random motion of solutes in the porous structure, the porosity, ϵ, has no intrinsic effect on the diffusion coefficient; it does, however, account for the decreased area through which solute flux occurs and must be included if quantities are defined per total material volume.

Experimentally, it is difficult to separate the effects of F and τ on the effective diffusivity. Often, empirical values of the product $F\tau$ are reported as the "tortuosity." This is particularly true in the literature of transport/reaction in porous catalyst pellets. For the large variety of catalysts, the "tortuosity"—equal to $1/h(\Re)$ in Equation 4-42—ranges from 1 to ~ 10 [103]. Although it has been difficult to correlate these "tortuosity" values with experimentally determined pore structure parameters, the "tortuosity" almost always decreases with increasing porosity.

Geometric analysis suggests that $1 < \tau < 3$. If the material is isotropic and the pores are distributed randomly, the tortuosity τ of the material can be determined geometrically [104]:

$$\frac{1}{\tau} = \frac{1}{2} \int_0^\pi \cos^2\theta \sin\theta \, d\theta = \frac{1}{3} \qquad (4\text{-}45)$$

This analysis assumes that the pores are infinitely narrow, so that solute diffusion occurs along the pore trajectory, θ, which is defined relative to a reference direction in the material. If the pore is wide enough to permit molecules to assume intermediate trajectories, the tortuosity is related to the porosity of the material:

---

3. Since the tortuosity is often interpreted as the increase in diffusional path length, and since the characteristic time for diffusion scales with the diffusional path length squared, an alternative definition for tortuosity is sometimes used:
$$\frac{D_{A,\text{eff}}}{D_{A,\text{pore}}} = \frac{1}{\tau^2}.$$
This text uses the definition provided above; where necessary, values from the literature have been converted to conform with this convention.

$$\frac{1}{\tau} = 1 - \frac{2}{3}(1 + \epsilon)(1 - \epsilon)^{2/3} \tag{4-46}$$

where ϵ is the volume fraction of the pore space or porosity. A similar analysis can be used to determine D_{eff} for porous materials composed of randomly arranged pore segments, each segment with a characteristic geometry [105]. Tortuosities determined by this expression, and Equation 4-43 for arrays of spheres where the diffusion coefficient within the sphere, $D_{\mathrm{A,s}}$, is equal to zero, are indicated in Figure 4.18. The prediction for Equation 4-43 is indicated only for values of ϵ greater than 0.5, or ϵ_s less than 0.5. Equation 4-43 applies for spheres that do not interact and, therefore, it should fail when the spheres touch. For spheres packed on a cubic lattice, this occurs at a porosity of $4\pi/24 \sim 0.5$.

Tortuosities have also been calculated for ensembles of cuboidal cells arranged in a regular or staggered lattice [106]; the ensembles were designed to mimic the arrangement of cells in tissues. The porosity was the only variable that led to substantial differences in the predicted tortuosity: no significant differences were found between regular or staggered arrays or arrays of elongated cells. These tortuosities are also indicated in Figure 4.18.

Geometric models of the pore structure can be used to determine values of the shape factor, F. This has been done by assuming capillaries with cross-sections that vary (i) as hyperbolas of revolution, (ii) as sinusoids, or (iii) as serial cylinders with differing radii [103]. For these geometric models, F is relatively insensitive to the geometric details, depending largely on the ratio of maximum to minimum pore cross-sectional area. Experimental values for the "tortuosity factor" (equal to $F\tau$) are as high as 10 in porous catalysts [103] and 10,000 for macromolecular diffusion in porous polymers [107]. In these cases the shape factor F must be significantly greater than 1. For porous catalysts, where F must vary between 1 and 4, simple models can account for the diffusive retardation. For diffusion in some types of porous polymers, where the apparent tortuosity $F\tau$, is much higher, more complicated models are required.

4.4.3 Percolation Descriptions of Porous Materials

Models of transport in environments as complex as the extracellular space of a tissue require a sophisticated description of the microstructure. In many tissues the extracellular space is stochastically arranged, consisting of an ensemble of individual spaces of random size that are distributed randomly. Useful models of transport in this situation can be based on percolation descriptions of porous media. Percolation descriptions have been used to describe fluid flow, electrical conduction, and phase transitions in random systems [108, 109].

Percolation Lattices. Percolation theory requires that space be represented as a lattice, often infinite in extent. In Figure 4.19, for example, two-dimensional space is discretized on to a square lattice. The points of

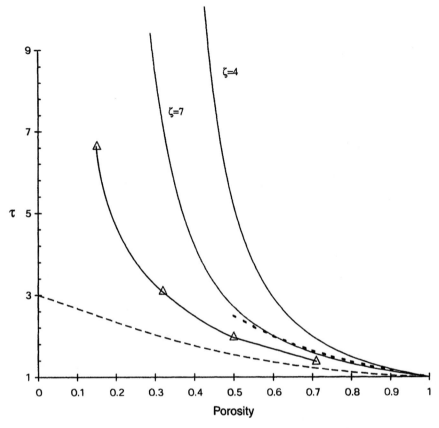

Figure 4.18 Tortuosity as a function of porosity for randomly oriented porous media. For diffusion in porous materials, the length of the diffusional path is increased. If the pores are randomly oriented, and large enough to permit random molecular trajectories, the tortuosity is a function of total porosity, Equation 4-46 (dashed line). The tortuosity predicted for diffusion around a lattice of sparsely populated spheres is obtained from Equation 4-43 assuming completely impermeable spheres (heavy dashed line). Tortuosities for ensembles of cuboidal cells are also included (triangles). The solid lines without symbols indicate tortuosity for a Bethe lattice of coordination number 4 or 7.

intersection on the lattice (i.e., the squares in Figure 4.19) are called sites and the connections between the sites (i.e., the edges of the squares) are called bonds. To make the example more concrete, assume that the lattice represents a tissue. A certain fraction of the available sites on the lattice corresponds to extracellular space; the remainder of the sites correspond to cells, or some other phase that cannot be penetrated by the diffusing molecule. A chaotically organized tissue can be simulated by randomly assigning a certain fraction of the sites to be extracellular; the porosity (or extracellular volume fraction) can then be prescribed by adjusting the

a

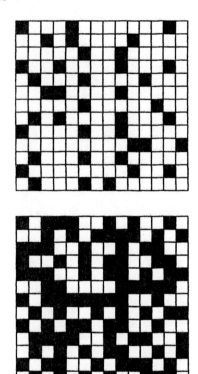

Figure 4.19
Typical lattice structures
used for percolation
theory. For percolation
lattices, like the square
lattices shown here,
there is a qualitative difference between lattice
properties at low and
high porosities. When
the lattice has a low
porosity, such as the
20% filled square lattice
(a), pores are isolated
from one another. At
high porosity, e.g., 60%
(b), pores form clusters
and connected pathways
throughout the macro-
scopic region.

b

probability, p, that a site will be outside of the cell. Figure 4.19 shows the square lattice at two different porosities, generated by selecting two different lattice probability values: 0.2 and 0.6. Black sites correspond to extracellular space or pores, and the white sites to the cellular space. If two adjacent sites are extracellular, the bond (or connection) between them is considered open and, therefore, molecules can move from one site to the other. Increasing the fraction of pores, p, increases the likelihood of finding two pores at adjacent sites and, therefore, increases the fraction of open bonds. Open bonds give the lattice conductance; in this case conductance is a measure of the ease with which molecules can move around in the lattice.

The preceding definitions are centered on lattice sites; a fraction of sites are randomly assigned to correspond to the extracellular space. A tissue could also be defined, based on lattice bonds by allowing the probability p to represent the fraction of open bonds in the lattice. Interconnected extracellular space then exists at all sites that are connected by open bonds. Bond percolation and site percolation are two distinct methods of describing space; each leads to quantitative predictions of material properties. The site percolation description corresponds more naturally to the porous materials: for example, in site percolation, the lattice probability p is exactly equal to the extracellular volume fraction.

As depicted on the square lattice in Figure 4.19, randomly assigning sites to each phase (extracellular or cellular) determines the structural properties of the material. Many important properties can be examined by considering the distribution of clusters in the lattice. A cluster is a group of connected pores; here, we focus on clusters of pores that are filled (i.e., black in Figure 4.19). The number of pores in a given cluster is the cluster size. At low filling probabilities most of the pores are isolated, only a few pores are connected, and the mean cluster size is close to unity. At a higher filling probability the lattice becomes more connected and the average size of a cluster increases. For lattices where each site has a probability p of being filled, there exists a critical probability p_c. When $p < p_c$ all clusters are of finite extent; when $p > p_c$ an infinite cluster exists. This can be observed in the finite square lattice of Figure 4.19: at low p ($< p_c$) all of the filled pores are isolated (panel A); at high p ($> p_c$) lattice-spanning clusters appear (panel B).

The square lattice is only one possible representation of space (Figure 4.20). Every lattice has an associated coordination number, z, which describes the number of bonds emanating from each site: for example, the square lattice in Figure 4.19 has a coordination number of 4. In addition there are lattices that have no obvious dimensionality, like the Bethe lattice (Figure 4.20). The Bethe lattice is a homogeneous tree-like structure, in which the number of sites

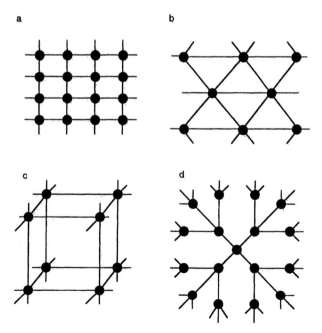

Figure 4.20 Types of lattice that can be used to represent space. (a) Square lattice with coordination number 4. (b) Triangular lattice with coordination number 6. (c) Cubic lattice with coordination number 6. (d) Bethe lattice with coordination number 4.

on the surface of the tree increases without bound as the size of the tree grows. The coordination number of the Bethe lattice can be from 2 to ∞. There are also lattice representations that are irregular; each site does not have the same characteristic shape. Voronoi lattices, both two- and three-dimensional, are constructed by placing points randomly in space and tesselating around these points to construct an internal surface [110]. Some relevant properties of each lattice—dimensionality D, coordination number z, critical probability p_c for site and bond percolation—are listed in Table 4.6.

The critical probability of a lattice depends on the dimensionality D and the coordination number z. Percolation lattices with the same dimensionality and coordination number have similar quantitative properties [108, 109]. The properties of many of these lattices (some are listed in Table 4.6) have been estimated, usually by numerical simulation. Few analytical expressions exist for real lattice parameters.

By contrast, since Bethe lattices are tree structures—containing no loops or closed paths within them—they are more easily analyzed than other, less regular, lattice structures. Analytical expressions have been derived for the percolation probability, cluster size distribution function [112], and effective conductivity [113] of Bethe lattices. The properties of these special Bethe lattices are quantitatively similar to regular tesselations (e.g., the square lattice) or irregular tesselations (e.g., the Voronoi tesselation) [114, 115]. Analysis of a given pore structure can proceed by selecting an effective Bethe coordination number, ζ, and using the properties of the Bethe lattice to describe the real system. The Bethe coordination number which best describes a given real structure may be different from the actual coordination number of the pore

Table 4.6   Percolation properties of various lattices

Lattice	D	z	p_c (site)	p_c (bond)
Honeycomb	2	3	0.70	0.65
Kagome	2	4	0.65	0.44
Square	2	4	0.59	0.50
Triangular	2	6	0.50	0.35
Voronoi polygon	2	6[a]	0.5	—
Diamond	3	4	0.43	0.39
Simple cubic	3	6	0.31	0.25
Body-centered cubic	3	8	0.24	0.18
Face-centered cubic	3	12	0.20	0.12
Hexagonal	3	12	0.20	0.12
Tetrakaidecahedron	3	14	0.18	—
Voronoi polyhedron	3	15.54[a]	0.16	—
Bethe		z	$1/(z-1)$	$1/(z-1)$

[a] Average coordination number.

Values reported in the literature [108, 111, 112]. D is the dimensionality of the lattice, z is the coordination number, and p_c is the critical probability.

space. For example, the two-dimensional triangular lattice or Voronoi polygon lattice both have p_c equal to 0.5. Quantitative predictions for a Bethe lattice with $\zeta = 3$, so that $1/(\zeta - 1) = 0.5$, can be employed for these two-dimensional lattices. For a three-dimensional lattice like the Voronoi polyhedron, a Bethe lattice with $\zeta = 7$ can be assumed. By varying the coordination number of the Bethe lattice over a small range, the properties of a number of different lattices can be accurately estimated.

Effective Transport Coefficients. Analytical expressions are available for effective diffusion on a Bethe lattice [113]. For a Bethe lattice with coordination number ζ, the effective diffusion coefficient is found from:

$$h(\mathfrak{R}) = -\left(\frac{\zeta - 1}{\zeta - 2}\right)\frac{C'(0)}{D_0} \tag{4-47}$$

where $C'(0)$ is defined by the integral equation:

$$\int_0^\infty e^{-lx}C(x)dx = \int_0^\infty G(L(r))$$

$$\left(\frac{1}{t + L(r)} + \frac{L^2(r)}{(t + L(r))^2}\int_0^\infty \exp\left(-\frac{L(r)t}{L(r) - t}x\right)C(x)^{\zeta - 1}dx\right)dL(r) \tag{4-48}$$

where t is a transform variable and with the condition $C(0) = 1$ [113]. $L(r)$, where r is the characteristic size of a pore, defines the distribution of transport coefficients on the lattice for all sites and $G(L(r))$ is a normalized probability density function for $L(r)$ over all possible lattice sites:

$$G(L(r)) = (1 - \varepsilon)\delta(L(r)) + \varepsilon F(L(r)) \tag{4-49}$$

where δ is the Dirac delta function and $F(L(r))$ is a probability density function describing the distribution of transport coefficients on the open sites in the lattice: $F(L(r))\, dL(r)$ is the probability that site-to-site diffusion for a given bond has a coefficient $L(r)$. For example, if distribution of pore sizes in a material is described by the function $f(r)$, then:

$$F(L(r)) = f(r)\frac{dr}{dL(r)} \tag{4-50}$$

A method of solution for Equations 4-47 to 4-50, involving power series expansion of $C'(0)$ to order $(\zeta - 1)^{-4}$, has been presented previously [114].

The effective diffusion coefficient depends on porosity, ϵ, and lattice coordination number, ζ, as shown in Figure 4.21. Since diffusion between any two lattice sites is assumed to depend only on the molecular diffusion coefficient, $h(\mathfrak{R})$ is equal to $1/\tau$. The tortuosity depends on both porosity and lattice coordination number (Figures 4.18 and 4.21). For a given coordination number, the tortuosity increases without bound at the critical porosity.

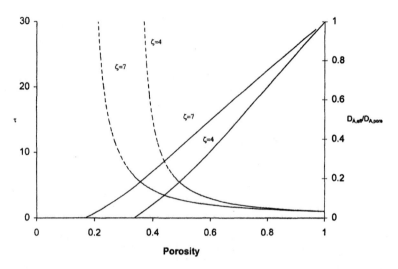

Figure 4.21 Effective diffusion coefficients and tortuosity on Bethe lattices. Effective diffusion coefficient (solid lines) and tortuosity (broken lines) on Bethe lattices with $\zeta = 4$ and 7. The effective diffusion coefficient (right-hand axis) and tortuosity (left-hand axis) are plotted versus the total porosity. Site-to-site diffusion coefficients have two possible values: a fraction of the sites (numerically equal to the porosity) have a diffusivity equal to the molecular diffusivity, the remainder of the sites have a diffusivity of 0. The effective diffusion coefficient and the tortuosity were obtained by the method of Stinchcombe [113] (details on the implementation of this technique are available elsewhere [116].

Prediction of effective diffusion coefficients, or more generally effective conductivities, on other lattices usually involves a finite element or finite difference approximation to the diffusion equation [115]. In three-dimensional media, effective diffusion coefficients for the cubic [117] and Voronoi polyhedron [115] lattices have been calculated. In general, lattices with the same dimensionality and coordination number, e.g., the hexagonal and Voronoi polygon, exhibit similar diffusion behavior [115]. Importantly, the effective diffusion coefficient on these lattice structures can be closely approximated by selecting a Bethe lattice with an appropriate effective coordination number ζ. For example, three-dimensional cubic and Voronoi polyhedron lattices, with coordination numbers of 6 and 16, have the same effective diffusion coefficient behavior as Bethe lattices with coordination numbers of 5 and 7. So it is reasonable to expect that the effective diffusion coefficient and tortuosity predicted in Figure 4.21 represent lattices with different geometries. The variation in tortuosity with porosity predicted by the percolation model (Figure 4.21) is similar to the variation predicted by geometric models (Figure 4.18).

4.4.4 Predicting Effective Diffusion Coefficients
from Tissue Structure

Do these models of diffusion in complicated microenvironments help us predict rates of drug movement in tissues? Unfortunately, insufficient experimental measurements are available to test predictions rigorously, but the models compare favorably to the data that are available. Nicholson *et al.* [90] have measured the effective diffusion coefficient for a variety of compounds in the brain; tortuosity can be predicted from these measured values (Figure 4.22). Since these measurements were made in the same tissue, the porosity or extracellular volume fraction should be equal (the extracellular volume fraction of the brain is ∼ 20%). All of the measured tortuosities are in the range predicted by the various models for media with this porosity (∼ 2 to 30, cf. Figure 4.18).

However, the data clearly suggest that the effective diffusion coefficient depends on molecular weight; this effect is not predicted by these models of porous structure (i.e., the tortuosity models in Figure 4.18 do not depend on molecular size). The tortuosity for a small water-soluble molecule, the tracer ion TMA, is ∼ 2 (Figure 4.22). Large molecules have tortuosity values greater than 2 and the tortuosity increases with molecular size. For larger molecules, this "tortuosity"—which is estimated from the effective diffusion coefficient—must reflect a decrease in diffusion rate due to actual tortuosity in the extra-

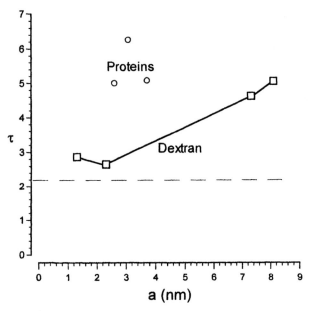

Figure 4.22 Tortuosity for size-fractionated dextrans and three albumins in the brain. Dashed line indicates the reduced diffusion coefficient for a small ion (TMA), and is probably a reasonable estimate for the intrinsic tortuosity of the diffusional path in the extracellular space.

cellular space, as well as a decrease in diffusion rate that results from hindered diffusion due to the polymer gel or solution in the extracellular fluid. These two effects can be represented independently:

$$\frac{D_{eff}}{D_{A,\infty}} = \frac{D_{eff}}{D_{A,pore}} \cdot \frac{D_{A,pore}}{D_{A,\infty}} = (\text{Actual tortuosity}) \cdot (\text{Additional tortuosity})$$

$$(4\text{-}51)$$

Since the tortuosity for TMA is ~ 2, it is reasonable to assume that this represents the intrinsic tortuosity for the extracellular space (i.e., $D_{eff}/D_{A,pore} = 2$). The additional "tortuosity" from the extracellular space should depend on molecular size, and can be estimated from models of diffusion in polymer solutions or gels (as shown in Figures 4.10 and 4.12).

4.5 DIFFUSION WITH BINDING IN TISSUES

Molecules diffusing through a tissue often interact with elements of the tissue. Some of these interactions are non-specific and some of these interactions—such as the receptor–ligand interactions introduced in Section 3.5.1—are highly specific. Specific interactions are of considerable consequence in drug delivery, since binding is essential for the action of many drugs (recall Table 2.2). In addition, binding can influence the rate of movement of a drug molecule through a tissue or—as we will see in Chapter 5—across a membrane barrier in tissues.

Antibody–antigen binding has been studied extensively. In addition, the diffusive migration of antibodies has been measured in the interstitial space of tumors that were growing within living animals [118]. When non-specific antibodies were tested, the diffusion coefficients for IgG and Fab′ fragments were nearly as rapid as in water (relative diffusion coefficients D_{tissue}/D_{water} were 0.3 and 0.4, respectively). Antibodies that were known to bind to an antigen on the cell surface, however, exhibited different diffusion characteristics (Figure 4.23). At low concentration, most of the antibody was immobile in the interstitial space (panel A); as concentration increased, so did the fraction of antibody that was free to diffuse. An apparent binding affinity, K_{app}, was calculated from the reduction in diffusion coefficient (panel B). At low doses, this apparent binding constant should be equal to the product of the intrinsic association constant and the concentration of binding sites in the tissue (see Equation 3-54).

The data in Figure 4.23 illustrate the importance of binding on the rate of diffusion through tissues. Diffusion can also influence the rate of binding. Consider a single cell suspended in a solution containing a ligand that binds to a receptor on the cell surface (Figure 4.24). Ligand binding to the cell surface must occur in two steps: (a) diffusion of ligand to the cell surface and (b) binding between ligand and receptor molecules. The binding kinetics are defined (as in Chapter 3):

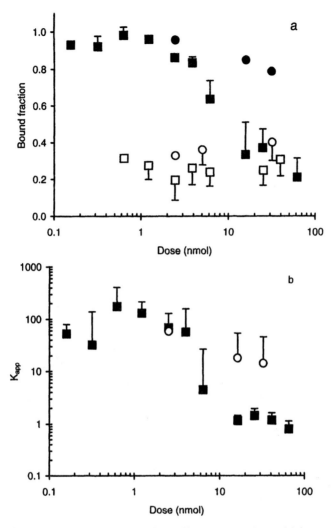

Figure 4.23 Binding of antibody during diffusion in the interstitial space of a tumor. Adapted from [118]. In panel a, the bound fraction is plotted versus the dose of antibody given to the animal. The solid symbols represent specific antibody and the open symbols represent non-specific antibody; squares indicate IgG and circles indicate Fab′ fragments. In panel b, an apparent binding constant for the specific antibody was calculated from the mobile fraction and the previously measured diffusion coefficient for non-specific IgG and Fab′. The diffusion coefficient in water was estimated, for IgG D_0 was 3.9×10^{-7} cm^2/s and for Fab′ D_{tissue} was 6.6×10^{-7} cm^2/s. The tumor diffusion coefficient was lower, for IgG D_0 was 1.3×10^{-7} cm^2/s and for Fab′ D_{tissue} was 2.7×10^{-7} cm^2/s. All diffusion coefficients were corrected to 20 °C using the Stokes–Einstein equation.

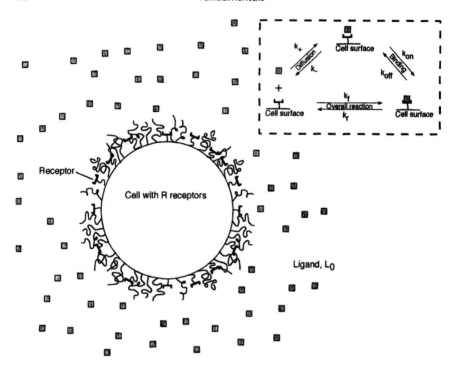

Figure 4.24 Model for ligand binding to cell surface receptors. A single cell with R receptors is held in medium containing ligand at bulk concentrations $c_{L,0}$. The system is characterized by rate constants for the intrinsic rate of association and dissociation (k_{on} and k_{off}), diffusion of ligand to the cell surface (D), and rate constants for the overall association and dissociation reactions (k_f and k_r).

$$L + R \underset{k_{off}}{\overset{k_{on}}{\rightleftarrows}} L - R$$

$$K_d = \frac{1}{K_a} = \frac{c_L c_R}{c_{L-R}} = \frac{k_{off}}{k_{on}}$$

(4-52)

where the intrinsic binding kinetics are now represented by rate constants k_{on} and k_{off}.

Ligand molecules must diffuse to the cell surface before they can participate in the reactions shown in Equation 4-52. In the previous chapter, the concentration gradient in the vicinity of a spherical cell was determined for the situation in which the solute concentration disappears rapidly at the surface (Equation 3-46); rapid disappearance provided a boundary condition ($c_A = 0$ at the cell surface). In the present situation, the reaction at the surface is not instantaneous; therefore, the rate of diffusion at the cell surface must be balanced by the rate of ligand disappearance due to the binding reaction:

$$4\pi a^2 D \frac{dc_L}{dr}\bigg|_{r=a} = k_{on} c_R c_L \qquad \text{at } r = a \qquad (4\text{-}53)$$

where c_R is the number of receptors on the cell surface (molecules/cell). The differential equation and second boundary condition remain the same:

$$D \frac{1}{r^2} \frac{d}{dr}\left(r^2 \frac{dc_L}{dr}\right) = 0 \qquad (4\text{-}54)$$

$$c_L = c_{L,0} \qquad \text{at } r = \infty \qquad (4\text{-}55)$$

The solution to Equation 4-54 subject to the two boundary conditions (Equations 4-53 and 4-55) is:

$$c_L(r) = -\frac{k_{on} c_R a c_{L,0}}{4\pi D a + k_{on} c_R} \frac{1}{r} + c_{L,0} \qquad (4\text{-}56)$$

The overall rate of binding at the cell surface—which accounts for a balance between diffusion of ligand to the surface and a finite rate of association with receptors at the surface—can be calculated from the flux at the cell surface:

$$\text{Rate} = -D \frac{dc_L}{dr}\bigg|_{r=A} (4\pi a^2) = \frac{k_{on} c_R a c_{L,0}}{4\pi D a + k_{on} c_R} \frac{4\pi a^2}{a^2} \qquad (4\text{-}57)$$

The overall rate of binding is also represented by an overall forward rate constant k_f (see Figure 4.24), which is evaluated from Equation 4-57:

$$k_f = \frac{\text{Rate}}{c_{L,0}} = \frac{(4\pi D a) c_R k_{on}}{4\pi D a + c_R k_{on}} \qquad (4\text{-}58)$$

In Chapter 3, the rate of diffusion of solute to a cell surface was determined for the special case of very rapid reaction at the surface. In that special case, the rate of diffusion determines the rate of reaction at the surface (recall Equation 3-48); the rate constant for the diffusion-limited reaction is:

$$k_+ = 4\pi D a \qquad (4\text{-}59)$$

The overall forward rate constant can, therefore, be written in terms of the rate constant for the diffusion-limited reaction (k_+) and the rate constant for the intrinsic binding reaction $(c_R k_{on})$:

$$k_f = \frac{k_+ c_R k_{on}}{k_+ + c_R k_{on}} \qquad (4\text{-}60)$$

The maximum rate of the overall reaction occurs when the diffusion-limited rate constant is very large $(k_+ \gg c_R k_{on})$; $k_{f,max} = c_R k_{on}$.[4] A similar analysis yields the overall reverse rate constant, k_r, which describes the dissociation of ligand molecules from the cell:

---

4. Note that the rate constants k_f and k_r are defined on a per cell basis, not a per receptor molecule basis. A more detailed description of alternative conventions is available [119].

$$k_r = \frac{k_+ c_R k_{off}}{k_+ + c_R k_{on}} \tag{4-61}$$

These overall rate constants depend on the number of receptors per cell, c_R, as shown for k_f in Figure 4.25.

4.6 DIFFUSION WITHIN CELLS

For many drug molecules, binding to a receptor on the cell surface is required for activity. Drug binding to a cell receptor can activate intracellular signaling pathways or transport processes that eventually bring the drug molecule inside the cell (membrane transport mechanisms are discussed in the next chapter). If the drug acts within the cell, it must move from the point of entry to the site of action. For some drugs, particularly for protein- and gene-based agents, this active site is within a specialized compartment or organelle in the cell; rates of transport within the cell are therefore important.

4.6.1 Structure of the Cytoplasm

Diffusion is an essential mode of molecular transport within individual cells. In addition, the cytoplasm is relatively inaccessible, so intracellular diffusion has not been extensively studied. Consider a typical human cell, schematically depicted in Figure 4.26, in which a plasma membrane encompasses formed organelles suspended within a water-rich matrix; the relative volume of orga-

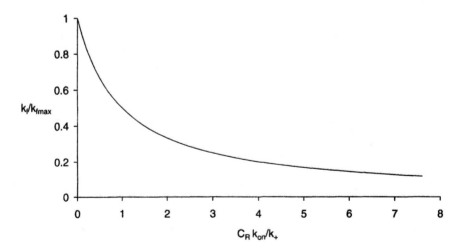

Figure 4.25   The overall rate of ligand binding to cell-surface receptors depends on receptor number. The relative rate of the overall association reaction is shown as a function of dimensionless receptor number ($c_R \times k_{on}/4\pi Da$). This calculation agrees with experimental data for ligand binding as a function of receptor number (see Chapter 4 of [119] for details and references).

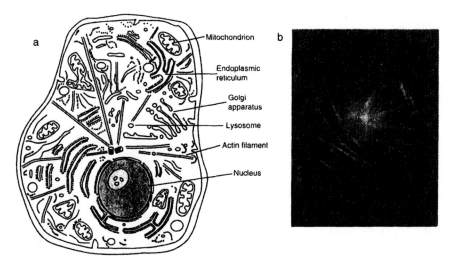

Figure 4.26   Intracellular structure. (a) Illustration of the internal structure of a typical human cell. (b) Fluorescence micrograph showing the organized actin filaments in a surface-attached fibroblast.

nelles varies among cell types (Table 4.7). The intercellular fluid that makes up the bulk of the cytoplasm is 75–80 weight % water, and ~ 20% protein. The structure and composition of this intracellular fluid is complex. The dry matter within the cell has been imaged using an elegant method that permits analysis of intracellular structure during cell movement [120]. Rheologically, the fluid behaves as a weakly viscoelastic gel; the overall rheological properties of the cytoplasm appear to be due to actin and actin-binding proteins [121].

The cytoskeleton is an organized, but dynamic, assembly of proteins that is responsible for cell shape and behavior. The cytoskeleton provides mechanical

Table 4.7   Relative volume of cell compartments

Relative volumes of cell compartments	Percentage	Number
Cytosol	54	1
Mitochondria	22	1,700
Rough ER cisternae	9	1
Smooth ER cisternae and Golgi cisernae	6	
Nucleus	6	1
Peroxisomes	1	400
Endosomes	1	200
Lysosomes	1	300

The total percentage volume and the total number of organelles is presented for a hepatocyte. ER = endoplasmic reticulum. Adapted from [122], p. 553.

strength, but it must also be able to rearrange to accommodate cell movement and cell–cell interactions. Diffusion and binding are both important in the formation and rearrangement of actin filaments within the cytoskeleton. Actin-binding proteins form reversible cross-links that cause actin filaments to organize into bundles. For actin-binding proteins with weak affinity constants, rearrangement can readily occur in the network; the actin fibers form bundles, which is the thermodynamically favored state. For actin-binding proteins of higher affinity, the network is stabilized more rapidly, so bundle formation cannot occur; therefore, the actin fibers are kinetically constrained in the network configuration [123].

4.6.2 Measurements of Intracellular Transport

Diffusion of macromolecules in the cytoplasm is slower than diffusion in water; this effect is more pronounced with larger molecules (Figure 4.27) [29, 30, 124, 125]. The functional dependence of diffusion coefficient on molecular size is similar to that observed for diffusion of proteins in concentrated polymer solutions or gels [126] (Equations 4-28 and 4-29 are frequently used to analyze diffusion in actin solutions and cytoplasm [127]). Globular proteins—lactalbumin, ovalbumin, and serum albumin—diffuse approximately five times slower in the cytoplasm of cultured neurons than in water [9]. When size-fractionated dextrans were microinjected into neuron processes, the reduction in diffusion coefficient was greater for larger molecules (Figure 4.27). The filamentous cytoskeleton appears to create this size-dependent reduction in the diffusion

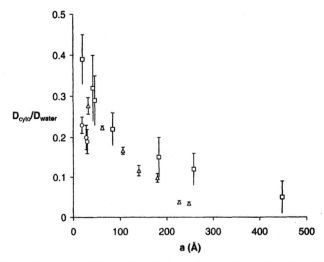

Figure 4.27 Diffusion coefficients for ficoll in the cytoplasm. Relative diffusion coefficient ($D_{A,cyto}/D_{A,water}$) for ficoll diffusion in the cytoplasm of cultured fibroblasts (triangles), albumin diffusion in neurite cytoplasm (circles), and dextran diffusion in neurite cytoplasm (squares). Data from [9, 29].

coefficient; the reduction of diffusion in cytoplasm is greater than the reduction measured in reconstituted F-actin solutions [127], suggesting that organized features of the cytoplasm also contribute to hindrance of diffusion.

The axons of neurons appear to have special transport processes for moving molecules into and out of the cell body (soma). The presence of special transport mechanisms in these cells is essential for their function, since neurons vary greatly in size, with some extending over 1 m, while protein synthesis occurs predominantly in the soma. Transport rates for different solutes occur in two categories: slow (1–10 mm/day) and fast (100–400 mm/day), and occur in the anterograde (away from the cell body) and retrograde (toward the cell body) directions. The speed of the slow processes is consistent with rates of diffusion, but the fast processes require additional mechanisms [9, 128].

The characteristic time for diffusion within cells or cell compartments of various sizes can be estimated from the diffusion coefficient (Table 4.8). Diffusion is an efficient method for distributing molecules throughout a small cell ($\sim 1\,\mu m$): even when the diffusion coefficient is low, diffusion times are less than 10 s. For larger cell compartments, diffusion is very slow, so that large cells ($\sim 100\,\mu m$) cannot rely on diffusion to distribute substrates or newly produced proteins.

Most organelles have a characteristic architecture that is important for function, but may serve as an additional impediment to diffusion within the cell compartment. Organelle membranes within the cell can serve as local barriers (Table 4.9), which affect the rate of diffusion of molecules locally. This effect has been estimated for diffusion in the mitochondrial matrix and the endoplasmic reticulum (Figure 4.28a and b). The mitochondrial matrix was modeled as a closed cylinder with multiple barriers occluding the lumen to simulate the mitochondrial cristae; the presence of multiple barriers produced a modest decrease in the rate of diffusion down the axis of the cylinder (Figure 4.28a); this calculation is consistent with recent measurements of green fluorescent protein (GFP) diffusion in the mitochondrial matrix, which was approximately three-fold slower than diffusion in saline. The endoplasmic reticulum was modeled as an array of interconnected cylinders with a continu-

Table 4.8 Characteristic times (in seconds) for diffusion in cells or cell compartments

L (μm)	Values of D (cm^2/s)				
	10^{-9}	10^{-8}	10^{-7}	10^{-6}	10^{-5}
1	10	1	0.1	0.01	0.001
10	1,000	100	10	1	0.1
100	10^5	10^4	1000	100	10

The characteristic time for diffusion across the compartment was determined for cells or cell compartments of 1, 10, 100 μm in typical dimensions, and for molecules diffusing within the cytoplasm with diffusion coefficients ranging from 10^{-9} to 10^{-5} cm^2/s: characteristic time $= L^2/D$

Table 4.9   Relative amounts of membranes for cell organelles

Membrane type	Hepatocyte volume $\sim 5,000\,\mu m^3$	Pancreatic exocrine cell volume $\sim 100\,\mu m^3$
Plasma membrane	2	5
Rough ER	35	60
Smooth ER	16	< 1
Golgi apparatus	7	10
Mitochondria		
Inner membrane	7	4
Outer membrane	32	17
Nucleus		
Inner membrane	0.2	0.7
Secretory vesicle	—	3
Lysosome	0.4	—
Peroxisome	0.4	—
Endosome	0.4	—
Total membrane area	$\sim 110,000\,\mu m^2$	$\sim 13,000\,\mu m^2$

The relative amount of membrane (%) is presented for a hepatocyte and a pancreatic exocrine cell. ER = endoplasmic reticulum. Adapted from [122], p. 553.

ously connected lumen; free diffusion in this structure was not significantly slower than diffusion in an unobstructed matrix. However, the rate of dispersion was substantially reduced in the presence of binding to the lumenal surface of the cylinder.

The structure of the nucleus, which may turn out to be the most complex (and most important) subcellular structure for drug targeting, is illustrated in Figure 4.28c. The outer nuclear membrane is continuous with the endoplasmic reticulum. Chromosomes are long, linear sequences of DNA and have associated proteins, including histone proteins which comprise the same mass as the DNA. Each chromosome has a centromere, which is necessary for segregation during mitosis, and two telomeres, which occur at each end of the chromosome and are essential for replication of the terminal sequences. The chromatin of each chromosome occupies a territory; the territories are separated by interchromosomal domains in which large molecular tracers (500-kDa dextran) can freely diffuse (crossing the $10\,\mu m$ nucleus in $\sim 7\,s$). Chromatin is mobile within the territory of a chromosome: labeled centromeres are observed to move at $\sim 10\,\mu m/h$, but movement occurs at infrequent intervals. Within each chromosome, regions of early replication tend to be located nearer the center of the nucleus. Chromatin can be indentified morphologically as heterochromatin, which is highly condensed and inactive, and euchromatin, which contains the active gene sequences. Active genes tend to occur more frequently at spatial locations in the chromosome territory far from the heterochromatin, which contains certain nuclear proteins that can suppress transcription.

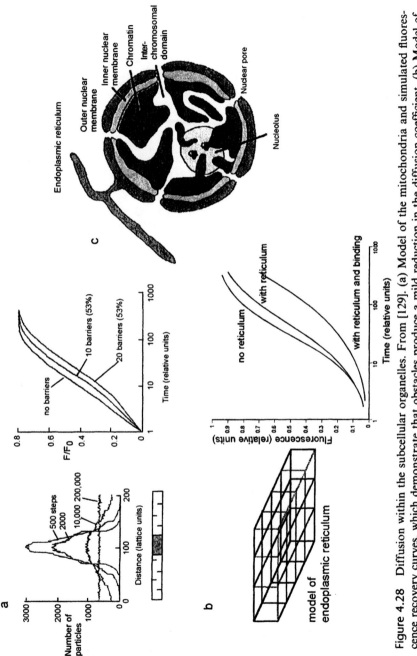

Figure 4.28 Diffusion within the subcellular organelles. From [129]. (a) Model of the mitochondria and simulated fluorescence recovery curves, which demonstrate that obstacles produce a mild reduction in the diffusion coefficient. (b) Model of the endoplasmic reticulum and simulated fluorescence recovery curves, which demonstrate a mild influence on diffusion and substantial reduction in diffusion with binding to the lumenal surface. (c) Structure of a typical interphase nucleus. Modified from Chapter 8 of [122] and [130].

4.7 DIFFUSION AND REACTION

Most biological systems are chemically and structurally complex. As a result, the rate of solute diffusion is often low (Table 4.8), because physical structures can impede diffusing molecules and because the diffusing molecules are often large. The processes of life (i.e., growth and metabolism) occur through an orderly—i.e., highly regulated and strongly coupled—array of biochemical reactions. The reaction between any two biochemical substrates, A and B, requires contact or collision between the two reactants. In biological systems, it is often the rate of diffusion that determines the frequency of collision and, therefore, the rate of reaction.

4.7.1 Rates of Enzymatic Reactions

Most cellular functions depend on enzymatic reactions. Turnover rates for intracellular enzymes are typically several hundred reactions per second, although the variation in catalytic power is high (Table 4.10). A typical intracellular compartment ($L \sim 1\,\mu\text{m}$) will be well mixed by diffusion: if $D \sim 10^{-6}\,\text{cm}^2/\text{s}$, with a diffusion time of $\sim 10\,\text{ms}$. In this micrometer-scale compartment, any two molecules will collide every 1 s [131]. Clearly, this rate of collision will have important implications for the overall rate of reaction for enzymes with high turnover number.

4.7.2 Diffusion-limited Biochemical Reactions

A commonly used model to analyze enzyme–substrate (or receptor–ligand) reactions is shown in Figure 4.29 (this model is similar to the one used to

Table 4.10   Maximum turnover numbers, and reaction time for enzymes

Enzyme	Turnover number (1/s)	Reaction time (ms)
Carbonic anhdrase	600,000	0.0017
3-Ketosteroid isomerase	280,000	0.0036
Acetylcholinesterase	25,000	0.040
Penicillinase	2,000	0.5
Lactate denhydrogenase	1,000	1
Chymotrypsin	100	10
DNA polymerase I	15	67
Tryptophan synthetase	2	500
Lysozyme	0.5	2,000

The turnover number, which is equal to the maximal reaction rate V_{max} divided by the concentration of active sites on the enzyme, is indicated for a variety of enzymes (from [58], p. 191). The reaction time was determined as 1/(turnover number), which estimates a characteristic time for a round of catalysis by the enzyme.

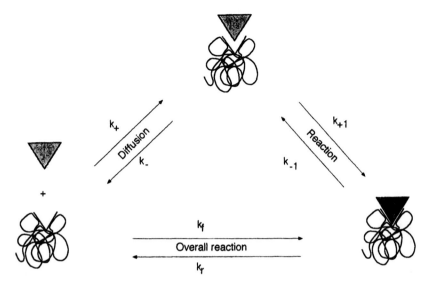

Figure 4.29 Schematic mechanism for diffusion-limited reactions. The overall rate of reaction is characterized by forward and reverse rate constants (k_f and k_r). The reaction occurs in two steps: diffusion brings the two reactants in proximity (rate constants k_+ and k_-) and then reaction occurs with (intrinsic rate constants (k_{+1} and k_{-1}).

evaluate binding at the cell surface in Section 4.5). Here, the overall reaction of solutes A and B occurs in two reversible steps: the first step involves "encounter" or collision—in which the two solutes come in close enough proximity to permit reaction—and the second step involves the intrinsic rates of reaction.

A mass balance equation for the reaction kinetics can be written for each of the species A, B, A—B, and A–B. The assumption that the intermediate species (A—B) reaches a steady-state concentration rapidly compared to the other species yields:

$$\frac{dc_{AB}}{dt} = k_f c_A c_B - k_r c_{AB}$$

$$\text{where } k_f = \frac{k_{+1}k_+}{k_- + k_1} \quad \text{and} \quad k_r = \frac{k_{-1}k_-}{k_- + k_1}$$

(4-62)

In this formulation, the overall apparent rates of the forward and reverse reaction (k_f and k_r) can be written in terms of rates of encounter (k_+ and k_-) and the intrinsic forward and reverse reaction rates (k_{+1} and k_{-1}).

The encounter step involves the simultaneous diffusion of two separated solute species that collide during random motion. The kinetics of this event (k_+ and k_-) can be determined by analysis of the diffusion equation. The relevant diffusion equation was provided in the previous chapter, in the context of the diffusion of solutes to the surface of a cell. If the cell is replaced with one of the solute molecules, A, which is assumed to be fixed in space and surrounded by solute B molecules that can diffuse with diffusion coefficient D equal to $D_A + D_B$, the rate of collision is equal to the steady-state flux of B molecules to the surface of the A molecule:

$$k_+ c_{B,\infty} = -D(4\pi a^2)\frac{\partial c_B}{\partial t} \tag{4-63}$$

where $c_{B,\infty}$ is the bulk concentration of solute B and a is the radius of the encounter complex (A—B) that is formed upon collision of B with A. Substituting the steady-state concentration profile for c_B (identical to Equation 4-59) yields:

$$k_+ = 4\pi a D \tag{4-64}$$

Remember that this reaction-rate constant was derived for a single molecule of A. A similar procedure can be used to estimate the reverse-rate constant, except that the boundary conditions on the diffusion equation must be modified: instead of $c_B = 0$ at the complex surface and $c_B = c_{B,\infty}$ far from that surface (as was used to find Equation 4-64), the reverse reaction starts with a single B molecule "bound" in the encounter complex (of volume $4\pi a^3/3$), which must subsequently diffuse into an unbounded fluid in which its concentration is negligible. The rate of dissociation of the complex is equal to the rate of diffusion of B molecules away from the complex surface:

$$k_- \left[\frac{1}{\frac{4}{3}\pi a^3}\right] = -D(4\pi a^2)\frac{\partial c_B}{\partial t} \tag{4-65}$$

which, on substitution of c_B that is found by solution of the diffusion equation, yields:

$$k_- = \frac{3D}{a^2} \tag{4-66}$$

The encounter-rate constants k_+ and k_- can usually be estimated from physical properties of the two reactants (i.e., the diffusion coefficient and encounter radius), using Equations 4-64 and 4-66. Therefore, the relative importance of diffusion vs reaction in determining the overall rate of reaction can be estimated by using the definitions of k_f and k_r in Equation 4-62.

4.7.3 Example of DNA-binding Proteins

The relative rates of diffusion and reaction are important in the regulation of biochemical pathways. Transcription factors are proteins that bind to specialized regions of DNA and thereby facilitate transcription by RNA polymerase. The binding of one class of transcription factors, the basic leucine zipper (bZIP) proteins, is illustrated in Figure 4.30. The bZIP factors have a C-terminal leucine zipper domain, a basic region that binds to DNA, and a domain that is important for transcriptional regulation. The bZIP factors must dimerize for transcriptional activity and, therefore, it is usually assumed that dimerization occurs before DNA binding (dimer pathway). A recent study shows that monomer and dimer binding (reactions 3 and 2 in Figure 4.30) are very rapid (i.e., diffusion limited), so that the dimerization reaction (reaction 1) is not necessary for function of bZIP factors [132]. In fact, the monomer pathway may provide kinetic advantages in transcriptional regulation. Monomeric bZIP can bind to DNA more rapidly than dimeric bZIP (because of its smaller size and faster diffusion). In addition, because the monomer bZIP–DNA interaction is probably weaker than the dimer bZIP–DNA interaction (binding is due to electrostatic interactions between the basic protein and the charged polyphosphate backbone; the monomer has fewer basic residues to interact), monomers can "slide" along the DNA backbone. "Sliding" represents an opportunity for bZIP to perform a one-dimensional search of the DNA strand. Faster searching allows the transcription factor to find its target rapidly; binding to the target is stabilized by dimerization of bZIP at the target site.

The importance of one-dimensional diffusion in the function of DNA-binding proteins has been best studied with the restriction endonuclease *Eco*RV, which cleaves the DNA polymer after recognizing a specific restriction site on the DNA strand.

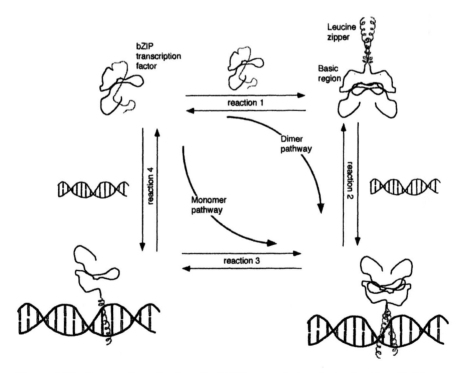

Figure 4.30 Proposed mechanisms for bZIP transcription factor function. Stable association of the bZIP dimer with DNA can occur via a monomer or dimer pathway. Binding of both monomer and dimer is diffusion limited. Analysis of the kinetics of both pathways suggests that the monomer pathway may have an overall kinetic advantage.

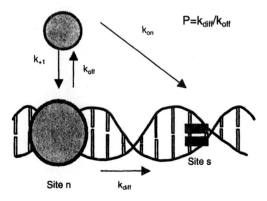

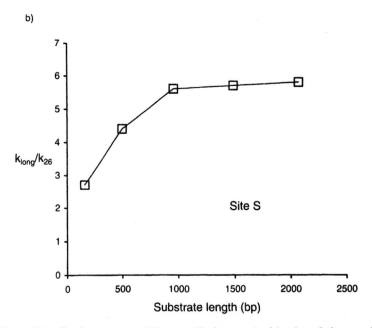

Figure 4.31   The importance of linear diffusion on the kinetics of cleavage by EcoRV. EcoRV recognizes and cleaves GAT↓ATC sites on DNA. The overall rate of cleavage was determined for PCR-amplified DNA sequences of different length, but each containing one restriction site. (a) Model for association and one-dimensional diffusion. (b) Experimental data for relative rate of cleavage for long DNA vs short DNA (26 bp) [133].

The role of linear diffusion in the overall kinetics of cleavage can be analyzed by the model shown in Figure 4.31 [134]:

$$\frac{k_{on}}{k_{+1}} = \sum_{n=1}^{L} \exp\left[-(n-s)^2/P\right] \tag{4-67}$$

where k_{on} is the overall rate of association with the target site (at position s), k_{+1} is the rate of association of $EcoRV$ with a site n, and P is the probability that the enzyme will diffuse one base pair along the DNA rather than dissociate from the DNA polymer; P is equal to the ratio of the rate constant for diffusion (k_{diff}) to the rate constant for dissociation (k_{off}). For the example shown in Figure 4.31, a value of $P = 5 \times 10^5$ provides the best fit of Equation 4-67 to the experimental data. This value of P suggests that $EcoRV$ takes 5×10^5 random steps along the DNA chain for each association event; since diffusive spread in a random walk increases with the square root of the number of steps (recall Equation 3-8), the restriction enzyme surveys ~ 707 base pairs with each visit to the chain.

SUMMARY

- Diffusion coefficients in aqueous environments are usually measured by observing changes in concentration over time.

- Diffusion coefficients in water depend on molecular size and shape, as well as concentration.

- Diffusion in membranes, polymer solutions, and gels can be described adequately by simple models.

- Overall rates of diffusion through tissues—or within cells—depend on tissue composition, local architecture, and the extent of binding,

- Diffusion can influence the rate of biochemical reactions, because reactions occur after random collisions of reactants in diffusive motion.

REFERENCES

1. Levin, V., J. Fenstermacher, and C. Patlak, Sucrose and inulin space measurements of cerebral cortex in four mammalian species. *American Journal of Physiology*, 1970, **219**, 1528–1533.
2. Dykstra, K.H., et al., Quantitative examination of tissue concentration profiles associated with microdialysis. *Journal of Neurochemistry*, 1992, **58**, 931–940.
3. Strasser, J.F., et al., Distribution of 1,3-bis(2-chloroethyl)-1-nitrosourea (BCNU) and tracers in the rabbit brain following interstitial delivery by biodegradable polymer implants. *Journal of Pharmacology and Experimental Therapeutics*, 1995, **275**(3), 1647–1655.
4. Dang, W. and W.M. Saltzman, Dextran retention in the rat brain following controlled release from a polymer. *Biotechnology Progress*, 1992, **8**, 527–532.
5. Saltzman, W.M., et al., Antibody diffusion in human cervical mucus. *Biophysical Journal*, 1994, **66**, 508–515.

6. Radomsky, M.L., *et al.*, Macromolecules released from polymers: diffusion into unstirred fluids. *Biomaterials*, 1990, **11**, 619–624.

7. Groebe, K., S. Erz, and W. Mueller-Klieser, Glucose diffusion coefficients determined from concentration profiles in EMT6 tumor spheroids incubated in radioactively labeled L-glucose. *Advances in Experimental Medicine and Biology*, 1994, **361**, 619–625.

8. Bjelke, B., *et al.*, Long distance pathways of diffusion for dextran along fibre bundles in brain. Relevance for volume transmission. *NeuroReport*, 1995, **6**, 1005–1009.

9. Popov, S. and M.M. Poo, Diffusional transport of macromolecules in developing nerve processes. *Journal of Neuroscience*, 1992, **12**(1), 77–85.

10. Terada, S., *et al.*, Visualization of Slow Axonal Transport *in Vivo*. *Science*, 1996, **273**, 784–788.

11. Johanson, S., M. Crouch, and I. Hendry, Retrograde axonal transport of signal transduction proteins in rat sciatic nerve. *Brain Research*, 1995, **690**, 55–63.

12. Tao, L. and C. Nicholson, Diffusion of albumins in rat cortical slices and relevance to volume transmission. *Neuroscience*, 1996, **75**(3), 839–847.

13. Levin, V.A., J.D. Fenstermacher, and C.S. Patlak, Sucrose and inulin space measurements of cerebral cortex in four mammalian species. *American Journal of Physiology*, 1970, **219**(5), 1528–1533.

14. Fenstermacher, J. and T. Kaye, Drug "diffusion" within the brain. *Annals of the New York Academy of Sciences*, 1988, **531**, 29–39.

15. Schnitzer, J.J., *et al.*, Absolute quantitative autoradiography of low concentrations of 125I-labeled proteins in arterial tissue. *Journal of Histochemistry and Cytochemistry*, 1987, **35**(12), 1439–1450.

16. Dykstra, K.H., *et al.*, Microdialysis study of zidovudine (AZT) transport in rat brain. *Journal of Pharmacology and Experimental Therapeutics*, 1993, **267**(3), 1227–1236.

17. Sung, C., *et al.*, The spatial distribution of immunotoxins in solid tumors: assessment by quantitative autoradiography. *Cancer Research*, 1993, **53**, 2092–2099.

18. Krewson, C.E., M. Klarman, and W.M. Saltzman, Distribution of nerve growth factor following direct delivery to brain interstitium. *Brain Research*, 1995, **680**, 196–206.

19. Kanekal, S., *et al.* Three-dimensional reconstruction of two-dimensional tumor autoradiograms obtained by storage-phosphor radioluminography, in *87th Annual Meeting of the American Association for Cancer Research*. Washington, DC: Matirix Pharmaceutical, Inc., 1996.

20. Nugent, L. and R. Jain, Extravascular Diffusion in Normal and Neoplastic Tissues. *Cancer Research*, 1984, **44**, 238–244.

21. Clauss, M.A. and R.K. Jain, Interstitial transport of rabbit and sheep antibodies in normal and neoplastic tissues. *Cancer Research*, 1990, **50**, 3487–3492.

22. Jain, R.K., Barriers to drug delivery in solid tumors. *Scientific American*, 1994, **271**(1), 58–65.

23. Williams, R.M., D.W. Piston, and W.W. Webb, Two-photon molecular excitation provides intrinsic 3-dimensional resolution for laser-based microscopy and microphotochemistry. *FASEB Journal*, 1994, **8**(11), 804–813.

24. Xu, C., *et al.*, Multiphoton fluorescence excitation: new spectral windows for biological nonlinear microscopy. *Proceedings of the National Academy of Sciences, USA*, 1996, **93**, 10763–10768.

25. Maiti, S., *et al.*, Measuring serotonin distribution in live cells with three-photon excitation. *Science*, 1997, **275**, 530–532.

26. Wolf, D.E., Designing, building, and using a fluorescence recovery after photobleaching instrument, in D.L. Taylor and Y. Wang, *Fluorescence Microscopy of Living Cells in Culture Part B*, San Diego, CA: Academic Press, 1989, pp. 271–306.

27. Edidin, M., S.C. Kuo, and M.P. Sheetz, Lateral movements of membrane glycoproteins restricted by dynamic cytoplasmic barriers. *Science*, 1991, **254**, 1379–1382.

28. Wier, M. and M. Edidin, Constraint of the translational diffusion of a membrane glycoprotein by its external domains. *Science*, 1988, **242**, 412–414.

29. Luby-Phelps, K., *et al.*, Hindered diffusion of inert tracer particles in the cytoplasm of mouse 3T3 cells. *Proceedings of the National Academy of Sciences USA*, 1987, **84**, 4910–4913.

30. Luby-Phelps, K., F. Lanni, and D. Taylor, The submicroscopic properties of cytoplasm as a determinant of cellular function. *Annual Review of Biophysics and Biophysical Chemistry*, 1988, **17**, 369–396.

31. Berk, D.A., *et al.*, Fluorescence photobleaching with spatial fourier analysis: Measurement of diffusion in light-scattering media. *Biophysical Journal*, 1993, **65**, 2428–2436.

32. Kaufman, E.N. and R.K. Jain, Measurement of mass transport and reaction parameters in bulk solution using photobleaching. *Biophysical Journal*, 1991, **60**, 596–616.

33. Berk, D., *et al.*, Fluorescence photobleaching with spatial fourier analysis: measurement of diffusion in light-scattering media. *Biophysical Journal*, 1993, **65**, 2428–2436.

34. Williams, R., D. Piston, and W. Webb, Two-photon molecular excitation provides intrinsic 3-dimensional resolution for laser-based microscopy and microphotochemistry. *FASEB Journal*, 1994, **8**(11), 804–13.

35. Axelrod, D., *et al.*, Mobility measurement by analysis of fluorescence photobleaching recovery kinetics. *Biophysical Journal*, 1976, **16**, 1055–1069.

36. Johnson, E.M., *et al.*, Hindered diffusion in agarose gels: test of the effective medium model. *Biophysical Journal*, 1996, **70**, 1017–1026.

37. Salmon, E.D., *et al.*, Diffusion coefficient of fluorescein-labeled tubulin in the cytoplasm of embryonic cells of a sea urchin: video image analysis of fluorescence redistribution after photobleaching. *Journal of Cell Biology*, 1984, **99**(6), 2157–2164.

38. Chary, S.R. and R.K. Jain, Direct measurement of interstitial convection and diffusion of albumin in normal and neoplastic tissues by fluorescence bleaching. *Proceedings of the National Academy of Sciences, USA*, 1989, **86**, 5385–5389.

39. Nugent, L.J. and R.K. Jain, Extravascular diffusion in normal and neoplastic tissues. *Cancer Research*, 1984, **44**, 238–244.

40. Nicholson, C., Ion-selective microelectrodes and diffusion measurements as tools to explore the brain cell microenvironment. *Journal of Neuroscience Methods*, 1993, **48**, 199–213.

41. Nicholson, C. and J.M. Phillips, Ion diffusion modified by tortuosity and volume fraction in the extracellular microenvironment of the rat cerebellum. *Journal of Physiology*, 1981, **321**, 225–257.

42. Lundbaek, J.A. and A.J. Hansen, Brain interstitial volume fraction and tortuosity in anoxia. Evaluation of the ion-selective micro-electrode method. *Acta Physiologica Scandinavica*, 1992, **146**(4), 473–484.

43. Rice, M.E. and C. Nicholson, Diffusion characteristics and extracellular volume fraction during normoxia and hypoxia in slices of rat neostriatum. *Journal of Neurophysiology*, 1991, **65**(2), 264–272.

44. Kelly, R.S. and R.M. Wightman, Detection of dopamine overflow and diffusion with voltammetry in slices of rat brain. *Brain Research*, 1987, **423**, 79–87.

45. Nicholson, C. and L. Tao, Diffusion properties of brain tissue measured with electrode methods and prospects for optical analysis, in J. Dirnagl, *Optical Imaging of Brain Function and Metabolism*, New York: Plenum Press, 1993.

46. Reisfeld, B., et al., Tracking and modeling controlled drug release and transport in the brain. *Magnetic Resonance Imaging*, 1993, **11**, 247–252.

47. Einstein, A., A new determination of molecular dimensions. *Annalen der Physik*, 1906, **19**, 289–306.

48. Bird, R.B., W.E. Stewart, and E.N. Lightfoot, *Transport Phenomena*. New York: John Wiley, 1960 780 pp.

49. Polson, A., *J. Phys. Colloid Chem.*, 1950, **54**, 649.

50. Tyn, M. and T. Gusek, Prediction of diffusion coefficients of proteins. *Biotechnology and Bioengineering*, 1990, **35**, 327–338.

51. Geankopolis, C.J., *Transport Processes and Unit Operations*, 3rd ed. Englewood Cliffs, NJ: Prentice-Hall, 1993.

52. Haller, M.F. and W.M. Saltzman, Localized delivery of proteins in the brain: Can transport be customized? *Pharmaceutical Research*, 1998, **15**, 377–385.

53. Sober, H.A., ed., *Handbook of Biochemistry: Selected Data for Molecular Biology*, 2nd ed. Boca Raton, FL: CRC Press, 1970.

54. Kosar, T.F. and R.J. Phillips, Measurement of protein diffusion in dextran solutions by holographic interferometry. *AIChE Journal*, 1995, **41**(3), 701–711.

55. Gosnell, D.L. and B.H. Zimm, Measurement of diffusion coefficient of DNA in agarose gel. *Macromolecules*, 1993, **26**, 1304–1308.

56. Lehninger, A.L., *Biochemistry*, 2nd ed. New York: Worth Publishers, 1975.

57. de la Torre, J.G., S. Navarro, and M.C.L. Martinez, Hydrodynamic properties of a double-helical model for DNA. *Biophysical Journal*, 1994, **66**, 1573–1579.

58. Stryer, L., *Biochemistry*, 2nd ed. New York: W.H. Freeman, 1988.

59. Eimer, W. and R. Pecora, Rotational and translational diffusion of short rodlike molecules in solution: oligonucleotides. *Journal of Chemical Physics*, 1991, **94**(3), 2324–2329.

60. Wilke, C.R. and P. Chang, Correlation of diffusion coefficients in dilute solutions. *AIChE Journal*, 1955, **1**, 264-270.

61. Cussler, E.L., *Diffusion: Mass Transfer in Fluid Systems*. Cambridge: Cambridge University Press, 1984, 525 pp.

62. Kausch, H. and M. Tiorrell, Polymer interdiffusion. *Annual Review of Material Science*, 1989, **19**, 341–377.

63. Anderson, J. and C. Reed, Diffusion of spherical macromolecules at finite concentration. *Journal of Chemical Physics*, 1976, **64**, 3240–3250.

64. Keller, K., E. Canales, and S. Yum, Tracer and mutual diffusion coefficients of proteins. *Journal of Physical Chemistry*, 1971, **75**, 379–387.

65. Phillies, G., G. Benedek, and N. Mazer, Diffusion in protein solutions at high concentrations: A study by quasielastic light scattering spectroscopy. *Journal of Chemical Physics*, 1976, **65**, 1883–1892.

66. Gibbs, S.J., E.N. Lightfoot, and T.W. Root, Protein diffusion in porous gel filtration chromatographic media studied by pulsed field gradient NMR-spectroscopy. *Journal of Physical Chemistry*, 1992, **96**, 7458–7462.

67. Sorlie, S.S. and R. Pecora, A dynamic light scattering study of a 2311 base pair DNA restriction fragment. *Macromolecules*, 1988, **21**, 1437–1449.
68. Batchelor, G.K., *Journal of Fluid Mechanics*, 1976, **74**, 1.
69. Deen, W.M., Hindered transport of large molecules in liquid filled pores. *AIChE Journal*, 1987, **33**, 1409–1425.
70. Renkin, E.M., Filtration, diffusion, and molecular sieving through porous cellulose membranes. *Journal of General Physiology*, 1954, **38**, 225–243.
71. Anderson, J.L. and J.A. Quinn, Restricted transport in small pores: A model for steric exclusion and hindered particle motion. *Biophysical Journal*, 1974, **14**, 130–150.
72. Ogston, A., B. Preston, and J. Wells, On the transport of compact particles through solutions of chain polymers. *Proc. R. Soc. Lond*, 1973, **333**, 297–316.
73. Phillies, G., The hydrodynamic scaling model for polymer self-diffusion. *Journal of Physical Chemistry*, 1989, **93**, 5029–5039.
74. Yam, K.L., D.K. Anderson, and R.E. Buxbaum, Diffusion of small solutes in poymer-containing solutions. *Science*, 1988, **241**, 330–332.
75. Phillies, G., *et al.*, Probe diffusion in solutions of low molecular weight polyelectrolytes. *Macromolecules*, 1989, **22**, 4068–4075.
76. Peppas, N., P. Hansen, and P. Buri, A theory of molecular diffusion in the intestinal mucus. *International Journal of Pharmaceutics*, 1984, **20**, 107–118.
77. Phillips, R.J., W.M. Deen, and J.F. Brady, Hindered transport of spherical macromolecules in fibrous membranes and gels. *AIChE Journal*, 1989, **35**, 1761–1769.
78. Brady, J. Hindered diffusion, in AIChE Annual Meeting. San Francisco, CA, 1994.
79. Clague, D. and R. Phillips, Hindered diffusion of spherical macromolecules through dilute fibrous media. *Phys. Fluids*, 1996, **8**(7), 1720–1731.
80. Perrins, W., D. McKenzie, and R. McPhedran, Transport properties of regular arrays of cylinders. *Proc. R. Soc. Lond.*, 1979, **369**, 207–225.
81. Johansson, L. and J.-E. Lofroth, Diffusion and interaction in gels and solutions. 4. Hard sphere Brownian dynamics simulations. *Journal of Chemical Physics*, 1993, **98**(9), 7471–7479.
82. Johnson, E.M., *et al.*, Diffusion and partitioning of proteins in charged agarose gels. *Biophysical Journal*, 1995, **68**(4), 1561–1568.
83. de Gennes, P.-G., *Journal of Chemical Physics*, 1971, **55**, 572.
84. Perkins, T.T., D.E. Smith, and S. Chu, Direct observation of tube-like motion of a single polymer chain. *Science*, 1994, **264**, 819–822.
85. Amsden, B., Solute diffusion within hydrogels. Mechanisms and models. *Macromolecules*, 1998, **31**, 8382–8395.
86. Reinhart, C. and N. Peppas, Solute diffusion in swollen membranes. Part II. Influence of crosslinking on diffusive properties. *Journal of Membrane Science*, 1984, **18**, 227–239.
87. Tao, L. and C. Nicholson, *Diffusion of Albumins in Rat Cortical Slices and Relevance to Volume Transmission*. New York: New York University Medical Center, 1996.
88. Curran, R.E., *et al.*, Cerebrospinal fluid production rates determined by simultaneous albumin and inulin perfusion. *Experimental Neurology*, 1970. **29**, 546–553.
89. Lundbaek, J.A. and A.J. Hansen, Brain interstitial volume fractin and tortuosity in anoxia. Evaluation of the ion-selective microelectrode method. *Acta Physiol. Scand.*, 1992, **146**, 473–484.

90. Nicholson, C., J. Phillips, and A. Gardner-Medwin, Diffusion from an iontophoretic point source in the brain: role of tortuosity and volume fraction. *Brain Research*, 1979, **169**, 580–584.

91. Eis, M., *et al.*, Quantitative diffusion MR imaging of cerebral tumor and edema. *Acta Neurochir*, 1994, **60**, 344–346.

92. Freeman, C.L., *et al.*, A study of the mass transport resistance of glucose across rat capsular membranes. *Materials Research Society Symposium Proceedings*, 1989, **110**, 773–778.

93. Sharkawy, A.A., *et al.*, Engineering the tissue which encapsulates subcutaneous implants. I. Diffusion properties. *Journal of Biomedical Materials Research*, 1997, **37**, 401–412.

94. Maxwell, J.C., *A Treatise on Electricity and Magnetism*. Vol. 1. 1Oxford: Clarendon Press, 1873, 365 pp.

95. Crank, J., *The Mathematics of Diffusion*. 2nd ed. Oxford: Oxford University Press, 1975, 414 pp.

96. Kozeny, J., *Hydralik*. Wien: Springer-Verlag, 1927.

97. Dullien, V., Single phase flow through porous media and pore structure. *Chemical Engineering Journal*, 1975, **10**, 1–34.

98. Greenkorn, R., Steady flow through porous media. *AIChE Journal*, 1981, **27**, 529–545.

99. Brakel, J., Pore space models for transport phenomena in porous media: review and evaluation with special emphasis on capillary liquids transport. *Powder Technology*, 1975, **11**, 205–236.

100. Scheidegger, A.E., *The Physics of Flow Through Porous Media*. Toronto: University of Toronto Press, 1974.

101. Cunningham, R. and R. Williams, *Diffusion in Gases and Porous Media*. New York: Plenum Press, 1980, pp. 129–264.

102. Greenkorn, R.A., *Flow Phenomena in Porous Media*. New York: Marcel Dekker, 1983.

103. Satterfield, C.N., *Mass Transport in Heterogeneous Catalysis*. 1970, Cambridge, MA: MIT Press, 1970.

104. Pismen, L., Diffusion in porous media of a random structure. *Chemical Engineering Science*, 1974, **29**, 1227–1236.

105. Bhatia, S., Stochastic theory of transport in inhomogeneous media. *Chemical Engineering Science*, 1986, **41**, 1311–1324.

106. El-Kareh, A., S. Braunstein, and T. Secomb, Effect of cell arrangement and interstitial volume fraction on the diffusivity of monoclonal antibodies in tissue. *Biophysical Journal*, 1993, **64**, 1638–1646.

107. Saltzman, W.M. and R. Langer, Transport rates of proteins in porous polymers with known microgeometry. *Biophysical Journal*, 1989, **55**, 163–171.

108. Shante, V. and S. Kirkpatrick, An introduction to percolation theory. *Advances in Physics*, 1971, **20**, 325–357.

109. Stauffer, D., Scaling theory of percolation clusters. *Physics Reports*, 1979, **54**, 1–74.

110. Winterfeld, P., L. Scriven, and H. Davis, Percolation and conductivity of random two dimensional composites. *Journal of Physics C: Solid State Physics*, 1981, **14**: 2361–2376.

111. Mohanty, K., J. Ottino, and H. Davis, Reaction and transport in disordered composite media: Introduction of percolation concepts. *Chemical Engineering Science*, 1982, **37**, 905–924.

112. Fisher, M. and J. Essam, Some cluster size and percolation problems. *Journal of Mathematical Physics*, 1961, **2**, 609–619.

113. Stinchcombe, R., Conductivity and spin-wave stiffness in disordered systems—an exactly soluble model. *Journal of Physics C: Solid State Physics*, 1974, **7**, 179–203.

114. Reyes, S. and K. Jensen, Estimation of effective transport coefficient in porous solids based on percolation concepts. *Chemical Engineering Science*, 1985, **40**, 1723–1734.

115. Winterfeld, P.H., Percolation and conduction phenomena in disordered composite media, PhD thesis, University of Minnesota, 1981.

116. Saltzman, W.M., A microstructural approach for modelling diffusion of bioactive macromolecules in porous polymers, PhD thesis, Massachusetts Institute of Technology, 1987

117. Kirkpatrick, S., Percolation and conduction. *Reviews of Modern Physics*, 1973, **45**, 574–588.

118. Berk, D.A., F. Yuan, M. Leunig, and R. K. Jain, Direct in vivo measurement of targeted binding in a human tumor xenograft. *Proceedings of the National Academy of Sciences, USA*, 1997, **94**, 1785–1790.

119. Lauffenburger, D.A. and J.J. Linderman, *Receptors: Models for Binding, Trafficking, and Signaling*. New York: Oxford University Press, 1993, 365 pp.

120. Brown, A. and G. Dunn, Microinterferometry of the movement of dry matter in fibroblasts. *Journal of Cell Science*, 1989, **92**, 379–389.

121. Sato, M., W. Schwarz, and T. Pollard, Dependence of the mechanical properties of actin/α-actinin gels on deformation rate. *Nature*, 1987, **325**, 828–830.

122. Alberts, B., *et al.*, *Molecular Biology of the Cell*, 3rd ed. New York: Garland Publishing, 1994.

123. Wachsstock, D.H., W.H. Schwarz, and T.D. Pollard, Cross-linker dynamics determine the mechanical properties of actin gels. *Biophysical Journal*, 1994, **66**, 801–809.

124. Luby-Phelps, K., F. Lanni, and D. Taylor, Behavior of a fluorescent analogue of calmodulin in living 3T3 cells. *Journal of Cell Biology*, 1985, **101**, 1245–1256.

125. Luby-Phelps, K., D. Taylor, and F. Lanni, Probing the structure of cytoplasm. *Journal of Cell Biology*, 1986, **102**, 2015–2022.

126. Hou, L., F. Lanni, and K. Luby-Phelps, Tracer diffusion in F-actin and Ficoll mixtures. Toward a model for cytoplasm. *Biophysical Journal*, 1990, **58**, 31–43.

127. Jones, J.D. and K. Luby-Phelps, Tracer diffusion through F-actin: Effect of filament length and cross-linking. *Biophysical Journal*, 1996, **71**, 2742–2750.

128. Glass, J.D. and J.W. Griffin, Retrograde Transport of Radiolabeled Cytoskeletal Proteins in Transected Nerves. *Journal of Neuroscience*, 1994, **14**, 3915–3921.

129. Olveczky, B.P. and A.S. Verkman, Monte Carlo analysis of obstructed diffusion in three dimensions: application to molecular diffusion in organelles. *Biophysical Journal*, 1998, **74**, 2722–2730.

130. Lamond, A.I. and W.C. Earnshaw, Structure and function in the nucleus. *Science*, 1998, **280**, 547–553.

131. Hess, B. and A. Mikhailov, Self-organization in living cells. *Science*, 1994, **264**, 223–224.

132. Berger, C., *et al.*, Diffusion-controlled DNA recognition by an unfolded, monomeric bZIP transcription factor. *FEBS Letters*, 1998, **425**, 14–18.

133. Jeltsch, A., *et al.*, Linear diffusion of the restriction endonuclease EcoRV on DNA is essential for the *in vivo* function of the enzyme. *EMBO Journal*, 1996, **15**(18), 5104–5111.

134. Jeltsch, A. and A. Pingoud, Kinetic characterization of linear diffusion of restriction endonuclease EcoRV on DNA. *Biochemistry*, 1998, **37**, 2160–2169.
135. Fair, B., D. Chao, and A. Jamieson, Mutual translational diffusion coefficients in bovine serum albumin solutions measured by quasielastic laser light scattering. *Journal of Colloid and Interface Science*, 1978, **66**, 324–330.
136. Ottewill, R. and N. Williams, Study of particle motion in concentrated dispersions by tracer diffusion. *Nature*, 1987, **325**, 232–234.
137. Anderson, J., F. Rauh, and A. Morales, Particle diffusion as a function of concentration and ionic strength. *Journal of Physical Chemistry*, 1978, **82**, 608–616.
138. Muramatsu, N. and A. Minton, Tracer diffusion of globular proteins in concentrated protein solutions. *Proceedings of the National Academy of Sciences, USA*, 1988, **85**, 2984–2988.
139. Muramatsu, N. and A. Minton, An automated method for rapid determination of diffusion coefficients via measurements of boundary spreading. *Analytical Biochemistry*, 1988, **168**, 345–351.
140. Bohrer, M.P., G.D. Patterson, and P.J. Carroll, Hindered diffusion of dextran and ficoll in microporous membranes. *Macromolecules*, 1984, **17**, 1170–1173.
141. Composto, R., E. Kramer, and D. White, Fast macromolecules control mutual diffusion in polymer blends. *Nature*, 1987, **328**, 234–236.
142. Barisas, B.G. and M.D. Leuther, Fluorescence photobleaching recovery measurement of protein absolute diffusion constants. *Biophysical Chemistry*, 1977, **10**, 221–229.
143. Blum, F., S. Pickup, and R. Waggoner, NMR measurements of solvent self-diffusion coefficients in polymer solutions. *Abstracts PAP Am. Chem. Soc.* 1990, **199**, 125–126.
144. Conrath, G., *et al.*, *In situ* determination of the diffusion coefficient of a solute in a gel system using a radiotracer. *Journal of Controlled Release*, 1989, **9**, 159–168.
145. Park, I., C. Johnson, and D. Gabriel, Probe diffusion in polyacrylamide gels as observed by means of holographic relaxation methods: search for a universal equation. *Macromolecules*, 1990, **23**, 1548–1553.
146. Soda, K. and A. Wada, Dynamic light-scattering studies on thermal motions of native DNAs in solution. *Biophysical Chemistry*, 1984, **20**, 185–200.
147. Voordouw, G., *et al.*, Isolation and physical studies of the intact supercoiled, the open circular and the linear forms of ColE 1-PLASMID DNA. *Biophysical Chemistry*, 1978, **8**, 171–189.
148. Newman, J., *et al.*, Hydrodynamic properties and molecular weight of fd bacteriophage DNA. *Biochemistry*, 1974, **13**, 4832–4838.
149. Wattenbarger, M.R., *et al.*, Tracer diffusion of proteins in DNA solutions. *Macromolecules*, 1992, **25**, 5263–5265.

5

Drug Permeation through Biological Barriers

...Nothing so aggravates an earnest person as a passive resistance.
Herman Melville, *Bartleby, the Scrivener* (1853)

In multicellular organisms, thin lipid membranes serve as semipermeable barriers between aqueous compartments (Figure 5.1). The plasma membrane of the cell separates the cytoplasm from the extracellular space; endothelial cell membranes separate the blood within the vascular space from the rest of the tissue. Properties of the lipid membrane are critically important in regulating the movement of molecules between these aqueous spaces. While certain barrier properties of membranes can be attributed to the lipid components, accessory molecules within the cell membrane—particularly transport proteins and ion channels—control the rate of permeation of many solutes. Transport proteins permit the cell to regulate the composition of its intracellular environment in response to extracellular conditions.

The relationship between membrane structure, membrane function, and cell physiology is an area of active, ongoing study. Our interest here is practical: what are the basic mechanisms of drug movement through membranes and how can one best predict the rate of permeation of an agent through a membrane barrier? To answer that question, this section presents rates of permeation measured in some common experimental systems and models of membrane permeation that can be used for prediction.

5.1 MOBILITY OF LIPIDS AND PROTEINS IN THE MEMBRANE

The external surface of the plasma membrane carries a carbohydrate-rich coat called the glycocalyx; charged groups in the glycocalyx, which are provided principally by carbohydrates containing sialic acid, cause the surface to be negatively charged. On average, the plasma membrane of human cells contains, by mass, 50% protein, 45% lipid, and 5% carbohydrate. Given the mass ratio

113

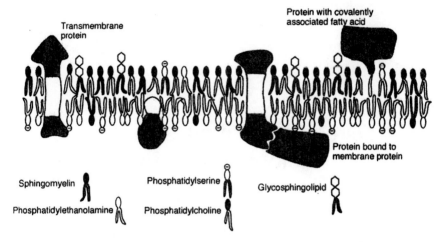

Figure 5.1   Model of cell membranes as lipid bilayers with associated proteins. Composite from diagrams in a variety of sources, including Chapter 6 of [1].

of protein to lipid is $\sim 1 : 1$, and assuming reasonable values for the average molecular weight and cross-sectional area for each type of molecule ($50 \times M_{W,lipid} = M_{W,protein}$; $A_{lipid} = 50\,\text{Å}^2$ and $A_{protein} = 1{,}000\,\text{Å}^2$), the area fraction of protein on a typical membrane is $\sim 33\%$. The lipid composition varies in membranes from different cells depending on the type of cell and its function (Table 5.1). In addition, the outermost monolayer of lipids, called the outer leaflet, has a different lipid composition from the inner leaflet (see erythrocyte membrane in Table 5.1).

Cell membranes are heterogeneous; lipids and accessory molecules are distributed throughout a phase that, because of its slight thickness, is roughly two-dimensional. Most lipid and protein molecules are mobile in the plane of the membrane. Protein mobility was first described by Frye and Edidin, who observed the redistribution of labeled proteins in the membranes of two cells after fusion [1] (Figure 5.2). Molecular mobility confers essential properties to the membrane.

A number of biophysical techniques have been used to measure rates and patterns of molecular motility. One of the most powerful of these approaches is fluorescence photobleaching recovery (FPR) (Figure 5.2). In FPR, a focused laser is used to photobleach molecules within a small, defined volume of tissue; fluorescent molecules from adjacent regions of tissue subsequently diffuse into the bleached volume, causing the fluorescence intensity to increase with time (Figure 5.3). The time interval for recovery depends on the diffusion coefficient and the size of the bleached region ($t = L^2/D$); the size is usually small ($< 1\,\mu\text{m}$) and adjusted so that recovery occurs within a fraction of a minute. Over this short interval, elimination can usually be neglected and D is obtained by adjusting solutions to the diffusion equation to fit the measured intensity recovery curve.

Table 5.1 Approximate lipid compositions of different cell membranes

	Liver plasma membrane	Erythrocyte plasma membrane		Myelin	Mitochondria (inner and outer membranes)	Endoplasmic reticulum	E. coli
		Total (%)	Outer: Inner				
Cholesterol	17	23		22	3	6	0
Phosphatidylethanolamine (PE)	7	18	20:80	15	35	17	70
Phosphatidylserine (PS)	4	7	0:100	9	2	5	trace
Phosphatidylcholine (PC)	24	17	75:25	10	39	40	0
Sphingomyelin (SM)	19	18	80:20	8	0	5	0
Glycolipids	7	3		28	trace	trace	0
Others	22	13		8	21	27	30

Adapted from [1]. Membranes differ significantly in their composition, suggesting that permeabilities may also differ substantially at different sites.

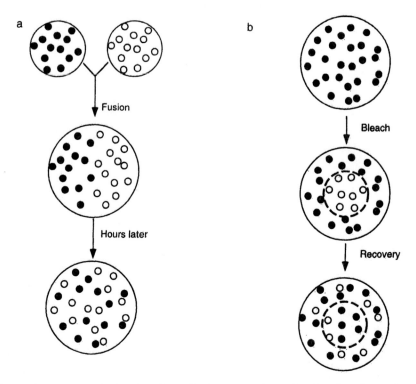

Figure 5.2 Schematic showing the redistribution of labeled molecules after fusion and after photobleaching. (a) Redistribution of membrane proteins that had different fluorescent labels (shown as black or white circles) after cell fusion. (b) Fluorescent recovery after photobleaching, in which the fluorescent molecules (black circles) in an isolated section of the membrane are bleached (white circles). The recovery of fluorescence within the isolated region is an indicator of mobility in the membrane.

5.2 PERMEATION THROUGH LIPID MEMBRANES

The permeation of low molecular weight non-electrolytes through lipid membranes has been studied by a number of investigators. In a typical experiment, a tracer molecule is added to a cell suspension and the rate of accumulation of tracer within the cell is measured. Using this technique, and tracer solutes with a range of physicochemical properties, the permeability, k_s (cm/s), of non-electrolytes was determined as a function of the oil/water partition coefficient. The oil/water partition coefficient is a convenient measure of the relative solubility of a solute within a membrane [2]. A reasonable estimate of the permeability of a solute can frequently be obtained by comparison with experimental measurements for solutes that cover a range of partition coefficients (Figure 5.4). Of the physical properties that have been correlated with permeability, the partition coefficient appears to be the single best indicator for permeability of non-electrolytes. However, it is an imperfect predictor; there is considerable varia-

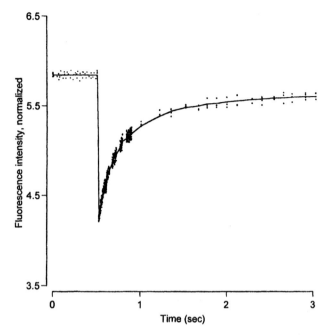

Figure 5.3 Typical trace from a FPR measurement. Representative of measurements acquired by this technique. See [2] for first quantitative description of this approach.

bility in the permeability of compounds with similar partition coefficients (Figure 5.4).

A closer inspection of the definition of membrane permeability suggests additional methods to correlate the permeability with other properties of the diffusing species. The permeability, called k_s or P, is related to the equilibrium partition coefficient, K, the diffusion coefficient in the membrane, D_m, and the thickness of the membrane, L:

$$P = k_s = \frac{KD_m}{L} \tag{5-1}$$

This definition of the permeability arises from the equation for steady-state flux through the membrane, N:

$$N = P(c_0 - c_L) \tag{5-2}$$

where c_0 and c_L are concentrations of the solute in the external phase on either side of the membrane. Diffusion through polymers is often empirically correlated by power law expressions (see Chapter 4), such as:

$$D_m = D_m^0(M_w)^{-s_m} \tag{5-3}$$

where M_w is the molecular weight of the diffusing species and the coefficients D_m^0 and s_m are determined by the characteristics of the polymer solution. When Equation 5-3 is substituted into Equation 5-1, the following correlation for

Fundamentals

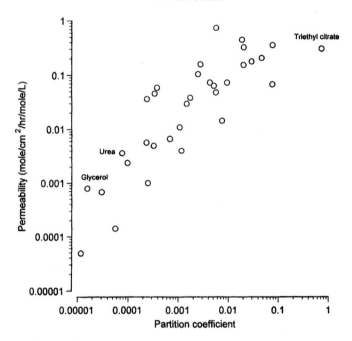

Figure 5.4 Permeability through a cell membrane as a function of olive oil/water partition coefficient. Adapted from [3], which was adapted from the original results of Collander and Barland.

permeability as a function of partition coefficient and molecular weight is obtained:

$$P = P_0 K M_w^{-s_m} \tag{5-4}$$

where P_0 is equal to D_m^0/L. This expression suggests that non-electrolytes of various molecular weight diffusing through cell membranes behave like solutes diffusing through a polymer film [3]; this correlation agrees well with experimental data (Figure 5.5). This analysis, which supports the concept of the plasma membrane as a fluid phase, was published several years before the presentation of the fluid mosaic model of cell membranes (Figure 5.1) [4].

The permeability of the erythrocyte plasma membrane has been measured by monitoring the rate of movement of a variety of solutes into erythrocytes (Table 5.2 and Figure 5.6). For non-electrolytes of molecular weight less than 200, permeability decreases with increasing molecular weight; compounds with molecular weight over 300 (such as sucrose) are essentially excluded from the membrane. Charged molecules do not partition into lipid bilayers; therefore, ions have a very low permeability in the erythrocyte membrane. The low intrinsic permeability of ions underlies the ability of membranes to support an electrical potential difference, which is discussed more fully in Section 5.4.3.

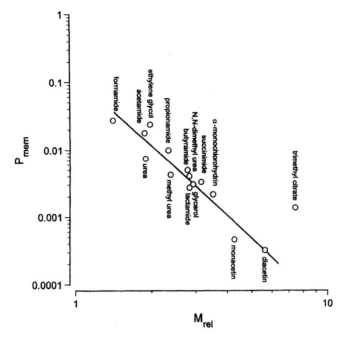

Figure 5.5 Membrane permeability versus molecular weight with correction for the partition coefficient. Figure 2 from [2], $\log_{10}(P/K)$ versus M_w. K was estimated from the olive oil/water partition coefficient.

Solute permeability is also a function of membrane composition. The rate of permeation of doxorubicin measured through artificial lipid bilayers composed of cholesterol, PC, PS, PE, and SM (abbreviations defined in Table 5.1) changes substantially with composition (Figure 5.7). This is an important finding, since cells have distinct membrane compositions. In addition, the composition of the individual leaflets of a membrane can vary (Table 5.1), suggesting the possibility of asymmetric passive transport (i.e., permeabilities that differ depending on the direction of transport, out-to-in or in-to-out).

Despite their large size, some oligonucleotides can permeate through membranes in artificial liposomes [8]. The half-time for efflux was measured for liposomes with a fixed composition (Table 5.3).

5.3 PERMEATION THROUGH POROUS MEMBRANES

In the previous section, the permeation of solutes through uniform lipid membranes was discussed; however, cell membranes and cellular barriers are not perfectly uniform (Figure 5.1). Proteins interrupt the continuous lipid membrane and provide an additional pathway for the diffusion of water-soluble molecules. Protein channels in the membrane, for example, permit the selective diffusion of certain ions. In the blood vessel wall, water-filled spaces between the adjacent endothelial cells provide an alternate path for transport.

Table 5.2 Permeability of human red blood cells

Solute	M_w	Permeability of erythrocyte membranes (cm/s)	Permeability of lipid vesicles (cm/s)
Carbon dioxide	34		0.2–0.6
Water	18	0.0088	0.005
Methanol	32	0.003	
Urea	60	4.0×10^{-4}	2×10^{-6}
Butanol	74		3.0×10^{-4}
Ethylene glycol	62	2.0×10^{-4}	
Ethanol	46	1.5×10^{-4}	
Methylurea	74		5.0×10^{-5}
Glycerol	92		1.5×10^{-6}
Creatinine	113	6×10^{-8}	
Indole	117		4×10^{-4}
Uric acid	158	2×10^{-8}	
Glucose	180	1.5×10^{-7}	6×10^{-8}
Fructose	180	1.5×10^{-9}	
Mannitol	180	6×10^{-6}	
Tryptophan	204		1×10^{-7}
Sucrose	342	0	
Na^+	23		1×10^{-12}
K^+	39		6×10^{-12}
Cl^-	35		7×10^{-11}

Permeabilities from [5] and [6]. Data are plotted in Figure 5.6, with filled circles representing permeation through lipid bilayers and open circles representing permeation into red blood cells.

Figure 5.8 illustrates a useful model for diffusion of water-soluble molecules through aqueous pores in a membrane. In this model, the pore space is assumed to consist of a monodisperse ensemble of straight cylindrical pores. The transport of solutes through this hypothetical porous material can be analyzed by consideration of the dynamics of a single spherical particle through this idealized cylindrical pore. In many systems of biological interest, the radius of the diffusing particle, r_s, is almost as large as the radius of the pore, r_0. The movement of the particle is therefore hindered in comparison with the movement of the particle in an unbounded fluid. Hindered transport in biological systems has been studied by a number of groups over the last few decades [9–12]. One of the major advantages associated with models of transport in this idealized geometry is that model predictions can be tested experimentally using membranes with straight cylindrical pores and size-fractionated spherical solutes. This discussion of the equations for hindered transport follows the development in an excellent review [9].

The key parameters used in describing hindered transport through cylindrical pores are illustrated in Figure 5.8. A critical parameter is the ratio of the solute radius to the pore radius, $\lambda = a/r_p$. As λ approaches 1, the pore walls become an increasingly important obstacle for particle transport. In the limit as λ approaches zero, the fluid within the pore can be considered unbounded,

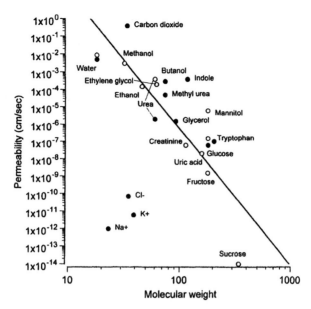

Figure 5.6 Permeability of membranes to small molecules. The figure shows permeability, P (cm/s), for a variety of molecules in red blood cells (open circles) and lipid vesicles (filled circles). The solid line is the best fit of P vs M_w^n for erthrocytes; the best value of n is -8. See legend to Table 5.2.

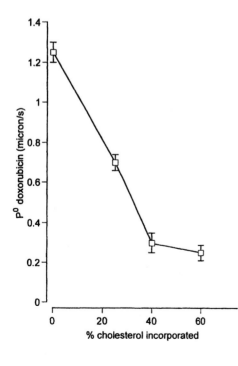

Figure 5.7
Permeability of doxorubicin in model membrane bilayers. From [7]. Permeability was measured in artificial unilamellar vesicles of known composition. Permeability varies with the addition of cholesterol (squares). Permeability also varies with change in phospholipid composition.

121

from the perspective of the particle. Between these two limiting cases, when $0 < \lambda < 1$, lie a number of situations of biological interest, such as permeation of water-soluble compounds through the aqueous pores in a capillary wall.

The driving force for particle movement through the pore, the gradient of chemical potential, is balanced by the drag force acting on the particle:

$$-\nabla \mu = F \tag{5-5}$$

Assuming that the activity coefficient is unity, so the chemical potential can be expressed in terms of the local solute concentration c, and considering only the axial gradient, Equation 5-5 becomes

$$-kT \frac{d \ln c}{dz} = F \tag{5-6}$$

where k_B is Boltzmann's constant and T is absolute temperature. The local net velocity of the particle with respect to the pore walls is U and the velocity of the fluid, v, is a well-developed, Pouseille flow with a parabolic profile that depends only on the dimensionless radius, $\beta = r/r_p$ (see Equation 6-5 and accompanying discussion):

$$v = 2v_0 (1 - \beta^2) \tag{5-7}$$

With these assumptions, the drag force on the particle can be written:

$$F = f_\infty K[U - Gv] \tag{5-8}$$

Table 5.3   The efflux of oligonucleotide analogs from liposomes

Compound	Size (number of bases)	Water/octanol partition coefficient	Efflux $t_{1/2}$ (days)
D-oligo	16	57,000	
D-oligo	14		7.2
D-oligo	7		9.1
D-oligo	4		8.7
S-oligo	14 (antisense to β-globulin gene site)	> 8,000 (no detectable partitioning into octanol)	9.3
Alt-MP-oligo	14		10.3
MP-oligo	12	24	
MP-oligo	14		7.2
MP-oligo	15	40	4.4, 6.2
Sucrose			10.1
Glucose			1.1

Data from [8]. Labeled methylphosphonate (MP-oligo), phosphorothioate (S-oligo), alternating methylphosphonate–phosphodiester (Alt-MP-oligo), and unmodified phosphodiester (D-oligo) oligodeoxynucleotides.

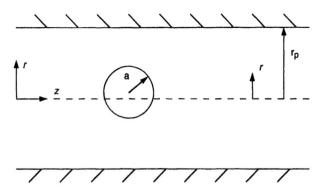

Figure 5.8 Hindered transport through porous membranes. Schematic diagram of model for hindered transport of a spherical particle in a cylindrical pore.

where f_∞ is the frictional drag coefficient on the particle in bulk solution, K is the enhanced friction due to the presence of the pore walls (which is equal to the pore friction coefficient/bulk friction coefficient), and G is the lag coefficient that accounts for the decreased approach velocity of the fluid due to the presence of the pore walls. In an unbounded fluid, $K = 1$ and $G = 1$ so that Equation 5-8 reduces to Equation 4-1. In the bounded fluid, with the pore walls near the particle, frictional drag will be enhanced ($K > 1$) and the approach velocity of the fluid will be reduced ($G < 1$)

Combining Equations 5-6 and 5-8 yields, with some rearrangement:

$$N_z = UC = -\frac{kT}{f_\infty}\frac{1}{K}\frac{dc}{dz} + Gvc \tag{5-9}$$

where N_z is the local solute flux in the z-direction, which has contributions due to solute diffusion (the first term on the right-hand side) and bulk fluid movement (second term on the right-hand side). Again, when the particle is diffusing in an unbounded ($K = G = 1$), stagnant ($v = 0$) fluid, this flux expression reduces to the familiar form presented in Table 3.2. The influence of the pore walls on solute flux is reflected in a decrease in the effective diffusion coefficient (i.e., $K > 1$), compared to diffusion in the bulk fluid, and a decrease in influence of bulk fluid velocity on particle convection (i.e., $G < 1$). Both of these factors depend on λ and β:

$$K^{-1} = \frac{D}{D_\infty} = f(\lambda, \beta); \, G = g(\lambda, \beta) \tag{5-10}$$

The local solute flux, N_z, can depend on both the axial and radial location of the diffusing particle. Equation 5-9 becomes more useful by averaging across the pore cross-section, which leads to an averaged flux $\langle N_z \rangle$ that depends only on the axial position, z:

$$\langle N_z \rangle = \frac{\int_0^{1-\lambda} N_z \beta d\beta}{\int_0^1 \beta d\beta} = -2D_\infty \int_0^{1-\lambda} K^{-1} \frac{dc}{dz} \beta d\beta + 4V_0 \int_0^{1-\lambda} Gc(1-\beta^2)\beta d\beta \quad (5\text{-}11)$$

where Equation 4-2 has been substituted for f_∞ and Equation 5-7 has been used to describe the radial velocity profile. The upper bound on several of the integrals is $1 - \lambda$, to reflect the fact that the center of the spherical particles can never get closer than one particle radius from the wall. Therefore, the flux of particles is zero within the annular region $1 - \lambda < \beta < 1$.

To complete the averaging operation in Equation 5-11, the radial dependence of c must be known. In the most general case, this expression should account for long-range (e.g., electrostatic) and short-range (e.g., steric) interactions between the pore walls and the particle [9]. Here, as in previous analyses [12], we assume that the average concentration in the pore, $\langle c \rangle$, is only a function of z, so that Equation 5-11 reduces to

$$\langle N_z \rangle = -K^d D_\infty \frac{d\langle c \rangle}{dz} + K^c v_0 \langle c \rangle \quad (5\text{-}12)$$

where K^d and K^c are defined:

$$K^d = 2 \int_0^{1-\lambda} K^{-1} \beta d\beta \quad \text{and} \quad K^c = 4 \int_0^{1-\lambda} G(1-\beta^2)\beta d\beta \quad (5\text{-}13)$$

To apply this local flux expression to the problem of solute transport through the characteristic pore shown in Figure 5.8, the solute concentration in the bulk fluids on either end of the pore, c_0 and c_L, must be related to the concentrations within the pore space, $\langle c \rangle_0$ and $\langle c \rangle_L$. For purely steric interactions between the pore wall and the solute, the relationship between the bulk and pore concentrations can be found by assuming local equilibrium between pore and bulk fluid, with an additional constraint that solutes are excluded from the annular region near the pore wall:

$$\Phi = \frac{\langle c \rangle}{c} = 2 \int_0^{1-\lambda} \beta d\beta = (1-\lambda)^2 \quad (5\text{-}14)$$

where Φ is the equilibrium partition coefficient between the bulk fluid and fluid in the pore.

The total steady-state flux through the pore can be determined by integration of Equation 5-12:

$$\langle N_z \rangle = \frac{Wv_0 c_0 \left[1 - \left(\frac{c_L}{c_0}\right) e^{-Pe} \right]}{1 - e^{-Pe}} \quad \text{where} \quad Pe = \frac{Wv_0 L}{HD_\infty}; \; W = \Phi K^c; \; H = \Phi K^d$$

$$(5\text{-}15)$$

which is obtained using the boundary conditions $\langle c \rangle_0 = \Phi c_0$ at $x = 0$ and $\langle c \rangle_L = \Phi c_L$ at $x = L$. This equation reduces to the following simple expressions in the limit of small and large Pe:

$$\langle N_z \rangle = \frac{HD_\infty}{L}(c_0 - c_L) \quad Pe \ll 1$$

$$\langle N_z \rangle = W v_0 c_0 \quad Pe \gg 1$$

(5-16)

When the Péclet number, Pe, is small, Equation 5-15 reduces to the familiar expression for flux as a function of membrane permeability (recall Equation 5-2), where the membrane permeability is now $P = HD_\infty/L$. When Pe is large, Equation 5-15 reduces to the expression for ultrafiltration, where the commonly used reflection coefficient is $\sigma = 1 - W$ (see Section 5.5.1).

The dependence of H and W on λ are shown in Figure 5.9. These hydrodynamic coefficients were calculated after determination of K and G, as in Equation 5-10. These coefficients can be determined by finding the drag on a sphere within a tube and the approach velocity for a sphere in parabolic flow, as described previously for spheres on the tube centerline [12] or distributed throughout the tube [11]. The solid lines in Figure 5.9 indicate values of H and W when spheres are confined to the tube axis (the "centerline approximation"):

$$H = (1 - \lambda)^2 [1 - 2.1044\lambda + 2.089\lambda^3 - 0.948\lambda^5]$$

$$W = (1 - \lambda)^2 (1 + \lambda)[1 - 2/3\lambda^2 - 0.163\lambda^3]$$

(5-17)

whereas the dashed lines indicate values of H and W over a more limited range of λ in the more general case. As is apparent from Figure 5.9, the "centerline approximation" overestimates H and W, but is sufficiently close to the general case that it is useful for many calculations.

5.4 PERMEATION IS ENHANCED BY MEMBRANE PROTEINS

Many molecules do not diffuse through lipid bilayers (see Figure 5.6 and notice that sucrose and ions do not permeate). One of the most important functions of accessory molecules in the membrane is regulation of transport of molecules that do not pass freely through the lipid bilayer. Several classes of accessory molecules are engaged in membrane transport, as described in the sections that follow.

5.4.1 Facilitated Diffusional Transport

Although the extracellular concentration of glucose is usually higher than the intracellular concentration, the low permeability of the lipid bilayer to glucose prevents passive transport of sufficient molecules of glucose to support metabolism (glucose permeates $\sim 1 \times 10^5$ times slower than water, see Figure 5.6). Glucose-transport proteins in the cell membrane solve this problem, by pro-

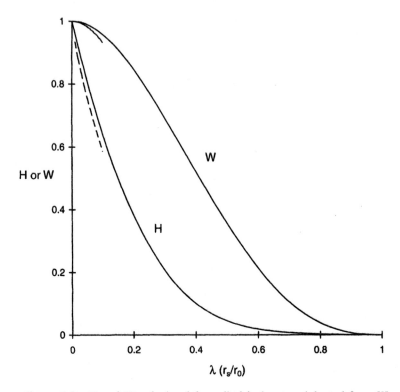

Figure 5.9   *H* and *W* calculated for cylindrical pores. Adapted from [9].

viding aqueous pathways that shuttle glucose through the hydrophobic bilayer. The glucose transporter facilitates glucose permeation by periodic changes in conformation: in one conformation, a glucose-binding site is exposed on the extracellular face, while in another conformation, the binding site is exposed to the intracellular face (Figure 5.10). Conformational changes occur due to natural thermal fluctuations in the membrane; conformational changes occur whether the glucose binding site is occupied or vacant. As a result of this periodic change in structure, the transport protein permits the passage of glucose (or another molecule that is able to bind to the transporter binding site), without the addition of any additional energy. Glucose molecules can move in either direction across the bilayer; the net flux will occur from the region of high to low concentration.

Facilitated transport proteins, such as the glucose transporter, are present in the cell membrane in limited number. As the concentration of the solute increases at the external membrane surface, the transporter binding sites become saturated, and the net rate of solute transport across the membrane approaches a maximal value (Figure 5.11). The rate of transport via the facilitated transport mechanism can be analyzed by considering binding of the transported solute (S) to the transmembrane carrier protein (C_p) to form a carrier–solute complex ($S-C_p$):

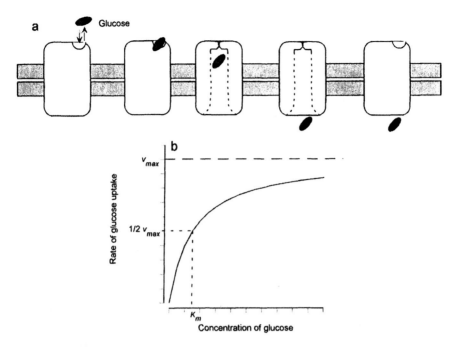

Figure 5.10   Illustration of the mechanism of action of the glucose transport protein. The rate of glucose uptake varies with extracellular glucose concentration(b). Modified from [13].

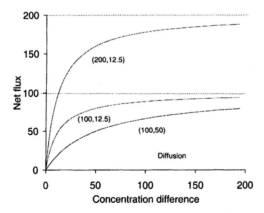

Figure 5.11   Facilitated transport proteins in cell membranes. Unlike simple diffusion through the membrane bilayer, facilitated transport systems become saturated as the solute concentration difference increases. In this hypothetical example, the permeability of the membrane in simple diffusion is 0.4 (units of flux/concentration). V_{max} (units of flux) and K_m (units of concentration) for facilitated transport are (200, 12.5), (100, 12.5), and (100, 50).

$$S + C_p \underset{k_{-1}}{\overset{k_1}{\rightleftarrows}} S - C_p \xrightarrow{k_2} S^* - C_p \xrightarrow{k_3} S^* + C_p \tag{5-18}$$

The conformational change of the carrier protein moves the solute from one side of the membrane (a state designated S) to the opposite side (designated S^*), from which it is released. If the rate of release of transported solute is rapid compared to the rate of conformational change ($k_3 \gg k_2$), the concentration of solute on the opposite side of the membrane is negligible ($S^* \sim 0$), and binding of solute to carrier is at equilibrium. The flux of solute across the membrane is given by:

$$N_{S^*} = \frac{1}{A} \frac{d[S^*]}{dt} = k_2[SC_p] \tag{5-19}$$

The concentration of the solute–carrier complex is assumed constant (i.e., the rate of formation of the complex is equal to the rate of dissociation), so that:

$$k_1[S][C_P] = k_{-1}[SC_P] + k_2[SC_P] \tag{5-20}$$

and the total number of carrier proteins is also assumed constant, C_{TOT}:

$$C_{TOT} = [C_P] + [SC_P] \tag{5-21}$$

Substitution of Equations 5-20 and 5-21 into Equation 5-19 yields:

$$N_{S^*} = \frac{k_2 C_{TOT}}{\left(1 + \dfrac{k_{-1} + k_2}{k_1} \dfrac{1}{[S]}\right)} \tag{5-22}$$

which can be simplified by definition of the lumped constants $V_{max} = k_2 C_{TOT}$ and $K_m = (k_{-1} + k_2)/k_1$:

$$N_{S^*} = \frac{V_{max}[S]}{([S] + K_m)} \tag{5-23}$$

This derivation is commonly used to describe the kinetics of product formation in enzyme-catalyzed reactions (substitute enzyme for carrier protein and product formation for the conformational change of the carrier protein). Under the assumptions of this simple model, carrier conformational change is the rate-limiting step, so it is reasonable to assume further that k_2 is much less than k_{-1}. In this case, the constant K_m is approximately equal to $K_d = k_{-1}/k_1$, the dissociation constant for the binding of solute to carrier. For this reason, it is common to refer to K_m as the "affinity" of the solute for the carrier (note the analogy to Equation 3-53). V_{max} is the maximum flux due to this carrier-mediated transport, which occurs when all of the carrier-binding sites are occupied (Figure 5.11).

Both facilitated and simple diffusion depend on concentration gradients: net solute transport always occurs from high to low concentration. Unlike diffusion, facilitated transport systems are specific, since they depend on binding of the solute to a site on the transport protein. For example, the D-glucose

Table 5.4 Specificity of the D-glucose transporter protein for various solutes

	K_m (mM)
D-Glucose	1.5
L-Glucose	> 3,000
D-Mannose	20
D-Galactose	30

transporter is very inefficient at transporting L-glucose, but it can accommodate D-mannose and D-galactose (Table 5.4).

The neutral amino acid transporter is responsible for movement of neutral amino acids from the blood into the brain; these compounds are not soluble in membranes and, therefore, would not diffuse into the brain in the absence of a transport system. This transporter is very efficient in brain capillaries, but it is found in other tissues as well (Table 5.5). The concentrations of amino acids in the blood are close to the values of K_m for brain capillaries, suggesting that the transporter is near saturation under normal conditions. Because of this, and because the transporter is equally efficient with a number of amino acids, changes in the blood concentration of one neutral amino acid can influence the rate of transport of all the other neutral amino acids.

5.4.2 Active Transport

Active transport systems are similar to facilitated transport systems; both involve the participation of transmembrane proteins that bind a specific solute (Figure 5.12). In active transport, however, energy is required to drive the conformational change that leads to solute transport. The Na^+/K^+ pump is the most well-characterized active transport system (Figure 5.13). The pump is an assembly of membrane proteins, with three Na^+-binding sites on the cytoplasmic surface, two K^+-binding sites on the extracellular surface, and a region of ATPase activity on the cytoplasmic surface, near the Na^+-binding sites.

Table 5.5 K_m values for transport of neutral amino acids in tissues

Amino acid	Intestinal epithelia	Renal tubule	Exocrine pancreas	Red blood cells	Liver	Blood–brain barrier
Phenylalanine	1			4	4	0.032
Leucine	2			2	6	0.087
Tryptophan		4				0.052
Methionine	5		3	5		0.083
Histidine	6	5				0.16
Valine	3		7	7		0.17

Values are given in mM. From [15].

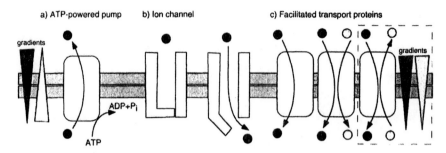

Figure 5.12 Transport proteins in cell membranes. (a) Energy-dependent, ATP-powered ion pumps such as the Na$^+$/K$^+$ exchange ATPase; (b) channels, gated or non-gated, which permit diffusion through an aqueous pathway, such as voltage-gated Na$^+$ channels; (c) passive, facilitated transport systems, which can act in uniport, symport, or antiport modes, such as the glucose transporter. The filled arrow indicates gradient of molecules indicated by the filled symbol.

Both facilitated transport systems and active transport mechanisms can be saturated, and the rates of transport are similar. In addition, both carrier-mediated transport systems have less transport capacity than channel-mediated transport or simple diffusion (Table 5.6).

5.4.3 Ion Transport, Membrane Potentials, and Action Potentials

Lipid bilayers are impermeable to ions and other charged species (Figure 5.6), but ions such as Na$^+$ and K$^+$ are found in abundance in both the extracellular and intracellular environment. Active transport proteins can move ions across membranes, even moving molecules against a concentration

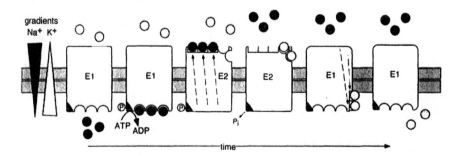

Figure 5.13 Na$^+$/K$^+$–ATPase moves ion across the plasma membrane. The conformational change is triggered by ATPase activity, which phosphorylates an aspartate residue of the protein. Formation of a high-energy acyl phosphate bond drives the E1 → E2 conformational change, which reverses when the phosphorus is eventually released. Na$^+$ and K$^+$ binding sites cycle between high affinity (semicircles in the schematic diagram) and low affinity (rectangles) states that are coordinated with the conformational change.

Table 5.6 Characteristics of membrane transport mechanisms

	Restrictions	Capacity (molecules/s)	Example
Simple diffusion	Solubility in membrane With a concentration gradient	a	See Figures 5.4 and 5.6
Channel-mediated transport	Specific channel required With a concentration gradient Gating is possible	10^7-10^8	Na^+ and K^+ channels
Facilitated transport	Specific transport proteins required With a concentration gradient[b]	10^2-10^4	Glucose transporter
Active transport	Specific transport proteins required Energy required	$1-10^3$	Na^+/K^+ pump

[a] Capacity is limited only by the solubility of solute in the membrane.

[b] Symport and antiport systems can move one of the transported solutes against a gradient.

gradient. Cells and microorganisms are extremely sensitive to changes in their ionic environment, making ion movement across membranes profoundly important in physiology.

Ion movement across membranes is regulated by specialized membrane proteins called ion channels (Figures 5.12 and 5.14). Some ion channels are selective, permitting the permeation of only specific ions. Selectivity is accomplished by a combination of channel characteristics including molecular sieving (which selects for ions of certain size), binding to the protein

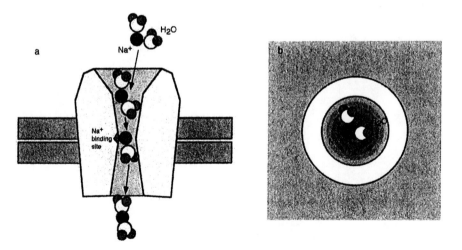

Figure 5.14 Mechanism for selectivity of ion channels. Modified from [13, 14]. (a) Cross-sectional side view of hypothesized mechanism for selectivity in a sodium channel. A weak binding interaction with the wall of the channel permits temporary release of water for hydration. (b) Top view of same structure.

surface, and stabilization of the non-hydrated ion. Some channels are gated; gated channels exist in multiple states that either permit or exclude ion movement across the membrane (Figure 5.12). In the open state, conductance (or permeability) of the ion rapidly increases (Figure 5.15). The closed/open state of the ion channel is regulated by extracellular and intracellular conditions. Some channels are voltage regulated; others are regulated by binding of a chemical ligand or mechanical stretching of the membrane.

Most channels behave as simple resistors to current flow: when the channel is open, the current through the channel varies linearly with membrane potential ($\Delta V = IR$). The antibiotic polypeptide gramicidin A forms a cylindrical channel by end-to-end dimerization of two peptides, one in each of the bilayer leaflets. Individual gramicidin channels exhibit a resistance of $8 \times 10^{10}\ \Omega$ over

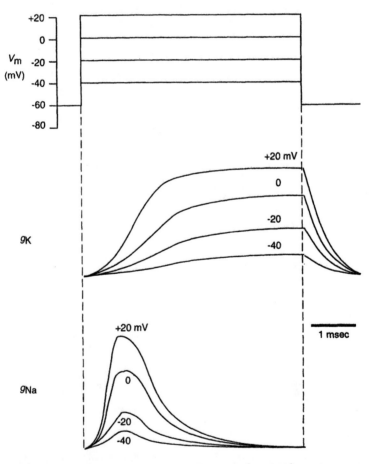

Figure 5.15 Kinetics of ion-channel permeability. Na^+ and K^+ conductances were measured during a step change in voltage. Measurements were accomplished by voltage-clamp techniques, modified from [16].

membrane potentials from -75 to $+75\,\mathrm{mV}$ (by standard convention, the membrane potential is positive if the cell interior is positively charged with respect to the exterior). This electrical resistance can also be expressed as a conductance, the reciprocal of resistance, of 12×10^{-12} siemens (S, where $1\,S = 1\,\Omega^{-1}$). In other channels, the relationship between voltage and current is non-linear, such that resistance varies with imposed potential. These channels are rectifying, in that current flow is unequal for positive or negative potentials.

The overall potential of a cell membrane depends on the concentrations of ions on either side of the membrane (Figure 5.16) and the permeability of the membrane to each ion. Membrane potential is generated by gradient-driven movement of ions across the membrane. Most cells have intracellular K^+ concentrations that are much higher than the extracellular K^+ concentration. If the membrane contains channels that permit K^+ permeation, ions will diffuse out from the cell into the extracellular environment. This selective permeation creates a separation of charge, because the negatively charged counterions for K^+ do not cross the membrane (Figure 5.16a). Therefore, an electrochemical potential develops and increases until, eventually, it balances the concentration driving force, at which point K^+ diffusion will cease. For a

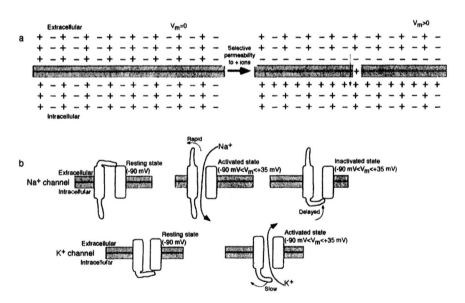

Figure 5.16 Development of membrane potential from unequal ion concentrations. (a) The presence of ion gradients across a membrane leads to a resting membrane potential. The value of the potential depends on the ion gradient and the permeability of the membrane to each ion. In the situation depicted here, positively charged ions are permeating through the membrane without their counterions, thereby generating a positive potential. (b) Voltage-gated Na^+ and K^+ channels change states as the overall membrane potential changes. The responses of these two classes of channels generate an action potential.

membrane that is permeable to a single ion, the equilibrium membrane potential, E_{ion}, is given by the Nernst equation:

$$E_{ion} = -\frac{RT}{z_{ion}F}\ln\left(\frac{C_{in}}{C_{out}}\right) \tag{5-24}$$

where R is the gas constant, F is Faraday's constant, T is absolute temperature, z is the valence of the ion, and C_{in} and C_{out} are the intracellular and extracellular ion concentrations. The membrane potential is defined so that a positive-ion gradient from inside to out results in a negative potential across the membrane. Equilibrium potentials for K^+ and Na^+ at 37 °C are −94 and +61 mV, respectively.

When a membrane is permeable to more than one ionic species, the net membrane potential, V_m, depends on concentration and permeability:

$$V_m = -\frac{RT}{F}\ln\left(\frac{P_K C_{K^+,in} + P_{Na} C_{Na^+,in} + P_{Cl} C_{Cl^-,in}}{P_K C_{K^+,out} + P_{Na} C_{Na^+,out} + P_{Cl} C_{Cl^-,out}}\right) \tag{5-25}$$

Na^+, K^+, and Cl^- are the most important ions for determining membrane potential in human cells. The importance of each ion in determining the resting membrane potential depends on the permeability of each ion in the membrane; permeability depends on the number of non-gated channels available for that ion. For many mammalian cells, the net membrane potential is ~ -60 mV, which is closer to the equilibrium potential of potassium than sodium due to the greater permeability of resting membranes to potassium.

Excitable cells in the nervous system are capable of transmitting signals over long distances very rapidly, with no attenuation of the signal as it moves. This remarkable property enables us to respond rapidly to our environment, as when we reflexively withdraw our hand from a flame. These rapidly moving signals are transmitted by changes in membrane potential. Since membrane potential depends on membrane permeability to ions (Equation 5-25), signal movement through tissue occurs via spatial changes in ion conductance. This signaling can be controlled, because ion conductance is regulated by voltage-gated ion channels, which can open or close in response to changes in the local membrane potential. Voltage-clamp techniques, first developed in 1949 (see [16]), permit the measurement of ion conductance at a fixed membrane potential (Figure 5.15). When the local potential changes, ion channels open, creating more paths for ion diffusion. Although the potential may change rapidly, changes in conductance occur more slowly; the overall kinetics of conductance change reflect the contributions from many individual ion channels, which open and close at different times.

For a fixed change in membrane potential, voltage-gated Na^+ -channels produce more rapid changes in conductance than K^+ -channels (Figure 5.15). This difference in speed of response is essential for signal propagation, as first described by Hodgkins and Huxley (see [16]). Action potentials—sudden, reversible changes in overall membrane potential—are initiated by local depolarization of the membrane: depolarization occurs when the overall membrane potential becomes slightly less negative than at its resting (polarized) state. A

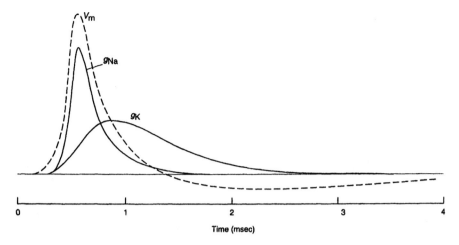

Figure 5.17 The action potential is the result of reversible changes in sodium and potassium conductance. Adapted from [16], which was adapted from Huxley (1964).

few voltage-gated Na^+ -channels open in response to this small depolarization; the opening of these few channels creates an increase in the inward Na^+ flux through the membrane, which causes further depolarization. This results in the opening of more Na^+ -channels, which causes further depolarization. As a result of this cascade, the initial small change in membrane potential suddenly becomes a large change; if enough channels open, the net membrane potential will approach the Na^+ equilibrium potential, given by Equation 5-24 (Figure 5.17). The Na^+ equilibrium potential is never achieved, because depolarization also opens voltage-gated K^+ -channels (Figure 5.15), which produce an outward flow of K^+ ions. This outward K^+ current tends to return the potential to its former value; since K^+ flux serves to re-establish the polarized state of the membrane, this process is called repolarization. The K^+ channels respond more slowly than Na^+ channels, but they remain open longer, resulting in a slight hyperpolarization (due to the residual permeability to K^+) at the end of the action potential. For a short period after the action potential, most membranes cannot be depolarized. This refractory state, which reflects the non-responsiveness of Na^+ channels for a brief period after opening, lasts for several milliseconds. Near the end of the refractory period, an action potential can be induced, but a greater threshold depolarization is required for initiation.

5.4.4 Other Transport Mechanisms

Passive diffusion, facilitated and active transport, and diffusion through channels account for most of the molecular transit across membranes. Several other mechanisms involving membrane-associated proteins are also important in the life of a cell, and represent important targets for drug therapy.

Receptor-Mediated Endocytosis and Signaling. Binding of some ligands to membrane receptor proteins can lead to rapid internalization of both receptor and ligand by a process called endocytosis (Figure 5.18). This process has been studied extensively for a variety of ligand–receptor systems including low-density lipoprotein (LDL), transferrin, and epidermal growth factor (EGF). Endocytosis frequently leads to accumulation of ligand in intracellular vesicles, including lysosomes, in which degradation of the ligand can occur. Ligand–receptor complexes have different fates after endocytosis. In the case of LDL and LDL receptor, the LDL dissociates from the LDL receptor within the endosome and the LDL receptor is recycled to the cell surface. LDL receptors are conserved in this process and an essential nutrient is brought into the cell without sacrificing the carrier protein. Transferrin, an iron-carrying serum protein, releases bound iron in the endosome (becoming apotransferrin, the iron-free form of the protein); the transferrin–receptor complex is recycled to the cell surface, after which transferrin is released into the extracellular environment. Both protein and

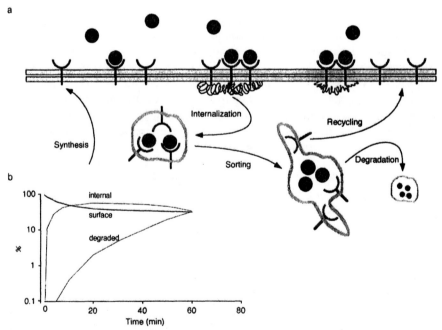

Figure 5.18   Receptor-mediated endocytosis. (a) Schematic diagram of the main steps in the endocytotic pathway. Binding at the cell surface induces clustering of receptor–ligand complexes in specialized regions of the membrane. Endocytotic vesicles form in these regions; separation of receptor and ligand occurs as the endosome matures. Separated receptors can recycle to the surface, while remaining material is degraded within lysosomes. (b) Typical time course of internalization and degradation.

receptor are conserved in this process. EGF and EGF–receptor are both degraded within the lysosomes. Since the receptor is degraded, EGF binding eventually leads to a decrease in the number of EGF receptors at the surface, a process called receptor down-regulation.

An agent that binds to and activates a specific receptor system is called an agonist; an agent that opposes activation is called an antagonist. Considerable effort in the pharmaceutical industry is devoted to finding agonists or antagonists that activate particular biochemical pathways with defined characteristics: for example,

- agonists with extended duration of action (which results from altered metabolism of the agonist);
- agonists with fewer side-effects (which often results from an increased binding specificity);
- agonists that can be orally administered (which is achieved by changing the physical properties of the compound).

As more information is collected about receptor–ligand interactions, receptor trafficking, and signal transduction, new opportunities for drug design are emerging. These new opportunities often target novel or unique steps in the ligand–receptor binding and signaling process. Examples of membrane receptor binding and signal transduction are illustrated in Figures 5.19 and 5.20. In both of these examples, the reaction network is complex and not completely understood, although certain aspects are yielding to analysis [18].

Activation of protein kinase activity is often involved in signal transduction. Binding of polypeptide hormones (such as insulin) and extracellular matrix ligands (such as fibronectin) activate kinase activity. Kinases are enzymes that catalyze the transfer of a phosphoryl group from ATP (or some other nucleoside triphosphate) to an acceptor molecule. For example, tyrosine kinases, when activated, phosphorylate tyrosine residues on proteins. Not uncommonly, the tyrosine residue that becomes phosphorylated is on the same protein that carries the kinase catalytic domain; the event is then called autophosphorylation. The binding of fibronectin to integrin receptor activates a number of kinases including FAK, Src, and Csk (Figure 5.19). EGF receptors, platelet-derived growth factor (PDGF) receptors, and insulin receptors have regions of tyrosine protein kinase activity within their cytoplasmic domains. Binding of ligand to the extracellular domain leads to a conformational change, and this induces autophosphorylation within the cytoplasmic domain, and this reveals a new kinase activity for cytoplasmic proteins, which carry the signal further into the cell. In both of these examples, the overall outcome of binding depends on the activity of a network of intracellular enzymes (including kinases) and other messengers.

Trimeric GTP-binding regulatory proteins (G-proteins) are common participants in signaling pathways. Receptors that are associated with G-proteins share common biochemical mechanisms (Figure 5.20a), but this same basic mechanism can produce a variety of outcomes within the cell. This functional

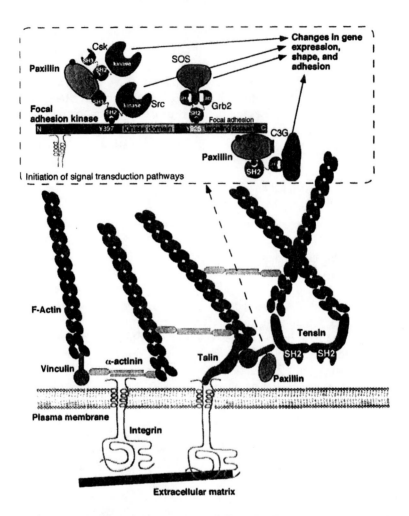

Figure 5.19   Signaling initiated by receptor binding. Binding of ligands to receptors can initiate biochemical signals within the cell; this concept is illustrated using integrin binding to extracellular matrix ligands. Binding at the cell surfaces causes association of cytoplasmic proteins (tensin, talin, vinculin, α-actinin, and F-actin) with the cytoplasmic domains of a transmembrane integrin receptor. The inset shows a hypothetical model for signal transduction based on focal adhesion kinase (FAK). The model includes tyrosine kinases (Src and Csk), adapter proteins (Grb2 and Crk), and guanine nucleotide exchange factors (SOS and C3G). Binding occurs through SH2 (Src homology 2) domains, which bind to proteins containing a phosphotyrosine (filled circles), and SH3 (Src homology 3) domains, which bind to proteins containing a proline-rich peptide motif (filled rectangles). The diagram was adapted from a variety of sources [13, 17–19].

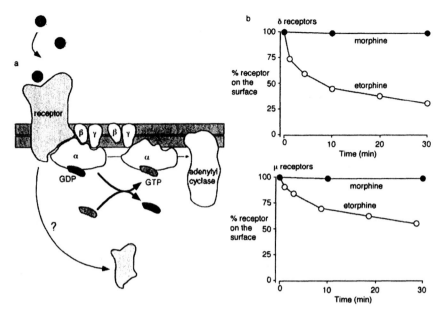

Figure 5.20 Agonists can have different effects on signal activation and internalization. (a) Binding of ligand activates this G-protein-linked receptor, which causes release of the α-subunit. (b) Morphine and etorphine are both agonists of μ- and δ-opioid receptors. The y-axis is an indication of the number of receptors remaining on the plasma membrane. Opioid receptors are rapidly internalized after etorphine treatment (with kinetics similar to internalization after enkephalin treatment), but not after morphine treatment. Adapted from [20].

diversity is due in part to the presence of different forms of the α, β, and γ subunits (e.g., the α subunit can be stimulatory, α_s, or inhibitory, α_i), but also reflects specific characteristics of the cell harboring the receptor. Binding of substrate induces a conformational change in the receptor that leads to binding of the trimeric G-protein. Binding causes release of GDP, with binding of GTP in its place, and dissociation of the α subunit from its partners. Now free in the membrane, the α subunit binds to another enzyme, such as adenylyl cyclase, and alters the activity of that enzyme. Since adenylyl cyclase catalyzes the conversion of ATP to cAMP, activation of the G-protein leads to a change in intracellular cAMP concentration; cAMP is an important secondary messenger. Reassociation of GDP with the α subunit reverses the process.

The versatility of receptor-ligand interactions in signal activation is apparent in the activity of the endorphin (endogenous opioid) family of naturally occurring peptides and proteins. When opioid peptides bind to opioid receptors, the receptor is internalized and signal transduction is initiated; binding activates a G-protein, which subsequently inhibits the activity of adenylyl cyclase, a membrane-associated enzyme. The alkaloid drug morphine is a potent agonist of opioid receptors, a family of receptors that respond to endo-

Table 5.7   Binding to opioid receptors

Agent	δ Receptor	μ Receptor
Morphine	9.2 μM	16 nM
Etorphine	1.8 nM	0.37 nM
Enkephalin analog peptides	2–9 nM	1 to 8 nM

From [20, 21]. The table displays the concentration of drug that is required to inhibit adenylyl cyclase activity by 50% (IC50).

genously produced opioid peptides. Morphine and the related compound etorphine bind to both μ and δ opioid receptors in the brain and in cultured cells (Table 5.7). However, unlike opioid peptides and etorphine, morphine is not internalized after binding (Figure 5.20b). Morphine activates G-protein signals without activating endocytosis. This example suggests that binding to a receptor can produce multiple activated states which initiate different signaling and trafficking events; in this case, receptor-mediated signal transduction can occur in modes with and without receptor internalization.

Changes in receptor activation result from changes in ligand concentration in the extracellular environment. Consider the situation in which a cell is initially surrounded by a uniform concentration of ligand, which is produced in the extracellular environment at some steady rate, $\dot{C}_0$. Although the ligand is continuously produced, its concentration remains constant because it is degraded by some natural processes (mechanisms of drug degradation and elimination are discussed in Chapter 8). For the signal to change, the rate of ligand generation must either increase or decrease. Modification of the rate of ligand production will eventually lead to a change in its concentration; a change in ligand concentration will alter receptor–ligand binding and, therefore, signal activation. The ligand concentration at the cell surface will change according to the differential equation (ligand accumulation = ligand production − ligand consumption):

$$\frac{dc}{dt} = \dot{C} - kc \tag{5-26}$$

At steady-state, dc/dt is equal to zero, so the ligand concentration is equal to the ratio of $\dot{C}/k$, where $\dot{C}$ is the rate of ligand production and k is the first-order rate constant for ligand degradation. If the rate of ligand production is suddenly increased to 10 times its previous level ($\dot{C} = 10\dot{C}_0$), then the concentration of ligand in the extracellular fluid will increase. The kinetics of ligand concentration change depend on the degradation rate (which can also be designated by the half-life for degradation, $t_{1/2} = \ln(2)/k$):

$$\frac{c}{c_0} = \frac{R}{R_0} + \left(1 - \frac{R}{R_0}\right)e^{-kt} \tag{5-27}$$

as illustrated in Figure 5.21. Rapid changes in ligand concentration, i.e., rapid changes in signal activation, can only occur for compounds that are rapidly degraded. For this reason, naturally occurring signaling molecules often have short half-lives in the body. Rapid degradation allows cells rapidly to regulate signal activation by increasing or decreasing the ligand production rate.

P-glycoprotein and Multidrug Resistance.   Tumor cells can become resistant to the anticancer drugs that are used to kill them. A common mechanism of drug resistance involves a 1.7×10^5 Da membrane protein called P-glycoprotein (Figure 5.22), which is capable of active transport of a wide variety of compounds from the interior of the cell to the exterior. In comparison with most other transport proteins, which are selective for the solutes they transport (Figure 5.12), P-glycoprotein is promiscuous. The same membrane protein appears to be capable of active transport of a wide variety of chemicals including anticancer drugs (doxorubicin, vincristine), steroid hormones (aldosterone, hydrocortisone), and many others.

How can this single membrane protein recognize and transport so many different compounds? Recently, the characteristics of 100 P-glycoprotein substrates were compiled and examined. P-glycoprotein substrates share certain structural characteristics: most have either two electron-donor groups separated by 2.5 ± 0.3 Å, two electron-donor groups separated by 4.6 ± 0.6 Å, or three electron-donor groups with the outermost groups separated by 4.6 ± 0.6 Å (Figure 5.22). This finding suggests that P-glycoprotein recognizes chemicals that have well-defined structural characteristics.

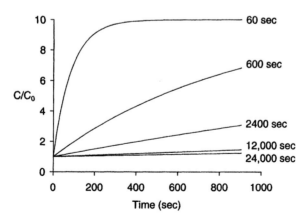

Figure 5.21   Influence of degradation on the kinetics of ligand change. Signaling molecules must have a rapid turnover in order to produce rapid changes in signal. Each curve represents the concentration for a ligand with a particular turnover rate; the half-life is indicated on the chart and varies between 60 and 24,000 s.

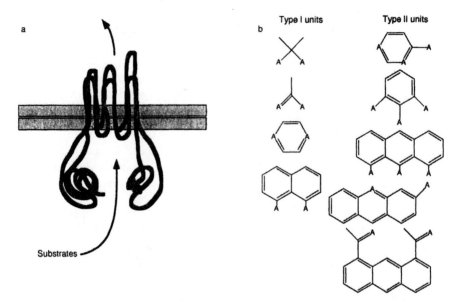

Figure 5.22 Structure of the P-glycoprotein pump, which confers multidrug resistance to tumor cells. (a) Schematic structure of the P-glycoprotein pump.
(b) Chemicals that are able to interact with the P-glycoprotein have type I and type II structure, where A is an electron-donor group. Type I substrates have two electron-donor groups with a separation of 2.5 ± 0.3 Å and type II substrates have two or three electron-donor groups with a separation (between the outermost groups) of 4.6 ± 0.6 Å. Adapted from [22].

5.5 PERMEATION THROUGH CELL LAYERS

5.5.1 Hydraulic Permeability

The movement of water through membranes and cell layers is critical for cell and tissue function. In its simplest formulation, the transmembrane volumetric fluid flux, V_w, is given by:

$$V_w = k_m(\Delta p - \Delta \pi) \tag{5-28}$$

where Δp and $\Delta \pi$ represent hydrostatic $(\Delta p = p_{in} - p_{out})$ and osmotic $(\Delta \pi = \pi_{in} - \pi_{out})$ pressure gradients across the membrane. In Equation 5-28, k_m represents the hydraulic permeability of the barrier; it is therefore similar to a mass transfer coefficient for fluid movement across the layer. This equation assumes that the membrane barrier is perfectly impermeable to the solutes that are generating the osmotic driving force. In many cases, the membrane barrier is imperfect; therefore, the osmotic driving force must be adjusted. The reflection coefficient, σ_i is used to make this correction:

$$\sigma_i = \begin{cases} 1 & \text{membrane is a perfect barrier} \\ (1 - v_i/v_w)_{\sigma=0} & \text{membrane is semipermeable} \\ 0 & \text{membrane is very permeable} \end{cases} \quad (5\text{-}29)$$

where v_i and v_w are the velocities of solute and water through the barrier. With this definition, the overall fluid flux can be written:

$$V_w = k_m \left(\Delta p - \sum_{i=1}^{n} \sigma_i RT \Delta c_i \right) \quad (5\text{-}30)$$

in which the van't Hoff equation for ideal, dilute solutions ($\pi = RTc$) has also been used.

5.5.2 Solute Permeation across Tissue Barriers

An alternative definition of solvent and solute flow is based on the irreversible thermodynamics and was initially derived by Kedem and Katchalsky (see [23] for a recent review). J_v is the solvent flow:

$$J_v = L_p S(\Delta p - \sigma_i \Delta \pi) \quad (5\text{-}31)$$

where L_p is the hydraulic conductance and S is the surface area. Correspondingly, the flow of solute i can be expressed as:

$$J_s = J_v(1 - \sigma_i)\bar{c}_i + PS\Delta c_i \quad (5\text{-}32)$$

where $\bar{c}_i$ is the mean concentration within the membrane and P is the permeability of the barrier to solute i. Equation 5-32 accounts for both diffusive transport (as defined in Equation 5-2) and convective transport due to solute movement across the barrier. The permeability P is related to the diffusion coefficient (see Equation 5-1) and the thickness of the barrier; therefore, Equation 5-32 can also be written in differential form:

$$J_s = J_v(1 - \sigma_i)c_i + D_i S \frac{dc_i}{dx} \quad (5\text{-}33)$$

and integrated from $x = 0$ to $x = L$ (the thickness of the barrier) to yield:

$$\text{Clearance} \, (Cl) = \frac{J_s}{c_{i,\text{plasma}}} = J_v(1 - \sigma_i) \frac{1 - \dfrac{c_{i,\text{interstitial}}}{c_{i,\text{plasma}}} e^{-Pe}}{1 - e^{-Pe}} \quad (5\text{-}34)$$

where $c_{i,\text{plasma}}$ and $c_{i,\text{interstitial}}$ are solute concentrations on the plasma and interstitial side of the barrier, respectively, Cl is the solute clearance, and Pe is a modified Péclet number, defined as $J_v(1 - \sigma_i)/PS$. The relationship between filtration rate (J_v) and clearance is shown in Figure 5.23: when the Pe is low ($Pe \sim 0$), diffusion of solute across the barrier is the primary mechanism of transport. When the Pe is high ($Pe > 3$), fluid convection is the dominant mechanism of transport, and transport is proportional to the product $J_v(1 - \sigma_i)$ (dashed lines in Figure 5.23).

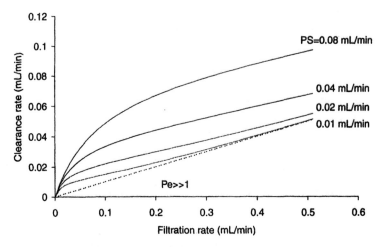

Figure 5.23 Clearance versus filtration rate for simple homogeneous transport through a tissue barrier. Redrawn from [23]. The solid lines indicate clearance for different values of the *PS* product. The dashed lines indicate transport in the absence of diffusion ($Pe \gg 1$). Equation 5-34 is plotted for $\sigma_i = 0.9$ at steady state (i.e., lymphatic clearance is equal to solute flux, see [23]).

5.5.3 Permeation through Endothelial Layers

General Concepts.   The capillary endothelium is the major site of exchange for nutrients, hormones, and proteins between blood and tissues. When the solute is a macromolecule, transport occurs through pores in the endothelium. Capillaries in the general circulation are highly permeable to water and glucose, and nearly impermeable to proteins larger than albumin (Table 5.8). Capillary permeability is qualitatively different from cell

Table 5.8 Relative permeability of muscle capillaries to solutes of various molecular weight

	Molecular weight	Relative permeability
Water	18	1
NaCl	59	0.96
Urea	60	0.8
Glucose	180	0.6
Sucrose	342	0.4
Inulin	5,000	0.2
Myoglobin	17,600	0.03
Hemoglobin	68,000	0.01
Albumin	69,000	0.001

Adapted from [24], p. 186.

membrane permeability: the ratio of glucose to water permeability is 0.6 for a muscle capillary (Table 5.8) and 0.0001 for a cell membrane (Figure 5.6). The enhanced permeability of water-soluble molecules through capillary walls is due to the presence of water-filled pores that are sufficiently large to accommodate the movement of glucose and sucrose. As the size of the solute increases, permeability through the capillary wall decreases, suggesting that the pores are only slightly larger than albumin, which has a radius of $\sim 35\,\text{Å}$.

Experimental measurements of rates of protein transport suggest that convection and diffusion through large pores (~ 200–$400\,\text{Å}$ radius) is the dominant mechanism of movement from blood into tissue [23]. However, the pores in the endothelium are not uniform in size. Pores of different radii present different resistances to diffusive and convective transport of macromolecules (see Figure 5.9); the overall clearance is influenced by these different transport resistances. For a typical protein, such as albumin, the overall clearance can be written as the sum of clearance from two different pore populations:

$$Cl_{\text{SMALL}} = PS_{\text{SMALL}} + \alpha_{\text{LARGE}}(1 - \alpha_{\text{LARGE}})L_pS(\sigma_{\text{SMALL},i} - \sigma_{\text{LARGE},i})$$
$$\Delta\pi(1 - \sigma_{\text{SMALL},i}) + \alpha_{\text{SMALL}}J_v(1 - \sigma_{\text{SMALL},i}) \tag{5-35}$$

$$Cl_{\text{LARGE}} = PS_{\text{LARGE}} + \alpha_{\text{LARGE}}(1 - \alpha_{\text{LARGE}})L_pS(\sigma_{\text{SMALL},i} - \sigma_{\text{LARGE},i})$$
$$\Delta\pi(1 - \sigma_{\text{LARGE},i}) + \alpha_{\text{LARGE}}J_v(1 - \sigma_{\text{LARGE},i}) \tag{5-36}$$

where the subscripts SMALL and LARGE indicate the two different pore populations, which contribute fractions α_{SMALL} and α_{LARGE} to the hydraulic conductance. The clearance has contributions from diffusion (PS), convection ($J_v(1 - \sigma_i)$), and recirculation of volume through large pores (Starling flow). Each pore population has different reflection coefficients for the solute, σ_i, and different contributions to the overall surface area, S. The overall clearance is found by summing these two contributions:

$$Cl = PS_i + \alpha_{\text{LARGE}}(1 - \alpha_{\text{LARGE}})L_PS(\sigma_{\text{SMALL},i} - \sigma_{\text{LARGE},i})^2\Delta\pi + J_v(1 - \sigma_i) \tag{5-37}$$

Figure 5.24 shows calculated fluxes of albumin (radius $35\,\text{Å}$) for a two-pore system in which the radius of the small pore is varied from 40 to $70\,\text{Å}$.

It is possible that some substances move through the capillary wall by transcytosis (Figure 5.25). The importance of this pathway is not clear, but capillaries in some tissues contain numerous vesicles.

Capillaries in different tissues vary tremendously in their permeability characteristics, as described in the sections that follow.

Capillaries in the Kidney. Filtration in the kidney occurs across a specialized capillary in the glomerulus. The glomerular capillary has several layers: a monolayer of endothelial cells that interface with the flowing blood, an acellular basement membrane, and a filtration layer formed by glomerular epithelial cells. The endothelium is fenestrated; that is, there are numerous

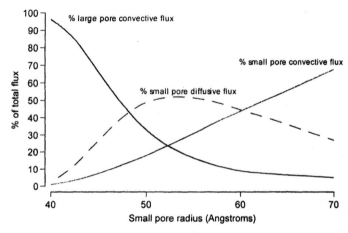

Figure 5.24   Mechanism of albumin transport as a function of small pore radius. Redrawn from [23]. The following parameters were used: $A_0/\Delta x$ (ratio of pore area to pore length) $= 300\,\text{m}/100\,\text{g},$, $\alpha_{\text{LARGE}} = 0.05$, $L_p S = 0.008\,\text{mL}/\text{min} \cdot \text{mmHg}$; and $C_i/C_p = 0.5$. Convection through large pores is most important at low small-pore radius; convection through small pores is most important at high small-pore radius; and diffusion through small pores is most important in the intermediate regime.

small holes or fenestrae throughout the continuous endothelium. Fenestrations can be important in overall transport through the capillary: fenestrated capillaries in other tissues, most notably the liver, permit almost free transit of proteins. These fenestrations in the glomerular capillary endothelium are large enough for proteins to pass; this characteristic contrasts with most capillaries, such as muscle capillaries, which are not fenestrated, and do not permit protein passage.

Proteins cannot, however, pass easily through the basement membrane, which is an entangled gel of proteins and proteoglycans; protein permeability is low because of the fine mesh within the basement membrane gel and because of

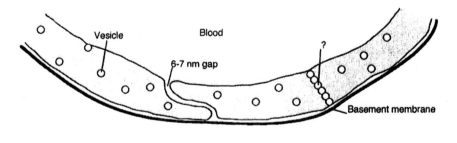

Figure 5.25   Vesicular transport through the capillary wall. Capillaries are capable of transcytosis, the directed transport of vesicles (containing extracellular fluid engulfed at one of the capillary surfaces).

the dense negative charge provided by the proteoglycans. The epithelial layer contains slits filled with porous diaphragms that permit the free passage of water; the structure of the diaphragm is still in debate, although the open gaps in the diaphragm are approximately the same size as albumin. The glomerular capillary, which is designed to permit filtration of large volumes of water, still allows for slow diffusion of macromolecules (Figure 5.26) [25]. Most of the diffusive resistance is due to the presence of small fenestrae and slits in the cellular layers ($\sim$ 80%), but the basement membrane provides a substantial resistance that increases with molecular size of the solute. The reduced diffusion coefficient (D/D_∞) decreases from $\sim$ 0.01 to $\sim$ 0.001 over the range of molecular radii shown in Figure 5.26.

Capillaries in the Brain (the Blood–Brain Barrier). Capillaries in the brain are less permeable than capillaries in other tissues. This limited permeability, which is frequently called the blood–brain barrier, is essential for brain function. Reduced permeation provides a buffer that maintains a constant brain extracellular environment, even at times when blood chemistry is changing. The basis for this lower permeability is the relative paucity of pores in the brain endothelium. Therefore, molecules that move from blood to brain must diffuse through the endothelial cell membranes. As expected from this observation, the permeability of brain capillaries depends on the size and lipid solubility of the solute. In general, molecules that are larger than several hundred in molecular weight do not permeate into the brain. Empirical relationships between cerebrovascular permeability and the oil[1]/ water partition coefficient have been developed [26] (see Figure 5.27):

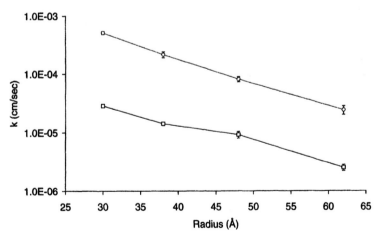

Figure 5.26 Diffusion of Ficoll through the glomerular capillary. Adapted from [25]. Diffusion of molecular weight fractions of Ficoll through isolated glomerular capillaries (squares) or capillaries with the cells removed (circles).

1. Octanol or olive oil is frequently used in experimental techniques for estimating the oil/water partition coefficient.

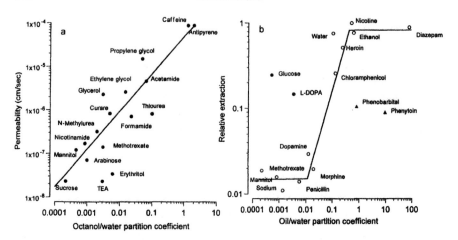

Figure 5.27 Correlation of cerebrovascular permeability with octanol/water partition coefficient. Redrawn from [26] and [16].

$$\log_{10} P = 4.30 + 0.866 \log_{10} K \qquad (5\text{-}38)$$

This correlation is useful for rapid estimation of the blood–brain barrier permeability. More accurate estimation can be obtained by examining other physical characteristics of the compound, such as cross-sectional area and critical micelle concentration [27] (Figure 5.28). Because of the extremely low permeability of brain capillaries to most molecules, brain endothelial cells have a variety of specialized transport systems—including transporters for glucose, amino acids, insulin, and transferrin—that enable essential molecules to move from the blood into the brain extracellular space. Useful reviews of the blood–brain barrier are available [28, 29].

Capillaries in Tumors. The blood vessels that develop in tumors are generally more permeable than vessels that develop in normal tissues. As a result, large molecules that would not penetrate through a normal vessel will permeate through a vessel in a tumor. The size cut-off for a vessel can be determined by injecting particles of a known size intravenously, then looking in the interstitial space of the tumor to see if the particles are extravasated. This approach was used in tumors that were grown subcutaneously or intracranially (Figure 5.29). Subcutaneous tumors were generally permeable to larger particles than intracranial tumors. The same tumor, when introduced subcutaneously or intracranially, had different permeability characteristics, highlighting the importance of local tissue environment in defining local vascular permeability.

5.5.4 Permeation through Epithelial Barriers

Skin and mucosal surfaces of the gut, respiratory, and reproductive tracts represent a principal barrier against entry of pathogens and molecules encountered in the environment. These multicellular structures are often the most

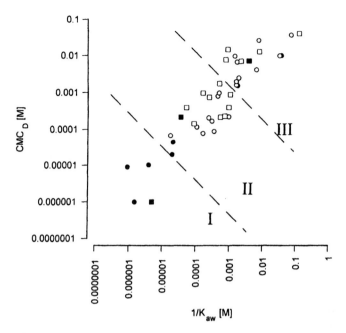

Figure 5.28 Correlation of cerebrovascular permeability with critical micelle concentration and air/water partition coefficient. Adapted from [27].

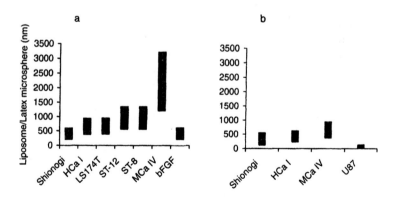

Figure 5.29 Extravasation of particles from vessels in subcutaneous and intracranial tumors. Adapted from [30]. In panel a, tumors were implanted subcutaneously and in panel b, tumors were implanted intracranially. The bars indicate the range of particle sites between noticeable extravasation (lower limit) and no extravasation (upper limit). The pore size depends on the tumor type (shown on x-axis) and the location of implantation.

Mucosal surface (apical, luminal)

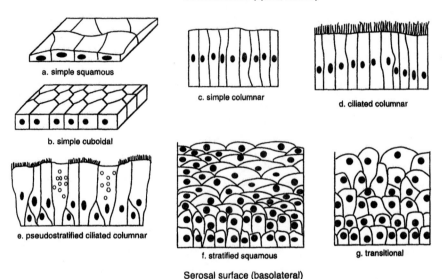

a. simple squamous

b. simple cuboidal

c. simple columnar

d. ciliated columnar

e. pseudostratified ciliated columnar

f. stratified squamous

g. transitional

Serosal surface (basolateral)

Figure 5.30   Illustration of the structure of epithelial tissue barriers in the body. Modified from [31] and Stedmann's Medical Dictionary. Epithelial tissues have a variety of histological patterns: (a) simple squamous as found in the initial segments of ducts in glands and the alveolus; (b) simple cuboidal as found in glandular ducts and ovary; (c) simple columnar as found in stomach, intestines, uterus, and cervix; (d) ciliated columnar as found in the small bronchi, intestines, and cervix; (e) pseudo-stratified ciliated columnar with mucus-secreting (goblet) cells as found in the trachea, large bronchi, and vas deferens; (f) stratified squamous as found in skin, esophagus, tongue, and vagina; and (g) transitional as found in the ureter and urinary bladder.

important barrier to the administration of drugs. For example, the oral route of drug administration, which is usually the preferred route, is limited by the extent to which drug molecules can penetrate the epithelial surface of the intestinal tract. The structure, and therefore the barrier characteristics, of epithelial tissues shows considerable variability; schematic diagrams of the major types of epithelium are shown in Figure 5.30.

Different mucosal surfaces differ in permeability. The permeability of nasal, rectal, and vaginal membranes from the rabbit was determined when $P_{nasal} > P_{vaginal} > P_{rectal}$ for mannitol and $P_{nasal} > P_{rectal} > P_{vaginal}$ for progesterone [32]. Permeability was assessed using electrical resistance and the permeability of epithelial barriers to a water-soluble fluorescent dye, 6-carboxyfluorescein. The rank order of permeabilities was intestinal $\sim$ nasal $\geq$ bronchial $\geq$ tracheal > vaginal $\geq$ rectal > corneal > buccal > skin [33].

Permeation through the Skin.   Our skin provides a mechanical barrier that prevents pathogen entry into our tissues and, therefore, protects us from infection. Skin also prevents excessive loss of fluid due to evaporation, by regulating the access of water to the body surface. The barrier properties of skin are impressive; although the average person has $1.8 \, \text{m}^2$ of exposed skin, insignificant quantities of most compounds enter the body through this surface. Because of its accessibility, the skin is also an attractive route for drug delivery, but to be a useful route of delivery, the formidable barrier properties of skin must be surmounted.

Skin is a keratinized, stratified, squamous epithelium and, therefore, is a complex, multilayer barrier (Figure 5.30f). The outermost layer of the skin, the stratum corneum, appears to be the major barrier to permeation of most compounds. Topical drug application is an effective and simple method for treatment of local skin diseases; drug molecules only need to penetrate locally in relatively small amounts. If systemic delivery is required, however, obstacles abound: overall bioavailability can be limited by the intrinsic permeability of the stratum corneum, by local degradation of drugs during permeation, or by low availability of blood flow to skin at the site of drug application. The overall flux of drug molecules through the skin, J, depends on concentration and permeability:

$$J = \frac{c_{\text{skin}} - c_{\text{blood}}}{(\Omega_{\text{blood}} + 1/P_{\text{sc}})} \tag{5-39}$$

where c_{skin} and c_{blood} are drug concentrations at the skin surface and in the blood, P_{sc} is the permeability of the stratum corneum, and Ω_{blood} is the effective resistance to transport through the tissue beneath the stratum corneum. This resistance includes the availability of blood flow, which serves as a reservoir for the drug molecules penetrating through the skin. For permeabilities of the stratum corneum that are less than $\sim 0.01 \, \text{cm/min}$, the resistance of the stratum corneum dominates the overall transport (Figure 5.31).

The permeation of compounds through excised skin correlates well with the lipid/water solubility ratio (Figure 5.32). The permeability measured

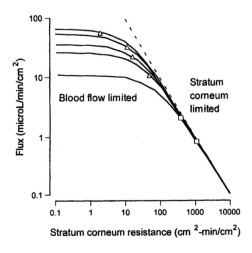

Figure 5.31
Transcutaneous flux as a function of resistance in the stratum corneum and blood flow. Adapted from [34].

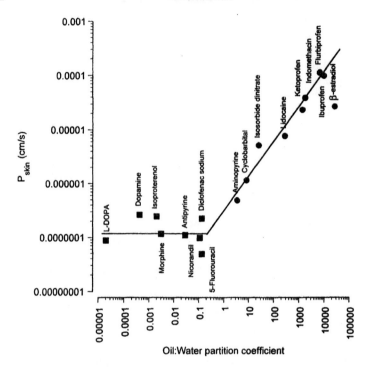

Figure 5.32 Permeability of compounds as a function of octanol/water partition coefficient. Adapted from Figure 13 of [35].

through excised hairless rat skin was represented well by the following expression:

$$P_{skin} = 4.78 \times 10^{-7} K_{o:w}^{0.589} + 8.33 \times 10^{-8} \qquad (5\text{-}40)$$

where the permeability is expressed in cm/s. For all of the compounds used in this study, the permeability was less than 6×10^{-3} cm/min, which suggests that overall transcutaneous flux is limited by the stratum corneum resistance, and not blood flow (see Figure 5.31).

Small molecules can diffuse in the lipid-rich environment of the stratum corneum, but rates of diffusion are slow for molecules greater than 200 in molecular weight (Figure 5.33).

The skin is readily accessible, but generally impermeable to most compounds. Therefore there has been considerable interest in finding non-destructive methods to enhance permeation through skin. Permeation can be enhanced through the action of chemical enhancers, which decrease the permeability of skin [37], although it has been difficult to identify agents that enhance permeability without damaging or irritating the skin. Alternative approaches, such as the use of ultrasound [38] or electrical fields [39] to enhance permeation, may prove useful in delivery of highly potent compounds that are needed chronically, such as insulin. These concepts are discussed further in Chapter 9.

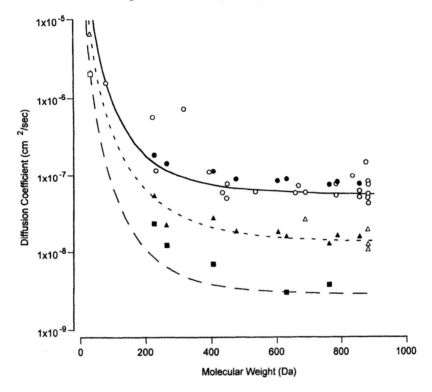

Figure 5.33 Lateral diffusion of small molecules in bilayers of stratum corneum. Adapted from Figure 4 of [36].

Permeation through Mucosal Layers

ALVEOLAR EPITHELIUM. Molecules move through the alveolar epithelium by transcytosis [40], where they eventually are deposited in the underlying interstitial space. The molecules may then enter the lung lymphatic pathways, or may be absorbed into local capillaries, a process that may also require either transcytosis or permeation through the capillary membrane. The extent and rate of absorption through the alveolar epithelium does not have a clear dependence on molecular weight, which is consistent with a transport process that depends on transcytosis (although some large molecules, such as ferritin (M_w 5×10^5) do not readily enter transcytotic vesicles). Rates of absorption can be estimated from the time to peak concentration in plasma [40]: 90 min in human for leuprolide acetate; 15–30 min in human for insulin; 30 min–4 h in rats for growth hormone; 4–12 h in sheep for α1-antitrypsin; 5 h in guinea pigs for albumin.

A thin layer of mucus (1–10 μm) and a thinner layer of surfactant (0.1–0.2 μm) protect the alveolar surface. Since particulates are rapidly taken up by macrophages, an agent delivered to the alveolus must quickly dissolve in the secretions to be absorbed. Still, even relatively large proteins enter the circulation if introduced into the lung properly. Bioavailability relative to subcutaneous injection has been reported to be ~ 30% for calcitonin (32 amino acids), 6–10% for insulin (51 amino acids), and 30% for interferon-α (165 amino acids) [40].

INTESTINAL EPITHELIA. Tight junctions in the intestinal epithelium provide a permeability barrier for molecules larger than 2.3 nm in diameter (Table 5.9). However, intestinal epithelia possess a variety of mechanisms for the transport of specific molecules. For example, maternal IgG antibodies are essential for the immune system of newborns. The movement of maternal IgG across the intestinal epithelium in mice occurs by a multistep process (Figure 5.34). The maternal antibodies, which are ingested by the newborn in mother's milk, bind, via their Fc region, to receptors on the apical surface of intestinal epithelial cells. This binding occurs readily at intestinal pH (6.0). Binding initiates vesicle formation and internalization of the receptor–IgG complex, which is shuttled through the cytoplasm as an endosome. The endosome fuses with the basolateral membrane of the epithelial cell, exposing the receptor–IgG complex to the subendothelial interstitial fluid, which has the pH of blood (7.0). Binding affinity is reduced at this pH, so the IgG is released from the transported receptor, which is now available for recycling to the apical surface. Transport via transcytosis occurs in other situations as well, including transport across the intact endothelium (Figure 5.34).

The intestinal epithelia also exhibit an interesting mechanism for the uptake of glucose, which utilizes several of the membrane transport mechanisms discussed in this chapter (Figure 5.35). Tight junctions between adjacent epithelial cells form a barrier to the movement of molecules, including glucose, in the intercellular space. Tight junctions are formed by interlocking protein particles, contributed by both of the neighbor cells. The junction prevents the diffusion of water-soluble molecules in the extracellular space, as well as the mobility of proteins in the membrane, creating segregated domains in each cell, with glucose-active transport proteins confined in the

Table 5.9 Permeability of some carbohydrates through isolated intestinal strips

Carbohydrate	Molecular weight	Radius of gyration (Å)	Intestinal permeability (μmol/g/h)
Inulin	5,000	14.8	0
Lactose	342	4.4	5
Mannose	180	3.6	19
Ribose	150	3.6	22
Glyceraldehyde	90	2.4	45

Permeability of golden hamster intestinal strips to passively diffusing carbohydrates. Original data from T.H. Wilson, *Intestinal Absorption*, Philadelphia: Saunders, 1962, Table 5, p. 4, as reported in [41].

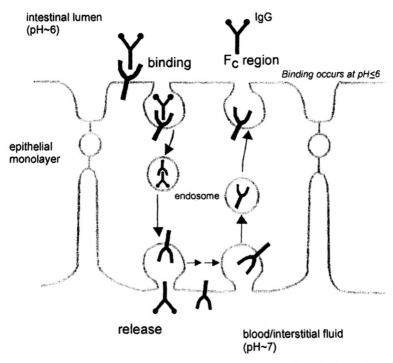

intestinal lumen
(pH~6)

IgG

binding

F$_C$ region

Binding occurs at pH≤6

epithelial
monolayer

endosome

release

blood/interstitial fluid
(pH~7)

Figure 5.34 Movement of maternal IgG across the intestinal epithelium of newborn mice. Adapted from [13, 14].

apical domain and glucose-facilitated transport channels confined to the basolateral domain. By maintaining this separation of domains, glucose can be transported up its concentration gradient. Active transport proteins move glucose molecules into the cell; glucose passively diffuses through facilitated transport channels into the subendothelial interstitial space; glucose is prevented from diffusing back down its gradient (into the intestinal lumen) by the junctional complexes.

SUMMARY

- Membranes form selectively permeable barriers that separate aqueous compartments in cells.

- Lipids and proteins are mobile within the membrane phospholipid bilayer.

- Non-electrolytes permeate membranes by dissolution and diffusion; permeation depends on both molecular size and solubility within the membrane phase.

- Solute permeation of porous membranes depends on the size of the solute relative to the pore.

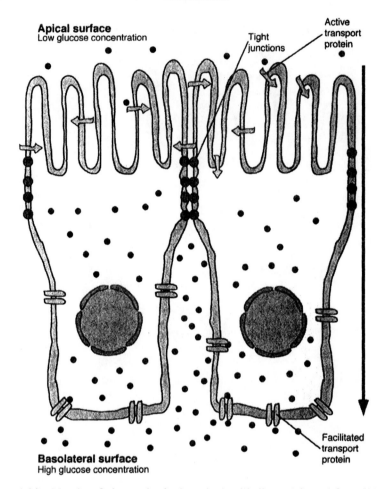

Apical surface
Low glucose concentration

Tight junctions

Active transport protein

Facilitated transport protein

Basolateral surface
High glucose concentration

Figure 5.35   Uptake of glucose in the intestinal epithelium. Adapted from [13, 14].

- Specialized proteins within the membrane provide mechanisms for facilitated transport, active transport, and ion transport.
- Movement of molecules across multicellular tissue barriers usually involves a combination of the mechanisms described in this chapter.

REFERENCES

1. Frye, L. and M. Edidin, The rapid intermixing of cell surface antigens after formation of mouse–human heterokaryons. *Journal of Cell Science*, 1970, **7**, 319–335.
2. Axelrod, D., *et al.*, Mobility measurement by analysis of fluorescence photobleaching recovery kinetics. *Biophysical Journal*, 1976, **16**, 1055–1069.
3. Diamond, J.M. and E.M. Wright, Biological membranes: The physical basis of ion and nonelectrolyte selectivity. *Annual Reviews of Physiology*, 1969, **31**, 581–646.

4. Singer, S.J. and G.L. Nicolson, The fluid mosaic model of the structure of cell membranes. *Science*, 1972, **175**, 720–731.
5. Colton, C., *et al.*, Diffusion of organic solutes in stagnant plasma and red cell suspensions. *Chemical Engineering, Progress Symposium Series*, 1970, **66**, 85–99.
6. Stryer, L., *Biochemistry*, 2nd ed. New York: W.H. Freeman, 1988.
7. Speelmans, G., *et al.*, Transport studies of doxorubicin in model membranes indicate a difference in passive diffusion across and binding at the outer and inner leaflets of the plasma membrane. *Biochemistry*, 1994, **46**, 13761–13768.
8. Akhtar, S., *et al.*, Interactions of antisense DNA oligonucleotide analogs with phospholipid membranes (liposomes). *Nucleic Acids Research*, 1991, **19**(20), 5551–5559.
9. Deen, W.M., Hindered transport of large molecules in liquid filled pores. *AIChE Journal*, 1987, **33**, 1409–1425.
10. Renkin, E.M., Filtration, diffusion, and molecular sieving through porous cellulose membranes. *Journal of General Physiology*, 1954, **38**, 225–243.
11. Brenner, H. and L. Gaydos, The constrained Brownian movement of spherical particles in cylindrical pores of comparable radius. Models of the diffusive and convective transport of solute molecules in membranes and porous media. *Journal of Colloid and Interface Science*, 1977, **58**, 312–356.
12. Anderson, J.L. and J.A. Quinn, Restricted transport in small pores: A model for steric exclusion and hindered particle motion. *Biophysical Journal*, 1974, **14**, 130–150.
13. Alberts, B., *et al.*, *Molecular Biology of the Cell*, 3rd ed. New York: Garland Publishing, 1994.
14. Lodish, H., *et al.*, *Molecular Cell Biology*. New York: W.H. Freeman, 1995.
15. Pardridge, W. and T. Choi, Neutral amino acid transport at the human blood–brain barrier. *Federation Proceedings*, 1986, **45**, 2073–2078.
16. Kandell, E.R., J.H. Schwartz, and T. Jessell, M., Eds., *Principles of Neural Science*, 3rd ed. Norwalk, CT: Appleton & Lange, 1991.
17. Clark, E.A. and J.S. Brugge, Integrins and signal transduction pathways: The road taken. *Science*, 1995, **268**, 233–239.
18. Lauffenburger, D.A. and J.J. Linderman, *Receptors: Models for Binding, Trafficking, and Signaling*. New York: Oxford University Press, 1993.
19. Longhurst, C.M. and L.K. Jennings, Integrin-mediated signal transduction. *Cellular and Molecular Life Sciences*, 1998, **54**, 514–526.
20. Keith, D.E., *et al.*, Morphine activates opioid receptors without causing their rapid internalization. *Journal of Biological Chemistry*, 1996, **271**, 19021–19024.
21. Keith, D.E., *et al.*, mu-Opioid receptor internalization: opiate drugs have differential effects on a conserved endocytic mechanism *in vitro* and in the mammalian brain. *Molecular Pharmacology*, 1998, **53**, 377–384.
22. Seelig, A., A general pattern for substrate recognition by P-glycoprotein. *Eur. J. Biochem.*, 1998, **251**, 252–261.
23. Rippe, B. and B. Haraldsson, Transport of macromolecules across microvessel walls: the two-pore theory. *Physiological Reviews*, 1994, **74**(1), 163–219.
24. Guyton, A.C. and J.E. Hall, *Textbook of Medical Physiology*. Philadelphia: W.B. Saunders, 1996.
25. Edwards, A., W.M. Deen, and B.S. Daniels, Hindered transport of macromolecules in isolated glomeruli. I Diffusion across intact and cell-free capillaries. *Biophysical Journal*, 1997, **72**, 204–213.

26. Rapoport, S., K. Ohno, and K. Pettigrew, Drug entry into the brain. *Brain Research*, 1979, **172**, 354–359.
27. Fischer, H., R. Gottschlich, and A. Seelig, Blood–brain barrier permeation: molecular parameters governing passive diffusion. *Journal of Membrane Biology*, 1998, **165**, 201–211.
28. Davson, H. and M.B. Segal, eds, *Physiology of the CSF and Blood–Brain Barrier*. Boca Raton, FL: CRC Press, 1996.
29. Pardridge, W.M., ed., *The Blood–Brain Barrier: Cellular and Molecular Biology*. New York: Raven Press, 1993.
30. Hobbs, S.K., *et al.*, Regulation of transport pathways in tumor vessels: role of tumor type and microenvironment. *Proccedings of the National Academy of Sciences*, 1998, **95**, 4607–4612.
31. Parkhurst, M.R., Leukocyte migration in mucus and interaction with mucosal epithelia: potential targets for immunoprotection and therapy, in *Chemical Engineering*. Baltimore, MD: The Johns Hopkins University Press, 1994.
32. Corbo, D.C., J.-C. Liu, and Y.W. Chien, Characterization of the barrier properties of mucosal membranes. *Journal of Pharmaceutical Sciences*, 1990, **79**(3), 202–206.
33. Rojanasakul, Y., *et al.*, The transport barrier of epithelia: a comparative study on membrane permeability and charge selectivity in the rabbit. *Pharmaceutical Research*, 1992, **9**(8), 1029–1034.
34. Whang, J., J. Quinn, and D. Graves, The skin as a barrier membrane: Relating *in vitro* measurements to *in vivo* behavior. *Journal of Membrane Science*, 1990, **52**, 379–392.
35. Hatanaka, T., *et al.*, Prediction of skin permeability of drugs. 1. Comparison with artificial membrane. *Chemical Pharmaceutical Bulletin*, 1990, **38**(12), 3452–3459.
36. Johnson, M.E., *et al.*, Lateral diffusion of small compounds in human stratum corneum and model lipid bilayer systems. *Biophysical Journal*, 1996, **71**, 2656–2668.
37. Williams, A.C. and B.W. Barry, Skin absorption enhancers. *Critical Reviews in Therapeutic Drug Carrier Systems*, 1992, **9**, 305–353.
38. Mitragotri, W., D. Blankschtein, and R. Langer, Ultrasound-mediated transdermal protein delivery. *Science*, 1995, **269**, 850–853.
39. Prausnitz, M.R., *et al.*, Transdermal delivery of heparin by skin electroporation. *Biotechnology*, 1995, **13**, 1205–1209.
40. Patton, J.S. and R.M. Platz, Pulmonary delivery of peptides and proteins for systemic action. *Advanced Drug Delivery Reviews*, 1992, **8**, 179–196.
41. Creasey, W.A., *Drug Disposition in Humans: The Basis of Clinical Pharmacology*. New York: Oxford University Press, 1979.

6

Drug Transport by Fluid Motion

> sensations sweet,
> Felt in the blood, and felt along the heart;
> And passing even into my purer mind
> William Wordsworth, *Lines composed a few miles*
> *above Tintern Abbey*, on revisiting the banks of
> the Wye during a tour, July 13, 1798

The rate of molecular movement by diffusion decreases dramatically with distance, and is generally inadequate for transport over distances greater than 100 μm (recall Table 4.8). The movement of molecules over distances greater than 100 μm occurs in specialized compartments in the body: blood circulates through arteries and veins; interstitial fluid collects in lymphatic vessels before returning to the blood; cerebrospinal fluid (CSF) percolates through the central nervous system (CNS) in the brain ventricles and subarachnoid space. In these systems, molecules move primarily by bulk flow, or convection.

Diffusive transport is driven by differences in concentration; convective transport is driven by differences in hydrostatic and osmotic pressure. This chapter introduces the principles of drug distribution by pressure-driven transport. The elaborate network of arteries, capillaries, and veins that carry blood throughout the body are described first in this chapter.[1] Hydrostatic pressure within the blood vasculature drives fluid through the vessel wall (recall Equation 5-28) and into the extravascular space of tissues. Fluid flow through the interstitial space is not well understood, although the importance of interstitial flows in moving drug molecules through tissue is beginning to be appreciated. Engineering approaches for analyzing fluid flows in the interstitium are described in the second section of the chapter. Finally, the specialized systems for returning interstitial fluid to the blood are essential for clearance of molecules from the interstitial space; therefore, the chapter also provides a description of the dynamics of lymph flow in the periphery and CSF production and circulation in the brain.

1. Biomedical engineers have long studied the circulatory system and made essential contributions to our understanding of cardiovascular physiology [1, 2].

6.1 BLOOD MOVEMENT IN THE CIRCULATORY SYSTEM

Our bodies appear, from the outside, to be solid masses that are slow to change, but, just beneath the surface, is a torrent of fluid motion. Blood moves at high velocity throughout the body within an interconnected and highly branched network of vessels. The cross-sectional area changes significantly along the network (Figure 6.1 and Table 6.1), and blood flow to the periphery emerges from the heart within a single vessel, which branches and rebranches to distribute blood to every tissue and organ. The network of branching vessels proceeds to vessels of ever decreasing diameter (aorta → artery → arterioles → capillaries; more detail is found in Tables B.3 and B.4). In the smallest vessels, the capillaries, vessel diameter is approximately the same as the diameter of cells within the flowing blood. One essential consequence of this highly branched structure is that every metabolically active cell in the body is within $100\,\mu$m of the nearest capillary, and is therefore assured adequate nutrients for metabolism.

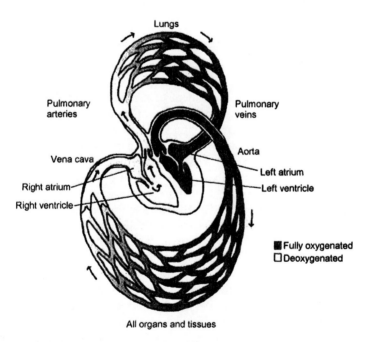

Figure 6.1 Diagram of the cardiovascular system. Blood flows through two connected circuits in the body. Starting from the left ventricle, blood is pumped through the aorta to the periphery, where it delivers oxygen to tissues. Deoxygenated blood is returned, through the veins to the right atrium and ventricle. Contraction of the right ventricle provides pressure for blood flow through the pulmonary vasculature, where oxygen is acquired during blood flow through the alveolar capillaries. Oxygenated blood returns through the pulmonary veins to the left atrium and ventricle. Adapted from Figure 14-7 of [3].

Table 6.1 Resistance to flow in vessels, hydraulic permeability of tissues, and typical fluid velocities

System	$\Omega/L\,(\mathrm{cm^4/dyn \cdot s})^a$	$k\,(\mathrm{cm^4/dyn \cdot s})^b$	$v\,(\mu\mathrm{m/min})$	Reference
Aorta	3.6–9.1		3.8×10^7 (63 cm/s)	Table B.3
Large artery	0.04–0.4		1.2–3.0×10^7 (20–50 cm/s)	
Capillary	2–9×10^7		3–6×10^4 (0.05–0.1 cm/s)	
Large vein	0.2–0.9		0.9–1.2×10^7 (15–20 cm/s)	
Vena cava	3.6		6.6–9.6×10^6 (11–16 cm/s)	
Mississippi River			1.7×10^7	d
Infusion into the brain		1.7×10^{-8}	14^c	[6]
Normal tissue		6.4×10^{-12}	0.0051^c	[5]
Tumor tissue		3.1×10^{-12}	0.0025^c	[5]
Polysaccharide gels		10^{-13} to 10^{-10}	0.08–0.00008^c	Figure 6.8
Tissue window			12–45	[10]
White matter in brain			10	[17]
UV-induced edema			4	[21]
Glioma			3–30	[22]

[a] From Equation 6-7.
[b] From Darcy's Law (Equation 6-9).
[c] This value was estimated from k by assuming $\Delta p = 10\,\mathrm{mmHg}$.
[d] Estimated from the average annual streamflow ($5 \times 10^4\,\mathrm{ft^3/s}$ according to U.S. Geological Survey records) and the average width and depth near Clinton, IA (500 and 5 m, respectively).

Hydrostatic pressure within the aorta is provided by contraction of heart muscle, which decreases left ventricular volume and, therefore, increases blood pressure within the chamber (Figure 6.2). Increasing pressure within the left ventricle eventually causes the aortic valve to open, ejecting blood into the elastic aorta, and initiating blood flow to the rest of the body. Rhythmic contraction of the heart repeats this ejection sequence; the presence of one-way valves (such as the aortic valve) assures blood flow in only one direction (ventricle $\rightarrow$ aorta).[2] A similar series of events drives blood from the right ventricle through the lung for oxygenation.

Flow through the vessels of the circulatory system is driven by the hydrostatic pressure created in the ventricles. The flow rate through a cylindrical vessel of radius R can be estimated from the equations for conservation of momentum (see [4]):

2. Without valves, the relaxation-contraction cycle of the ventricle would produce no net forward flow, but an endless repetition of blood ejection during the contraction phase and blood return during the relaxation phase.

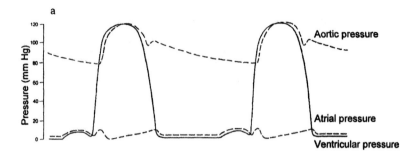

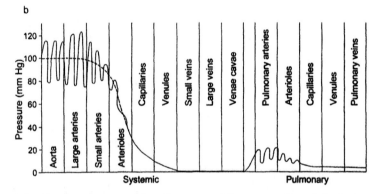

Figure 6.2 Blood pressure and flow is generated by contraction of the heart. (a) Pressure variations within the left ventricle, left atrium, and aorta during the cardiac cycle. (b) Pressure variations throughout the circulatory system. Adapted from [2].

$$\rho\frac{D\bar{v}}{Dt} = -\nabla p + \mu\nabla^2\bar{v} + \rho g \qquad (6\text{-}1)$$

where $\bar{v}$ is the fluid velocity, ρ is fluid density, μ is fluid viscosity, and p is hydrostatic pressure. Well-behaved flow through a cylindrical tube (Figure 6.3) occurs only in the axial direction ($\bar{v} = v_z(r)\hat{k}$), where v_z is a function of radial distance from the tube centerline, r. The z-component of Equation 6-1, in cylindrical coordinates, is:

$$\rho\left(\frac{\partial v_z}{\partial t} + v_r\frac{\partial v_z}{\partial r} + \frac{v_\theta}{r}\frac{\partial v_z}{\partial\theta} + v_z\frac{\partial v_z}{\partial z}\right) = -\frac{\partial p}{\partial z}$$

$$+ \mu\left[\frac{1}{r}\frac{\partial}{\partial r}\left(r\frac{\partial v_z}{\partial r}\right) + \frac{1}{r^2}\frac{\partial^2 v_z}{\partial^2 r} + \frac{\partial^2 v_z}{\partial^2 v}\right] + \rho g_z$$

$$(6\text{-}2)$$

This equation can be simplified by the following assumptions (which apply reasonably well to blood flows in most vessels in humans):

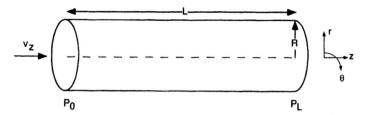

Figure 6.3 Geometry of cylindrical vessel. Many of the important characteristics of the flow of blood through the circulatory system can be modeled by a simple cylindrical geometry.

- steady flow $\left(\dfrac{\partial v_z}{\partial t} = 0\right)$;
- flow only in axial direction ($v_r = v_\theta = 0$);
- gravitational forces are negligible ($g_z \sim 0$).

These assumptions can be used to simplify Equation 6-2:

$$\frac{\partial p}{\partial z} = \mu \frac{1}{r} \frac{\partial}{\partial r}\left(r \frac{\partial v_z}{\partial r}\right) \tag{6-3}$$

which can be solved with the following conditions:

$$p = p_0 \text{ at } z = 0; \quad p = p_L \text{ at } z = L$$

$$v_z = 0 \text{ at } r = R; \quad \frac{\partial v_z}{\partial r} = 0 \text{ at } r = 0 \tag{6-4}$$

to yield:

$$v_z = \frac{(p_0 - p_L)R^2}{4\mu L}\left(1 - \frac{r^2}{R^2}\right) \tag{6-5}$$

This parabolic dependence of local velocity on radical position (Figure 6.4a) is characteristic of Hagen–Poiseuille flow.[3]

Equation 6-5 can be integrated over the vessel cross-section to obtain an overall rate of blood flow:

$$Q = \int_0^{2\pi} \int_0^R v_z(r) r\,dr\,d\theta = \frac{(p_0 - p_L)\pi R^4}{8\mu L} \tag{6-6}$$

which provides—by analogy to the resistance of electrical circuits ($\Delta V = i\Omega$)—the overall resistance of a cylindrical vessel to flow:

3. This low velocity flow was named in honor of Jean Leonard Marie Poiseuille and Gotthilf Heinrich Ludwig Hagen. Poiseuille (a physiologist) and Hagen (an engineer) independently published the first systematic measurements of pressure drop within flowing fluids in simple tubes in 1839 and 1840.

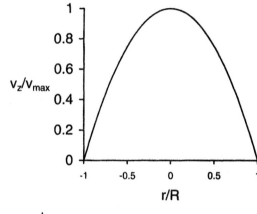

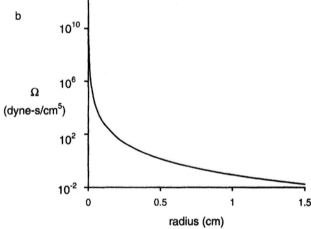

Figure 6.4 Velocity profile in low speed, steady flow through a cylindrical vessel. The velocity varies with the radial distance from the centerline squared, as in Equation 6-5.

$$\Omega = \frac{\Delta p}{Q} = \frac{8\mu L}{\pi R^4} \tag{6-7}$$

The resistance to flow is a strong function of vessel radius: $\Omega \propto R^{-4}$. As blood vessels become smaller, the resistance to flow increases dramatically (Figure 6.4b). Therefore, an additional consequence of the branching pattern of blood vessels is that the majority of the overall resistance to blood flow resides in the smallest vessels (Figure 6.2b); the majority of the pressure drop ($\sim 80\%$) occurs in arterioles and capillaries. This natural consequence of the physics of fluid flows is exploited in regulation of blood flow to organs of the body. Local blood flow to a tissue is controlled by constriction and dilation of

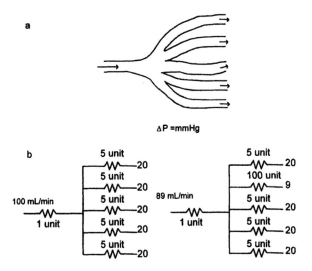

Figure 6.5 Flow distribution in branching networks can be regulated by resistance. (a) A schematic branching point, in which one large artery divides into five arterioles. The overall pressure drop across this structure is ΔP, such that the flow rate is 100 mL/min. (b) If the resistance of one of the branches changes (here, it is show to change from 5 to 100 units, which could be accomplished by $\sim$ 2-fold change in vessel diameter) and the pressure drop remains the same, the overall flow to this region of tissue will be reduced by 11%. In addition, the flow is no longer evenly distributed among the five branches.

the arterioles delivering blood to that tissue. Since the greatest overall resistance is provided by arterioles, and because individual arterioles have muscular walls which permit them to adjust their diameter—and hence resistance (Equation 6-7)—the proportion of blood flow arriving to the tissue served by an arteriole can be regulated with precision (Figure 6.5).

The circulatory system is also efficient at mixing. When drug molecules are injected into a vein, the concentration of that agent at the injection site is suddenly increased. The high velocity of local flow (Table 6.1), coupled with the extensive branching and rebranching of the cardiovascular circuit (Figure 6.1), produces rapid mixing of the injected agent. For agents confined to the plasma volume, mixing is complete in several minutes; that is, after several minutes, the blood in every vessel has an equal concentration of the drug.

6.2 INTERSTITIAL FLUID MOVEMENT

The importance of fluid flow in transporting drugs within the circulatory system is obvious. In the previous chapter, fluid flow was seen to be an important determinant of the overall rate of transport across porous membranes, as well. For some membranes, convection is more important than diffusion in determining the overall flux (recall Figure 5.24). This section illustrates other situations in which fluid flow can contribute to the movement of molecules in tissues.

6.2.1 Formulation of Transport Equations

The basic equations for mass transfer were derived in Chapter 3. The overall rate of change in local concentration depends on both the rate of diffusion and the rate of fluid flow (see Table 3.2):

$$D_A \nabla^2 c_A + \psi_A^m = \frac{\partial c_A}{\partial t} + \bar{v} \cdot \nabla c_A \tag{6-8}$$

In the presence of fluid flow ($\bar{v} \neq 0$), changes in bulk fluid velocity with position and time must be included in the description of mass transfer. This discussion follows the approaches used in previously published analyses of convection in tissues [5, 6].

Fluid flow through a porous medium (such as the interstitial space of a tissue) can be analyzed using Darcy's law:

$$\bar{v} = k \nabla p \tag{6-9}$$

where k is the hydraulic conductivity of the medium and p is the interstitial pressure. By defining a characteristic volume element in the tissue (as outlined in Chapter 3 for solute conservation), a mass balance on water can be obtained:

$$\nabla \cdot \bar{v} = \psi_w^V - \psi_w^L \tag{6-10}$$

where ψ_w^V, ψ_w^L are local rates of water (w) movement into the tissue from blood vessels (V) and out of tissue into the lymphatics (L). Rates of movement can be related to the hydraulic conductivity of the blood and lymphatic vessel walls, L_p and L_{pL}, respectively (see Section 5.5 for definition of L_p):

$$\nabla \cdot \bar{v} = L_p S[p_V - p_i - \sigma(\pi_V - \pi_i)] - L_{pL} S_L[p_i - p_L] \tag{6-11}$$

where S and S_L are the surface areas of the blood and lymphatic vessels, and p_v, p_i, and p_L are vascular, interstitial, and lymphatic pressures. In this model, water movement into the tissue interstitium occurs by pressure-driven flow through the capillary wall and water movement out of the tissue interstitium occurs by pressure-driven flow through the lymphatic vessel wall. The steady-state interstitial pressure, p_{iS}, at which rates of water inflow from the blood vessels and outflow to the lymphatics are balanced, can be defined:

$$p_{iS} = \frac{L_p S p_e + L_{pL} S_L p_L}{L_p S + L_{pL} S_L} \tag{6-12}$$

where p_e is an effective interstitial pressure that produces no net flux of water from the blood vessels into the interstitium, $p_e = p_v - \sigma(\pi_v - \pi_i)$.

Combining Equation 6-9 and Equation 6-11 yields:

$$\nabla \cdot k \nabla p_i = L_p S[p_V - p_i - \sigma(\pi_V - \pi_i)] - L_{pL} S_L[p_i - p_L] \tag{6-13}$$

which can be simplified with the use of Equation 6-12 and the assumption that hydraulic conductivity, osmotic pressures, and reflection coefficient are all constant:

$$\nabla^2 p_i = \frac{\phi^2}{R^2}(p_{is} - p_i) \qquad (6\text{-}14)$$

where $\phi = R\sqrt{(L_p S + L_{pL} S_L)/k}$ with R equal to a characteristic length. Equations 6-8, 6-9, and 6-14 provide a complete description of the distribution of solute A through the tissue in the presence of a convective flow provided that the resistance to flow, represented here by k, is known.

6.2.2 Fluid Flow in Porous Media and Gels

The structure of a tissue influences its resistance to the diffusional spread of molecules, as discussed previously (see Figure 4.18). Similarly, the structure of a tissue will influence its resistance to the flow of fluid. If Darcy's law is assumed, then the hydraulic conductivity, k, depends on tissue structure. Models of porous media are available; in the simplest model, the medium is modeled as a network of cylindrical pores of constant length, but variable diameter. This model produces a relationship between conductivity and geometry:

$$k = \frac{c\varepsilon^2}{\mu S_i^2} \qquad (6\text{-}15)$$

where ε is the interstitial volume fraction (porosity), S_i is the surface area of the interstitial space, c is a constant, and μ is the viscosity of the interstitial fluid. This expression was originally obtained by Kozeny [7]. A variety of other models have been used to model the geometry of complicated porous media, following the general approaches that are outlined in Section 4.4.

Water within the extracellular space of a tissue must flow through gels of macromolecules. Polysaccharides are a principal component of the extracellular space; they appear throughout tissues of animals in the form of glycosaminoglycans (GAGs) and proteoglycans. Most tissue polysaccharides are highly charged, due to the presence of sulfate and carboxyl groups (Figure 6.6a). In addition, GAGs and proteoglycans can form supramolecular assemblies within the tissue (Figure 6.6b). The molecules can become so large that they become entangled: overlap of domains appears to occur at ~ 2 mg/mL for aggregan monomers (Figure 6.7). Polysaccharides interact with some specificity with extracellular proteins, such as collagen, and appear to have an important role in the formation of structure within the extracellular space. The highly hydrophilic mesh of polysaccharides is essential for the mechanical properties of connective tissues: water is drawn into the polysaccharide-rich environment providing the tissue with resistance to compression. Polysaccharide composition of tissues can change significantly with disease, so understanding the

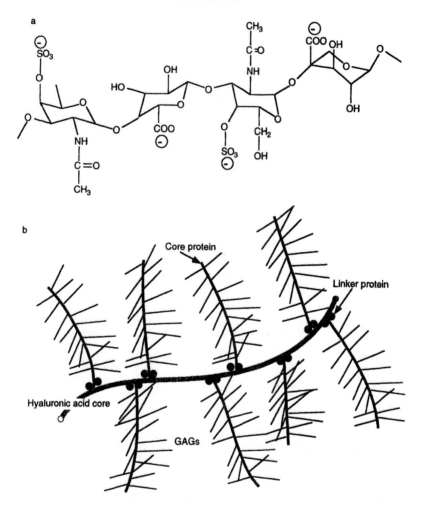

Figure 6.6 Chemical composition of the dermatan sulfate and structure of aggregan complexes. (a) Chemical structure of repeated disaccharide of the GAG dermatan sulfate (protons are not indicated in this schematic diagram). (b) Schematic diagram of the aggregan proteoglycan complex showing the relationship between the core and linker proteins, GAGs, and hyaluronic acid).

influence of polysaccharide content on drug movement through tissues is highly relevant for drug delivery.

The hydrodynamic resistance of solutions of connective-tissue polysaccharides has been measured by sedimentation techniques [9]; the resistance is expressed as the hydraulic conductivity (k), which was defined in Equation 6-9 (Figure 6.8). This analysis permitted the identification of two different groups of polysaccharides: those compounds containing $\beta_{1,4}$- and $\beta_{1,3}$-linkages (such as chondroitan sulfate, keratan sulfate, and hyaluronate) had lower hydraulic resistance than compounds containing some fraction of

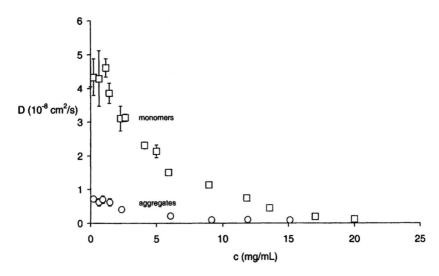

Figure 6.7 Diffusion coefficient as a function of concentration for aggregan. Adapted from [8]. Squares indicate diffusion coefficients for aggregan monomers; circles for aggregan aggregates.

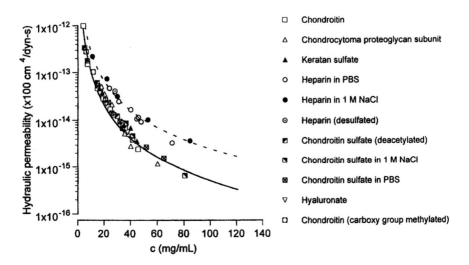

Figure 6.8 Hydraulic permeability of polysaccharide solutions. Adapted from [9]. Lines represent fits to the experimental data: the dashed line was fit to heparin in NaCl ($y = 3 \times 10^{-11} x^{-2.0}$) and the solid line was fit to chondroitin data ($y = 2 \times 10^{-11} x^{-2.3}$). Symbols represent different polysaccharides as indicated on the figure.

$\alpha_{1,4}$-linkages (such as heparin). Differences in hydraulic resistance were not correlated with any other characteristics of the polysaccharides (i.e., sulfation or osmotic activity), suggesting that the resistance to water movement is governed primarily by differences in the structure of the molecules that depend on linkage chemistry.

6.2.3 Interstitial Flow in Tissues and Tumors

Only a few experimental measurements of the rate of fluid movement in the interstitial space of tissues have been reported. Much is still to be discovered, but these initial studies are sufficient to provide guidance on the importance of convection in drug delivery to different tissues (Table 6.1).

Interstitial flows were measured by observing the motion of a spot bleached into fluorescent interstitial fluid in a tissue window preparation [10]. In this system, the fluid velocity was $0.2\,\mu m/s$ parallel to the vessel and $0.75\,\mu m/s$ perpendicular to the vessel. Convection was heterogeneous throughout the tissue region, but over a relatively small range $(0.1 < v < 1\,\mu m/s)$. This rate of interstitial flow corresponds to a Péclet number (Pe) for BSA and IgG of ~ 1, suggesting that convection and diffusion are equally important in the movement of macromolecules in the interstitial space.

Interstitial fluid pressures in normal tissues are approximately atmospheric or slightly sub-atmospheric, but pressures in tumors can exceed atmospheric by 10 to 30 mm Hg, increasing as the tumor grows. For 1-cm radius tumors, elevated interstitial pressures create an outward fluid flow of $\sim 0.1\,\mu m/s$ [11]. Tumors experience high interstitial pressures because: (i) they lack functional lymphatics, so that normal mechanisms for removal of interstitial fluid are not available, (ii) tumor vessels have increased permeability, and (iii) tumor cell proliferation within a confined volume leads to vascular collapse [12]. In both tissue-isolated and subcutaneous tumors, the interstitial pressure is nearly uniform in the center of the tumor and drops sharply at the tumor periphery [13]. Experimental data agree with mathematical models of pressure distribution within tumors, and indicate that two parameters are important determinants for interstitial pressure: the effective vascular pressure, p_e (defined in Section 6.2.1), and the hydraulic conductivity ratio, ϕ (also defined in Section 6.2.1) [14]. The pressure at the center of the tumor also increases with increasing tumor mass.

6.3 FLUID MOVEMENT IN THE LYMPHATIC CIRCULATION

Water moves from the blood, through the capillary wall and into the interstitial space. Some of this water is reabsorbed into distal regions of the capillary, or other vessels, but a substantial quantity of water is lost from the vascular into the interstitial volume. In humans, this loss amounts to $\sim 4\,L/day$.

The lymphatic vessels carry this interstitial fluid back to the circulatory system. Lymphatic vessels are structurally similar to capillaries; they have a wall formed by a single layer of endothelial cells. The lymph vessel wall, however, is more permeable than the capillary wall and not size selective. Therefore, the lymph vessels collect all of the molecules present in the interstitial space, including proteins and other macromolecules. In the absence of an effective lymphatic system, interstitial fluid accumulation in tissue produces swelling, which is known as edema.

Lymph vessels in tissues appear to be $\sim 40\,\mu$m in diameter and flattened. In the skin of the mouse tail, the lymph vessels form a uniform hexagonal network [15]. Local residence times in these lymphatic vessels are consistent with an average velocity of $2.7\,\mu$m/s for lymph flow. Local measurements of lymph flow using FPR indicate that lymph flow rates fluctuate over a range of values ($1.4 < v < 20\,\mu$m/s) with an average of $\sim 8\,\mu$m/s.

6.4 FLUID MOVEMENT IN THE BRAIN

6.4.1 Cerebrospinal Fluid

The tissue of the CNS is bathed in CSF, which circulates through the brain and spinal cord. CSF is confined to defined volumes of the nervous system, particularly the ventricles (Figure 6.9) and the subarachnoid space. Most of this fluid is produced by specialized cells of the ventricular lining (the ependyma) at regions called the choroid plexuses; about 500 mL are produced each day. The total volume of CSF is ~ 150 mL, which is substantially smaller than the daily rate of CSF production; therefore, the entire CSF is replaced several times per day. CSF circulates throughout an internal brain structure (an interconnected series of spaces called the ventricles), flows around the spinal cord and external surface of the brain, and returns to the circulatory system through valves leading to large veins of the head.

The composition of CSF is substantially different from plasma. Protein content is lower in CSF, but the extracellular fluid of the brain directly connects with the CSF flow, so proteins produced by cells in the brain can be found in CSF. Protein analysis of the CSF is a promising method for diagnosis of brain disease [16].

6.4.2 Interstitial Fluid Movement in the Brain

In the tissue of the CNS, under normal conditions, fluid flows within the gray matter are negligible; flows can be significant within white matter. For example, flows of $\sim 10\,\mu$m/ min were measured in white matter, but not gray matter, in the periventricular tissue of cats [17]. This extracellular fluid (ECF) flow, which moves towards the ventricular surface, is presumably the result of fluid leakage from the capillaries that moves predominantly through white matter tracts.

Surgical trauma and the introduction of osmotically active drug compounds may alter the normal fluid flow patterns by increasing the rate of

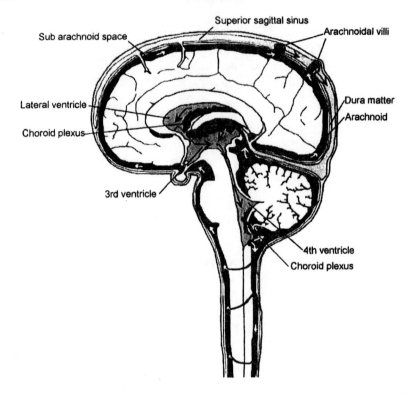

Figure 6.9   Cerebrospinal fluid circulation through the central nervous system. This schematic diagram illustrates the pattern of fluid flow from formation in the choroid plexus, movement through the internal ventricular system, flow around the external surfaces of the brain and spinal cord, and absorption into the venous system through the arachnoid villi. Redrawn from *The Ciba Collection of Medical Illustrations*, Vol. 1, *Nervous System*, Netter, F.H., Summit, NJ: CIBA (1953).

movement of water from the capillaries into the tissue extracellular space. For example, intravenous mannitol induced a flow of $\sim 3\,\mu\mathrm{m}/\mathrm{min}$ in periventricular gray matter away from the ventricle, presumably due to increased water flow into the capillaries [17]. Likewise, fluid flows occur around sites of injury and tumor (Table 6.1). Preliminary studies suggest that the direct injection of albumin may also induce ECF flows along white matter tracts [18]. In addition, there is some evidence that interstitial flows, particularly in the perivascular spaces, are important in the clearance of macromolecules from the brain [19, 20], since brains lack a lymphatic system.

SUMMARY

- Convective or pressure-driven fluid flow is remarkably efficient for carrying agents over large distances (i.e. > 1 cm) and for mixing agents within the human body.

- Fluid flow can be introduced into the equations that describe solute movement in tissues.

- The hydraulic permeability varies with tissue site and in conditions of injury or disease.

- Convection in tissues is likely to be most important for large molecules, for which the rate of diffusion is slow.

REFERENCES

1. Fung, Y.C., *Biodynamics: circulation*. New York: Springer-Verlag, 1984.
2. Berger, S., W. Goldsmith, and E. Lewis, eds., *Introduction to Bioengineering*. New York: Oxford University Press, 1996.
3. Vander, A., J. Sherman, and D. Luciano, *Human Physiology: the Mechanisms of Body Function*. Boston: WCB McGraw-Hill, 1998.
4. Bird, R.B., W.E. Stewart, and E.N. Lightfoot, *Transport Phenomena*. New York: John Wiley, 1960.
5. Baxter, L.T. and R.K. Jain, Transport of fluid and macromolecules in tumors I. Role of interstitial pressure and convection. *Microvascular Research*, 1989, **37**, 77–104.
6. Morrison, P.F., *et al.*, High-flow microinfusion: tissue penetration and pharmacodynamics. *American Journal of Physiology*, 1994, **266**, R292–R305.
7. Kozeny, J., *Hydralik*. Wien: Springer-Verlag, 1927.
8. Gribbon, P. and T.E. Hardingham, Macromolecular diffusion of biological polymers measured by confocal fluorescence recovery after photobleaching. *Biophysical Journal*, 1998, **75**, 1032–1039.
9. Comper, W.D. and O. Zamparo, Hydrodynamic properties of connective-tissue polysaccharides. *Biochemical Journal*, 1990, **269**(3), 561–564.
10. Jain, R.K., Barriers to drug delivery in solid tumors. *Scientific American*, 1994, **271**(1), 58–65.
11. Jain, R.K., Transport of molecules across tumor vasculature. *Cancer Metastasis Reviews*, 1987, **6**, 559–594.
12. Jain, R.K., Transport of molecules in the tumor interstitium: a review. *Cancer Research*, 1987, **47**, 3039–3051.
13. Boucher, Y., L.T. Baxter, and R.K. Jain, Interstitial pressure gradients in tissue-isolated and subcutaneous tumors: implications for therapy. *Cancer Research*, 1990, **50**, 4478–4484.
14. Jain, R.K. and L.T. Baxter, Mechanisms of heterogeneous distribution of monoclonal antibodies and other macromolecules in tumors: significance of interstitial pressure. *Cancer Research*, 1988, **48**, 7022–7032.
15. Leu, A., *et al.*, Flow velocity in the superficial lymphatic network of the mouse tail. *American Journal of Physiology*, 1994, H1507–H1513.
16. Hsich, G., *et al.*, The 14-3-3 brain protein in cerebrospinal fluid as a marker for transmissible spongiform encephalopathies. *New England Journal of Medicine*, 1996, **335**, 924–930.

17. Rosenberg, G., W. Kyner, and E. Estrada, Bulk flow of brain interstitial fluid under normal and hyperosmolar conditions. *American Journal of Physiology*, 1980, **238**, F42–F49.
18. Geer, C.P. and S.A. Grossman, Extracellular fluid flow along white matter tracts in brain: a potentially important mechanism for dissemination of primary brain tumors. *Proceedings of ASCO*, 1993, **12**, 177.
19. Cserr, H.F. and P.M. Knopf, Cervical lymphatics, the blood–brain barrier and the immunoreactivity of the brain. *Immunology Today*, 1992, **13**, 507–512.
20. Yamada, S., *et al.*, Albumin outflow into deep cervical lymph from different regions of rabbit brain. *American Journal of Physiology*, 1991, **261**, H1197–H1204.
21. Ferszt, R., *et al.*, The spreading of focal brain edema induced by ultraviolet irradiation. *Acta Neuropathologica*, 1978, **42**, 223–229.
22. Groger, U., P. Huber, and H.J. Reulen, Formation and resolution of human peritumoral brain edema. *Acta Neurochirurgica-Supplementum*, 1994, **60**, 373–374.

7

Pharmacokinetics of Drug Distribution

"I take it, Dr. Bauerstein, that strychnine, as a drug, acts quickly?"
Agatha Christie, *The Mysterious Affair at Styles*,
Chapter 11 (1920)

Pharmacology, the study of agents and their actions, can be divided into two branches. Pharmacodynamics is concerned with the effects of a drug on the body and, therefore, encompasses dose–response relationships as well as the molecular mechanisms of drug activity. Pharmacokinetics, on the other hand, is concerned with the effect of the body on the drug. Drug metabolism, transport, absorption, and elimination are components of pharmacokinetic analysis.

Physiology influences the distribution of drugs within the body; overall distribution depends on rates of drug uptake, rates of distribution between tissue compartments, and rates of drug elimination or biotransformation. Each of these phenomena potentially involves aspects of drug diffusion, permeation through membranes, and fluid movement that were introduced in the previous sections. The goal of pharmacokinetics is synthesis of these isolated basic mechanisms into a functional unit; this goal is most often achieved by development of a mathematical model that incorporates descriptions of the uptake, distribution, and elimination of a drug in humans or animals. This model can then be used to predict the outcome of different dosage regimens on the time course of drug concentrations in tissues. The development of a complete pharmacokinetic model for any given drug is a substantial undertaking, since the fate of any compound introduced into a whole organism is influenced by a variety of factors, and is usually complicated—in ways that are difficult to predict—by the presence of disease.

In this section, pharamacokinetics will be introduced by first considering the simplest situation: an agent is introduced into a single body compartment from which it is also eliminated. While quite sophisticated compartmental models can be developed from this basic construct, it is frequently difficult to relate model parameters (such as the volume of specific compartments or the rate of transfer between compartments) to physiological or anatomical parameters. To avoid this difficulty, physiological pharmacokinetic models are frequently employed; in these models, the kinetics of drug uptake, distribution, and elimination from local tissue sites are predicted by constructing anatomically and biochemically accurate models of the tissue environment. Physiological models are quite useful in the design of drug delivery systems, and will be described later in this section. Alternative modeling approaches, in which data are (1) interpreted without reference to any particular model or (2) interpreted with respect to variations across a large population of subjects, are also available [1], but are not considered here.

As will become apparent in this chapter, both compartmental and physiological approaches to pharmacokinetic modeling have important limitations. Compartmental models are empirical, so that great care must be exercised in extrapolating outside of the measured domain. On the other hand, physiological models often contain parameters that cannot be easily measured or accurately predicted. Our goal in this book is to develop pharmacokinetic methods that can aid in the rational design of drug delivery systems. The approach to pharmacokinetics is, therefore, pragmatic. Readers may also consult several other texts which introduce the fundamentals of pharmacokinetic modeling [1–3].

7.1 COMPARTMENTAL MODELS

7.1.1 One-compartment Models

The simplest model for distribution and elimination of an intravenously injected drug contains a single compartment representing the volume of distribution, V_d, of the compound (Figure 7.1a) [2]. Usually, the elimination of drug from this compartment is assumed to follow a simple kinetic expression; most commonly first-order kinetics are assumed. A mass balance on the single drug-containing compartment yields:

$$-kM = \frac{dM}{dt} \tag{7-1}$$

where M is the mass of drug within the compartment and k is a first-order elimination constant. Assuming that an initial mass of drug M_0 is rapidly introduced at time $t = 0$, Equation 7-1 is easily solved to produce:

$$M = M_0 e^{-kt} \tag{7-2}$$

The concentration of drug within the compartment, c, is related to the total mass within the compartment, $M = cV_d$, so that:

$$c = \frac{M_0}{V_d} e^{-kt} \tag{7-3}$$

the concentration of drug within the compartment decreases exponentially with time (Figure 7.2).

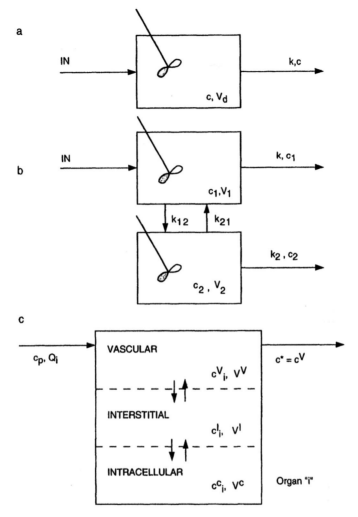

Figure 7.1 Compartmental models for drug distribution and clearance. Empirical pharmacokinetics models typically include (a) a single compartment or (b) multiple compartments. In physiological pharmacokinetic models, each organ is composed of multiple compartments, which reflect the anatomy of the organ, as shown in (c).

The half-life for drug residence within the compartment is related to the first-order rate constant:

$$t_{1/2} = \frac{\ln(2)}{k} \tag{7-4}$$

Simple pharmacokinetic models can be used to tailor therapies to the patient. After measuring the half-life of clearance in a patient, subsequent doses can be provided at intervals calculated to maintain the plasma concentration above a

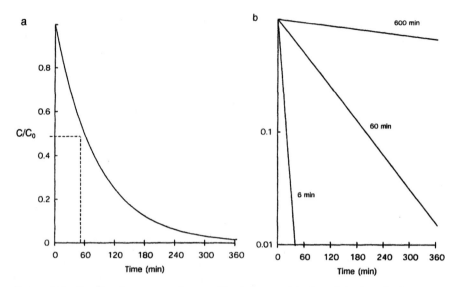

Figure 7.2   Predicted concentration profiles form empirical models. Results from simple pharmacokinetic models containing (a) a single compartment with $t_{1/2} = 60$ min ($k = 0.012$ min^{-1}) and (b) a semi-log plot illustrating clearance from a single compartment with $t_{1/2} = 6$, 60, or 600 min.

desired level. In a more complicated example of customized therapy, patients with B-cell lymphoma that expressed either the CD20 or CD37 antigen were given doses of ^{131}I-labeled murine monoclonal antibody (anti-CD20 or anti-CD37; 0.5, 2.5, or 10 mg/kg body weight, see Table 7.1) [4]. Pharmacokinetics and biodistribution were determined in patients after injection of radiolabeled antibody. Patients with favorable biodistribution were subsequently given therapeutic infusions of ^{131}I-labeled antibody, in doses calculated as appropriate

Table 7.1   Serum retention half-life for monoclonal antibodies in humans: effect of binding

Murine monoclonal antibody	Dose (mg/kg body weight)	Serum retention half-life (h)
MB-1 (anti-CD37, IgG1)	0.5	10.0 ± 5.4
	2.5	20.8 ± 7.6
	10	34.5 ± 9.8
B1 (anti-CD20, IgG2a)	0.5	35.5 ± 16.8
	2.5	48.2 ± 17
	10	48.1 ± 23.3
Non-binding Ab		40.7 ± 14.6

Data from [4].

for their particular tumor burden and biodistribution. The therapeutic antibodies, when followed by autologous bone marrow support, were effective only
in the subset of patients who demonstrated a favorable biodistribution.

7.1.2 One-compartment Model with Absorption

The one-compartment model can be extended to include slow absorption of
drug; the slow absorption step may represent entry of drug through the gastrointestinal tract or leakage into the circulation after subcutaneous injection.
Absorption is added by modifying the mass balance equation—Equation
7-1—to yield:

$$\frac{dM}{dt} = k_a D - kM \tag{7-5}$$

where D is the mass of the delivered dose that remains in the absorption
compartment and k_a is a first-order rate constant characterizing the absorption
step. A second mass balance must be added for the absorption compartment:

$$\frac{dD}{dt} = -k_a D \tag{7-6}$$

which is subject to the initial condition $D = D_0$ at $t = 0$. Equation 7-6 is easily
solved, and the result substituted into Equation 7-5 to obtain:

$$\frac{dM}{dt} = k_a D_0 e^{-k_a t} - kM \tag{7-7}$$

This differential equation can be solved subject to the boundary condition
$M = 0$ at $t = 0$:

$$\frac{M}{D_0} = \frac{k_a}{k - k_a} \left\{ e^{-k_a t} - e^{-kt} \right\} \tag{7-8}$$

Equation 7-8 is shown graphically in Figure 7.3a. Concentration in the
body compartment rises to some maximum value (which depends on D_0, k, and
k_a) and falls with a rate determined by k. Drugs administered in this manner
frequently require multiple doses, in an effort to maintain drug concentrations
within a therapeutic window for some prolonged time. The outcome of repe-
titive doses, with each dose assumed to influence the overall concentration
independently, is shown in Figure 7.3b.

7.1.3 Two-compartment Models

A two-compartment pharmacokinetic model can be used to describe more
complicated phenomena. Consider a drug that is rapidly injected into the
body, distributing instantaneously throughout one compartment and more
slowly throughout a second compartment (Figure 7.1b). This model might
be appropriate for describing the drug concentration in the plasma for a

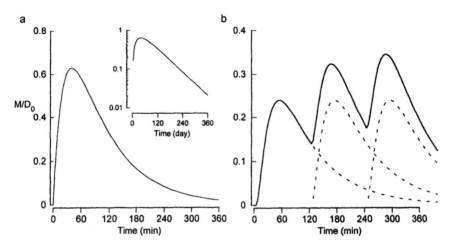

Figure 7.3 Predicted concentration profiles from models with slow absorption. Results from pharmacokinetic models containing a single compartment with elimination $t_{1/2} = 60$ min and absorption $t_{1/2} = 15$ min. (a) The concentration versus time curve for a single dose given at $t = 0$; the inset shows the curve calculation on a semi-log plot. (b) The solid line indicates the total concentration versus time for three consecutive doses given at $t = 0$, 120, and 240 min. The dashed lines indicate the concentration that would be produced by each individual dose.

drug that is intravenously injected, but distributes within both the plasma and an extravascular compartment.

Mass balances on compartments 1 and 2 yield:

$$V_2 k_{21} c_2 - V_1 k_{12} c_1 - V_1 k c_1 = V_1 \frac{dc_1}{dt} \tag{7-9}$$

$$V_1 k_{12} c_1 - V_2 k_{21} c_2 - V_2 k_2 c_2 = V_2 \frac{dc_2}{dt} \tag{7-10}$$

where k_{12} and k_{21} are transfer coefficients describing the rate of movement of compound from compartment 1 to 2 and 2 to 1, respectively, k is a first-order constant for elimination from compartment 1, k_2 is a first-order constant for elimination from compartment 2, V_1 and V_2 are the volumes of the two compartments, and c_1 and c_2 are concentrations of drug in each compartment. In most situations, k_{12} and k_{21} are equal (i.e., the resistance to drug permeation is the same in both directions) and equal to PA, where P is the permeability (defined in Chapter 5) and A is the surface area per volume for the interface between compartments 1 and 2.

Assume that at the time of injection ($t = 0$), the concentration in compartment 1 is $c_1^0 = M/V_1$, where M is the total mass of drug injected, and the concentration in compartment 2 is zero. Further assume that the drug is eliminated only from compartment 1; therefore, $k_2 = 0$. The two first-order differential equations can be combined to yield a second-order differential equation for c_1:

$$\frac{d^2 c_1}{dt^2} + (k_{12} + k + k_{21})\frac{dc_1}{dt} + k_{21}k c_1 = 0 \tag{7-11}$$

The general solution to this differential equation for concentration in compartment 1 (the central compartment) is:

$$c_1 = A_1 e^{-\alpha t} + A_2 e^{-\beta t} \tag{7-12}$$

where the parameters α and β were introduced to facilitate solution:

$$\alpha + \beta = k_{12} + k + k_{21}$$
$$\alpha\beta = k_{21}k \tag{7-13}$$

When Equation 7-12 is substituted into Equation 7-11, the concentration in compartment 2 can also be found:

$$c_2 = \frac{V_1}{V_2 k_{21}}\left[A_1(\beta - k_{21})e^{-\alpha t} + A_2(\alpha - k_{12})e^{-\beta t} \right] \tag{7-14}$$

which is identical in functional form to Equation 7-12, except that the values of the pre-exponential constants have changed.

Two-compartment models are frequently used when the disappearance of an intravenously injected agent follows a bi-exponential decay. The parameters A_1, A_2, α, and β are estimated from the data, using methods described elsewhere [2]. With the two-compartment model, elimination from the central compartment occurs in two phases: a fast phase with half-life $t_{1/2-\alpha} = \ln(2)/\alpha$, which is often attributed to drug distribution from the central compartment (compartment 1) to the peripheral compartment (compartment 2), and a slower phase with half-life $t_{1/2-\beta} = \ln(2)/\beta$, which is usually attributed to drug elimination from the central compartment. Since the initial concentration within the central compartment is known, c_1^0, it must be equal to the sum of the two constants, $A_1 + A_2$ (obtained from Equation 7-12 when $t = 0$). These constants, A_1 and A_2, indicate the fraction of the initial dose that is eliminated from the central compartment during the fast and slow phases, respectively (see example below for antibody kinetics). These parameters can be related to the transfer and elimination constants in the original model:

$$k_{21} = \frac{A_2\beta + A_2\alpha}{A_1 + A_2}$$
$$k = \frac{\alpha\beta}{k_{21}} \tag{7-15}$$
$$k_{12} = \alpha + \beta - k_{21} - k$$

Results from a typical two-compartment model are shown in Figure 7.4.

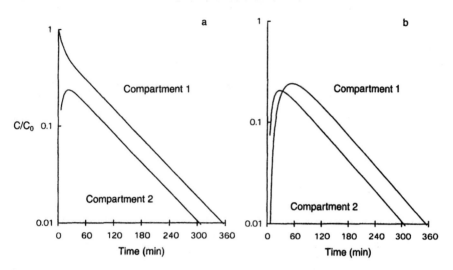

Figure 7.4   Predicted concentration profiles from a two-compartment model. Results from a two-compartment pharmacokinetic model with elimination only from the central compartment (compartment 1) with $t_{1/2} = 60$ min and distribution to a peripheral compartment (compartment 2) with $t_{1/2} = 6$ min. Equations 7-12 and 7-14 are both plotted. (a) Concentration after ingestion; (b) concentration with slow absorption.

7.1.4 Two-compartment Model with Absorption

This model can be used for additional situations—such as drug concentrations in the plasma following subcutaneous, intramuscular, or oral administration— if the kinetics of drug adsorption into the compartment are added. In this case, a mass balance on the central compartment (compartment 1) includes the slow absorption of drug:

$$V_a k_a c_a + V_2 k_{21} c_2 - V_1 k_{12} c_1 - V_1 k c_1 = V_1 \frac{dc_1}{dt} \qquad (7\text{-}16)$$

where k_a is the first-order rate constant for drug absorption into the central compartment, c_a is the drug concentration at the site of administration, V_a is the volume of the dosage compartment, and the other parameters were previously defined. Similar balance equations can also be written for the peripheral and absorption compartments. When solved, using a method similar to the one described above, the concentration in the central compartment is:

$$c_1 = A_1 e^{-\alpha t} + A_2 e^{-\beta t} - (A_1 + A_2) e^{-k_a t} \qquad (7\text{-}17)$$

The presence of a slow absorption phase often makes it difficult to observe the distribution of agent between the central and peripheral compartments (Figure 7.4b).

7.2 PHYSIOLOGICAL MODELS

Physiological pharmacokinetic models are based on actual anatomical compartments and physiological mechanisms of drug metabolism (Figure 7.5). In these detailed models, the fate of the administered compound is evaluated by explicitly considering the rates of compound entry, elimination, and efflux from every organ in the body. Only organs or tissues that contain negligible quantities of the compound are eliminated from investigation. This approach can lead to complex models, which require complex experimental evidence to

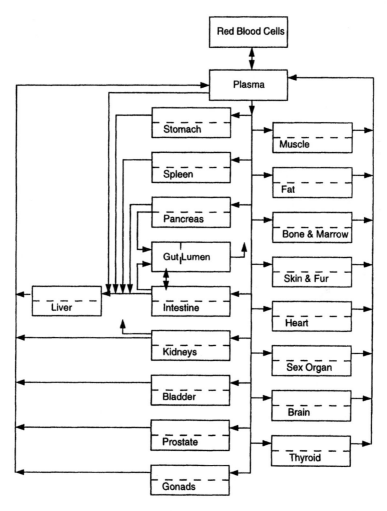

Figure 7.5 Schematic of a physiological pharmacokinetic model. Individual organs are represented by two-compartment models. The arrows indicate blood flow from organ to organ. Elimination is indicated in the kidney compartment only. The portal circulation is also explicitly included. Adapted from [5].

verify. Although cumbersome to execute, this level of detail provides several advantages over traditional, empirical pharmacokinetic models. Mechanisms of movement and elimination are used in the mathematical description. The molecular mechanisms are often similar in different species; therefore, physiological models can be extrapolated between species by inclusion of the appropriate geometry. Extension of the model to similar compounds is also possible. The basic components of a physiological pharmacokinetic model are presented here: more detailed reviews of the development and applications of physiological pharmacokinetic models are available [5].

A physiological pharmacokinetic model is constructed from models of individual organs; a basic element of the detailed model is shown in Figure 7.1c. The organ is divided into three anatomical compartments: the vascular, interstitial, and intracellular spaces. Concentrations (c) and volumes (V) within each compartment of each organ are identified by superscripts V, I, and C, indicating vascular, interstitial, and intracellular compartments. Subscripts indicate each individual organ in the model. Mass balances on each compartment of the individual organ model yield:

$$V_i^V \frac{dc_i^V}{dt} = Q_i c_p - Q_i c_i^v - N_i^{V \to I} - V_i^V q_{elim,i}^V$$

$$V_i^I \frac{dc_i^I}{dt} = N_i^{V \to I} - N_i^{I \to C} - V_i^I q_{elim,i}^I \qquad (7\text{-}18)$$

$$V_i^C \frac{dc_i^C}{dt} = N_i^{I \to C} - V_i^C q_{elim,i}^C$$

where Q_i is the volumetric flow rate of plasma into the organ, $N_i^{V \to I}$ and $N_i^{I \to C}$ indicate the net flux of drug from the vascular to interstitial space and the net flux from interstitial to intracellular space, respectively, and $q_{elim,i}^V$, $q_{elim,i}^I$ and $q_{elim,i}^C$ are the rates of drug elimination from each of the organ spaces. A set of equations similar to Equation 7-18 is written for each organ in the model. A separate balance equation for the plasma compartment is also required:

$$V_p \frac{dc_p}{dt} = \sum_{i=1}^k Q_i c_i^V - Q_p c_p + g(t) \qquad (7\text{-}19)$$

where V_p is the volume of the plasma compartment, k is the number of organs in the complete model, and $g(t)$ describes the introduction of drug into the plasma compartment. For a physiological model with k organs, Equations 7-18 and 7-19 provide $3k + 1$ equations in $3k + 1$ unknowns: $c_1^V, c_1^I, c_1^C, c_2^V, \ldots, c_k^V$, c_k^I, c_k^C, c_p.

In most experiments, it is not possible to measure the concentrations in the vascular, interstitial, and intracellular spaces separately. Usually, the total concentration within an organ, c_i is measured:

$$c_i = \frac{c_i^V V_i^V + c_i^I V_i^I + c_i^C V_i^C}{V_i^V + V_i^I + V_i^C} \qquad (7\text{-}20)$$

Obviously, a physiological model can be quite complex and is not easily applied in its full detail. Simplifications are helpful; some of the most common and useful simplifications are provided in the next subsections.

7.2.1 Flow-limited Compartments

Organs that are not well perfused are termed *flow limited*. In these cases, the intracompartmental fluxes, $N_i^{V \to I}$ and $N_i^{I \to C}$, are much greater than the flow rate of plasma through the organ, Q_i. In this limiting case, the three equations in Equation 7-18 reduce to:

$$V_i \frac{dc_i}{dt} = Q_i \left(c_p - \frac{c_i}{R_i} \right) \tag{7-21}$$

where R_i describes the equilibrium between the plasma and total tissue concentration:

$$R_i = \left(\frac{c_i}{c_p} \right)_{\text{equilibrium}} \tag{7-22}$$

The validity of this simplification can be justified by considering a simpler two-compartment model of an individual organ (Figure 7.6). A mass balance on the two-compartment organ gives:

$$V_1 \frac{dc_1}{dt} = Q_i (c_p - c_1) + PA \left(\frac{c_2}{K} - c_1 \right)$$
$$V_2 \frac{dc_2}{dt} = PA \left(c_1 - \frac{c_2}{K} \right) - fc_2 \tag{7-23}$$

where c_p is the concentration in the plasma, c_1 and c_2 are the concentrations in the two compartments with corresponding volumes V_1 and V_2, the permeability–area product PA is used to describe the exchange between the two compartments, f is a first-order rate constant for elimination or metabolism in the second compartment, and K describes the equilibrium between the two compartments ($= c_2/c_1$ at equilibrium). The assumption of flow limitation, for this

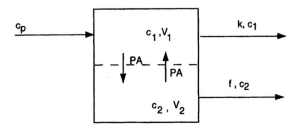

Figure 7.6 Flow-limited organ model. The three-compartment model of an individual organ (Figure 7.1c) can be reduced to this simpler two-compartment model if the rate of perfusion is slower than the rate of exchange across membranes in the organ.

particular case, implies rapid equilibration between the compartments, such that $c_2 \sim K c_1$. By solving the differential equations, Equation 7-23, it is possible to show that this is true if:

$$\frac{Q_i}{PA} \ll 1 + \frac{V_1}{KV_2} \quad \text{and} \quad \frac{K_f}{PA} \ll 1 \qquad (7\text{-}24)$$

7.2.2 Membrane-limited Compartments

If the organ is reasonably well perfused (i.e., the flow-limited conditions are not satisfied), the full physiological pharmacokinetic treatment may be reduced by assuming the organ is *membrane limited.* Here, the limitation on transport is assumed to occur at either the capillary membrane separating the vascular and interstitial compartments or the plasma membranes separating the interstitial and intracellular compartments. For example, when the net flux between the interstitial and intracellular compartments is much slower than the net flux between the vascular and interstitial compartments and the plasma flow rate, the three-compartment model can be reduced to a two-compartment model:

$$V_i^E \frac{dc_i^E}{dt} = Q_i(c_p - c_i^E) - N_i^{E \to C}$$

$$V_i^C \frac{dc_i^C}{dt} = N_i^{E \to C} \qquad (7\text{-}25)$$

where the superscript E represents the extracellular space, including both the vascular and interstitial volumes. A similar set of equations could be written for the case of capillary membrane limitation.

7.2.3 Drug Binding and Metabolism in Compartmental Models

Drug Binding. Many drugs will bind to proteins in the plasma (principally albumin) or within other compartments of the tissue. Binding can greatly influence the local rates of transport and elimination in individual organs, and hence can contribute to the observed pharmacokinetics. The extent of drug binding to proteins in plasma is discussed in Chapter 8; Table 8.1 provides the percentage of binding to proteins. Assuming that the percentage binding is constant over a concentration range of interest, the percentage binding ($100 \times f$, where f is the ratio of bound to total drug concentration) can be converted into a binding constant, K_b, an association constant, K_a, or a dissociation constant, K_d:

$$K_b = \frac{f}{1 - f}$$

$$K_b = \frac{n}{K_d} = n K_a \qquad (7\text{-}26)$$

where n is the number of available drug-binding sites per volume in the tissue compartment. (These binding constants were defined in Chapter 3.) If free and bound drug concentrations within organ i are distinguished by the variables $c_{A,i}$ and $c_{B,i}$ the following relationships can be used to calculate total concentration, c_i (Figure 7.7):

$$c_i = c_{A,i}(1 + K_b)$$
$$c_i = c_{B,i}\left(1 + \frac{1}{K_b}\right) \tag{7-27}$$

For drugs that bind to plasma proteins or proteins in other tissue compartments, the number of binding sites is limited. As the concentration of free drug in the compartment increases, the number of available binding sites will approach zero.. The definitions of f and K_b in Equation 7-26 are appropriate only within the range of drug concentrations for which n is constant. To estimate the extent of binding when binding sites are limited, a model for the equilibrium between bound and free drug must be assumed; a common model from physical chemistry is the Langmuir-type isotherm. The functional form of the Langmuir isotherm can be motivated from the definition of the dissociation constant (Equation 3-53):

$$K_d = \frac{n \times c_{B,i}}{c_{A,i}} = \frac{(n_0 - c_{B,i})c_{B,i}}{c_{A,i}} \tag{7-28}$$

where n_0 is the total number of binding sites per volume in the tissue (hence, n, the number of available sites, is equal to $n_0 - c_{B,i}$). With some rearrangement, the total concentration of drug in a tissue compartment, c_i, can be determined from the concentration of free drug, $c_{A,i}$:

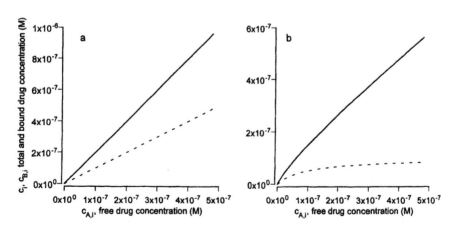

Figure 7.7 Total tissue concentration as a function of free drug concentration. Total drug concentration (solid lines) and bound drug concentration (dashed lines) are shown for two different models of binding: (a) linear binding with no saturation ($f = 0.5$, Equation 7-27) and (b) binding with limited binding sites ($n_0 = 10^{-7}$M, $K_d = 10^{-7}$ M, Equation 7-29).

$$c_i = c_{A,i} + \frac{n_0 c_{A,i}}{K_d + c_{A,i}} \qquad (7\text{-}29)$$

Drug Elimination. Excretion or metabolism can be added to an appropriate compartment, as indicated by the terms q_{elim} in Equation 7-18. A full description of enzyme-mediated metabolic pathways usually employs Michaelis–Menten kinetics:

$$q_{elim} = \frac{V_{max} c_{A,i}}{K_e + c_{A,i}} \qquad (7\text{-}30)$$

where it is assumed that only free drug molecules are available for metabolism. When the drug concentration is low ($c_{A,i} \ll K_e$), first-order kinetics can be assumed (with $k_1 = V_{max}/K_e$); if the drug concentration is high ($c_{A,i} \gg K_e$), zero-order kinetics can be assumed (with $k_0 = V_{max}$).

7.2.4 Applications of Physiological Pharmacokinetic Models

The literature contains many examples on the use of physiological pharmacokinetic models. An early, detailed, and influential example is the model for methotrexate [6]. This model used physiological parameters and was applied to mice, monkeys, dogs, and humans. Other papers, which provide good examples on the benefits of physiological modeling, describe the uptake of immunotoxins in solid tumors [7] and the biodistribution of monoclonal antibodies and antibody fragments [8].

7.3 PATTERNS OF MIXING IN TISSUES AND ORGANS

The compartments in pharmacokinetic models are "black boxes," which reflect the limits of our understanding of the physiological and metabolic processes impacting drug concentration. We can make different kinds of assumptions about the "black box" and these assumptions will influence the predictions of our model. In the models described above, the tissue compartment is assumed to be well mixed, and concentration differences within the compartment are neglected. However, the pattern of circulatory flow through a real region of tissue may be complex (Figure 7.8); perfect mixing within this compartment is not always assured.[1] This section considers the impact of alternative models of mixing on the predictions of pharmacokinetic models.

1. The pattern of blood flow through a region is the most obvious mechanism for mixing or distribution of drug molecules in the region, although the other mechanisms discussed in previous chapters (such as diffusion and lymphatic flow) also contribute. The pharmacokinetic compartment represents the limit of our understanding of the actual mechanisms of drug movement; "mixing," therefore, refers here to any process that influences the speed and extent of distribution of drug molecules within a tissue region.

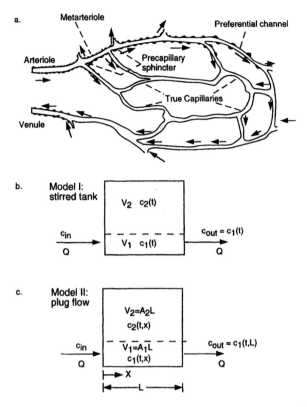

Figure 7.8 Models of capillary mixing with a tissue region. (a) Complex patterns of the capillary network in a region of tissue; (b) stirred-tank pharmacokinetic model; (c) plug-flow pharmacokinetic model.

7.3.1 Ideal Mixing Models

Engineers, who encounter "black boxes" frequently when analyzing processes, use two different ideal mixing models. The ideal stirred-tank model—which we have used throughout this chapter—assumes that molecules entering a compartment are instantaneously and uniformly distributed throughout the compartment (Figure 7.9a). Assume that an ideal stirred tank is operating with an inlet stream containing a dye molecule that does not react within the tank; if the tank has been operating for a long enough time, the concentration of the dye within the tank, the concentration of dye in the outlet stream, and the concentration in the inlet stream are all equal (c^*). If the inlet stream is suddenly changed, so that it now contains no dye molecules ($c_{in} = 0$), the concentration of dye in the tank and in the outlet stream will slowly decrease. Our basic assumption regarding the ideal stirred tank leads to several predictions: (1) the dye concentration throughout the tank is always uniform but changing with time, $c_{tank}(t)$; (2) the outlet concentration is equal to the concentration in the tank ($c_{tank} = c_{out}$); and (3) the tank/outlet concentration changes in a predictable way in response to the step change in inlet concentration:

$$c_{out} = c^* e^{-\frac{t}{\tau}} \qquad (7\text{-}31)$$

where t is the time after initiation of the step change in inlet concentration and τ is a characteristic time for the compartment ($= V_{tank}/Q$). For this example, the characteristic time is equal to the time required for the outlet concentration to decrease from c^* to $0.37c^*$.

The second model is ideal plug flow (Figure 7.9b). In ideal plug flow, molecules moving through the region do not mix, but rather travel in an orderly fashion from inlet to outlet. The simplest physical model of ideal plug flow is slow fluid flow through a long tube: each "packet" of fluid enters the tube just after some other "packet," and remains behind its neighbor from inlet to outlet. Each "packet" emerges from the tube in the order that they entered. The time required for a packet to move from inlet to outlet is $\tau (= V_{tank}/Q)$; a step change in inlet concentration produces a step change in outlet concentration that is offset by τ (Figure 7.9b).

These two ideal models display different behaviors in response to a step change in inlet concentration of a tracer. In principle, then, one can determine

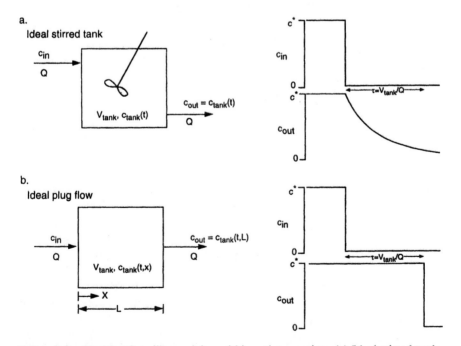

Figure 7.9   Models of capillary mixing within a tissue region. (a) Ideal stirred-tank model and response of the stirred tank to a step change in concentration of an inert tracer at the inlet. (b) Ideal plug-flow model and its response to a step change in concentration of tracer at the inlet.

whether a "black box" is more like a stirred or plug flow model by examining the response of that box to a step change in the inlet concentration of an inert tracer molecule. In physiology, this technique is called indicator dilution or dye dilution and it has been used extensively to evaluate the volume and flow patterns in regions of tissue [9].

7.3.2 Indicator Dilution Methods

Indicator kinetics or indicator dilution methods can be used to estimate the characteristics of tissues, organs, or whole animals [10]. Experimentally, a known amount of the tracer is introduced into an inlet artery that supplies the region or organ of interest, and the concentration of tracer that emerges from the region is measured. The inlet and outlet must be carefully selected, in order to examine the distribution within the intended region. For an individual organ or region of tissue, the mean transit time of an indicator through the region, $\bar{t}$, is related to the rate of blood flow to the organ, Q, and the volume of distribution in that organ or region, V_d:

$$Q = V_d \cdot \bar{t}^{-1} \tag{7-32}$$

In practice, the volume of distribution in the organ can be found using labeled inert compounds, which are allowed to reach equilibrium within the tissue:

$$V_d = \frac{M_\infty}{C_{\text{blood},\infty}} \tag{7-33}$$

where M_∞ is the total mass of compound within the system at equilibrium and $C_{\text{blood},\infty}$ is the concentration in the blood at equilibrium. A tissue-to-blood partition coefficient, λ (expressed in mL/mass of tissue), can be defined:

$$\lambda = \frac{C_{\text{tissue},\infty}}{C_{\text{blood},\infty}} = \frac{V_d}{W} \tag{7-34}$$

where W is the mass of the tissue under consideration. This partition coefficient can vary greatly for different compounds in different tissues; xenon, for example, distributes with $\lambda = 0.7\,\text{mL/g}$ in skeletal muscle and $10\,\text{mL/g}$ in adipose tissue. This partition coefficient provides a measure of the amount of compound in the entire tissue, including the associated blood volume. By dividing both sides of Equation 7-32 by the tissue mass, W, a relationship between the blood flow rate f (mL/g tissue/s) and the residence or transit time is found:

$$f = \lambda \cdot \bar{t}^{-1} \tag{7-35}$$

7.3.3 Microvascular Mixing in Tissue Compartments

Ideal mixing models can be used to evaluate the impact of mixing on the performance of a pharmacokinetic model. Consider the two-compartment organ models shown in Figure 7.8b and 7.8c. To simplify the calculations, flow-limited conditions are assumed, so that the concentration in compartment 2 is simply related to the concentration in compartment 1 through an equilibrium coefficient, K. Further, to find the behavior of each model to a concentration that changes with time, we assume that the concentration flowing into the organ is suddenly changed at time $t = 0$, decaying back to the initial concentration with a time constant $\tilde{t}$:

$$c_{in} = \begin{cases} 0; & t < 0 \\ c_0 e^{-t/\tilde{t}}; & t \geq 0 \end{cases} \tag{7-36}$$

Under what conditions do the two models give similar outlet concentrations? In model I (well stirred), a mass balance over the plasma compartment (i.e., compartment 1) gives:

$$(V_1 + V_2 K)\frac{dc_1}{dt} = Q\left[c_0 e^{-t/\tilde{t}} - c_1\right] \tag{7-37}$$

Defining a time constant for flow through the organ, τ, and a dimensionless time based on this time constant, T:

$$\tau = \frac{V_1 + V_2 K}{Q}; T = \frac{t}{\tau} \tag{7-38}$$

the solution to Equation 7-37 may be written:

$$c_1 = c_0 \left[\frac{\exp\left(-\dfrac{T\tau}{\tilde{t}}\right) - \dfrac{\tau}{\tilde{t}}\exp(-T)}{1 - \dfrac{\tau}{\tilde{t}}}\right] \tag{7-39}$$

For model II (plug flow), the concentration changes slowly throughout the tissue region. Since flow-limited behavior is assumed, a similar concentration gradient is slowly present within the peripheral compartment (compartment 2). Selecting a control volume, Δx, which includes the comparable regions of compartments 1 and 2, a mass balance on the control volume yields:

$$A_1 \Delta x \frac{\partial c_1}{\partial t} + A_2 \Delta x \frac{\partial c_2}{\partial t} = Q[c_1(x) - c_1(x + \Delta x)] \tag{7-40}$$

which reduces to the following differential equation, recognizing that the volume of each compartment, V, is equal to the cross-sectional area times the length of the compartment, AL:

$$\left(\frac{V_1 + V_2 K}{L}\right)\frac{\partial c_1}{\partial t} = -Q\frac{\partial c_1}{\partial x} \tag{7-41}$$

Introducing the non-dimensional variables, $X = x/L$ and $T = t/\tau$, where τ is defined as in Equation 7-38, Equation 7-41 reduces to:

$$\frac{\partial c_1}{\partial T} + \frac{\partial c_1}{\partial X} = 0 \tag{7-42}$$

which has the solution:

$$c_1 = c_0; \qquad\qquad T < X$$
$$c_1 = c_0 \exp\left[\frac{(X - T)\tau}{\tilde{t}}\right]; \quad T \geq X \qquad (7\text{-}43)$$

In this plug flow model, the concentration within the compartment is unchanged until the flow has reached a position within the compartment; likewise, the concentration at the outlet does not change until the flow has completely penetrated the compartment.

The two solutions for the outlet concentration can be condensed and compared more readily by defining $\epsilon = \tau/\tilde{t}$, so that:

$$\frac{c_{\text{out}}^{I}(t)}{c_0} = \frac{\exp(-\epsilon T) - \epsilon \exp(-T)}{1 - \epsilon}$$

$$\frac{c_{\text{out}}^{II}(L, t)}{c_0} = \exp[(1 - T)\epsilon] \qquad (7\text{-}44)$$

The results shown in Figure 7.10 can be compared to the characteristics of different tissues (Table 7.2). Assuming that the characteristic time for changes in the plasma concentration is $\tilde{t} \sim 30$ min, ϵ is approximately 0.1, 2, and 1 for viscera, adipose, and lean tissue, respectively; therefore, visceral tissue is more likely to experience concentration changes as in Figure 7.10a and adipose and lean tissues are more likely to respond as in Figure 7.10b. This comparison suggests that the choice of a mixing model is likely to be important when studying adipose or lean tissue, but not for visceral tissues.

7.4 THE PROBLEM OF SCALE

Biological systems vary tremendously in size, from unicellular organisms ($\sim 1\,\mu$m) to sequoias (~ 100 m), and in mass, they range from microbes

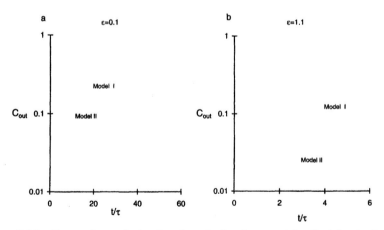

Figure 7.10 Comparison of stirred-tank and plug-flow models of mixing in tissue region. The figure shows comparison of plug flow and well-stirred flow for $\epsilon = 0.1$ and 1.1.

Table 7.2   Time constants for mixing in different tissues

Tissue	V (L)	Q (L/min)	τ (min)
Viscera	6.2	0.66	~ 2
Adipose	12.2	0.021	~ 50
Lean	39.2	0.033	~ 30

($\sim 10^{-13}$ g) to whales (10^8 g). Even within an individual species, biological variation is significant: height and weight vary noticeably among men and women. Can drug doses be adjusted to account for this variability?

Allometric scaling is commonly used to account for variations in biologic systems. If X is the variable of interest, the allometric equation assumes that this variable scales with a power of the mass M:

$$X = X_0 M^\alpha \tag{7-45}$$

where X_0 and α are parameters that fit the variation of X over a range of M. The allometric equation works well for many variables including surface area, total metabolic rate, and life span (Figure 7.11).

Interestingly, many variables scale with similar values of α (Table 7.3), suggesting that fundamental physical principles underlie the variation of biological organisms. Some of these scalings follow directly from Euclidian geometry; other scalings are more difficult to rationalize. Recent work suggests that the 1/4-power dependencies arise from the branching geometries that organisms use to distribute resources [11]. Branching architectures are usually quite efficient in design and behave as fractal networks. However, the 1/4-power dependence can be obtained by other arguments, as well, so the physical basis for allometric scaling principles remains controversial. These scalings can,

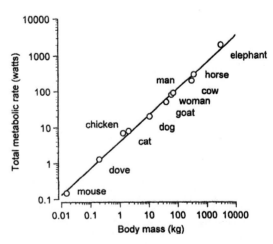

Figure 7.11
Correlation between total metabolic rate and body mass across species. The total metabolic rate varies with body mass according to Equation 7-45 with an exponent of $\frac{3}{4}$.

Table 7.3 Scaling dependencies for various variables

Variable	α
Diameters of tree trunks Diameters of aortas	3/8
Cellular metabolic rates Heart rate	−1/4
Blood circulation time Life span	1/4
Whole organism metabolic rate	3/4

From [11].

however, be useful for predicting the pharmacokinetics of a drug in one species based on pharmacokinetic data obtained in another.

SUMMARY

- Compartmental descriptions of the body can be used to model changes in drug concentration over time.

- Physiological pharmacokinetic models permit assignment of compartments to anatomical regions, and correlation of model parameters to meaningful physiological events.

- Physiological models can become quite complex, but certain features of drug transport can be used to simplify them.

REFERENCES

1. Welling, P.G., Pharmacokinetics: processes, mathematics, and applications. *ACS Professional Reference Book*. Washington, DC: American Chemical Society, 1997.
2. Creasey, W.A., *Drug Disposition in Humans: The Basis of Clinical Pharmacology*. New York: Oxford University Press, 1979.
3. Gibaldi, M., *Biopharmaceutics and clinical pharmacokinetics*. 4th ed. Malvern, PA: Lea & Febiger, 1990.
4. Press, O.W., *et al.*, Radiolabeled-antibody therapy of B-cell lymphoma with autologous bone marrow support. *New England Journal of Medicine*, 1993, **329**(17), 1219–1224.
5. Gerlowski, L.E. and R.K. Jain, Physiologically based pharmacokinetic modeling: principles and applications. *Journal of Pharmaceutical Sciences*, 1983. **72**(10): 1103–1127.
6. Bischoff, K.B., *et al.*, Methotrexate pharmacokinetics. *Journal of Pharmaceutical Sciences*, 1971, **60**, 1128–1133.
7. Sung, C., R. Youle, and R. Dedrick, Pharmacokinetic analysis of immunotoxin uptake in solid tumors: role of plasma kinetics, capillary permeability, and binding. *Cancer Research*, 1990, **50**, 7382–7392.

8. Baxter, L.T., *et al.*, Physiologically based pharmacokinetic model for specific and nonspecific monoclonal antibodies and fragments in normal tissues and human tumor xenografts in nude mice. *Cancer Research*, 1994, **54**, 1517–1528.
9. Seagrave, R.C., *Biomedical Applications of Heat and Mass Transfer*. Ames, IA: The Iowa State University Press, 1971.
10. Paaske, W.P., Into the black box: flows, fluxes, distribution volumes, and interstitial diffusion. *Acta Physiol. Scand.*, 1992, **143**, 109–113.
11. West, G., J. Brown, and B. Enquist, The fourth dimension of life: fractal geometry and allometric scaling of organisms. *Science*, 1999, **284**, 1677–1679.

III

DRUG DELIVERY SYSTEMS

> ...how can we intercede and not interfere?
> A.R. Ammons, *Garbage* (1993)

8

Drug Modification

> Now the drugs don't work
> They just make you worse...
> Richard Ashcroft, *Urban Hymns* (1997)

Previous chapters present the characteristics of drug movement through the body. Diffusion is an essential mode of transport at the microscopic scale; concentration gradients drive a substantial fraction of the molecular movements within cells and the extracellular space. The confinement and regulated passage of molecules within compartments of a tissue or cell is also essential for function; membranes confine molecules to spatial locations and regulate transport between these isolated spaces (Chapter 5). Membranes frequently are the major obstacles to the entry or distribution of therapeutic compounds (Chapter 7). Therefore, much of the effort in drug design and drug delivery is devoted to overcoming these diffusional or membrane barriers.

This chapter describes strategies for manipulating agents in order to increase their biological activity. The sections orbit a central assumption: i.e., agents can be modified to make analogous agents (analogs), which are chemically distinct from the original compound, but produce a similar biological effect. Nature uses a similar strategy, called "biotransformation" to assure elimination of many toxic compounds and drugs. Substantial chemical modification is often needed in order to impact physical properties that influence drug distribution such as stability or solubility; the challenge of drug modification is to identify chemical features that can be changed without sacrificing biological activity. Often, our understanding of the relationship between chemical structure and biological function for an agent is incomplete, making the rational production of analogs difficult.

Drug modifications are frequently directed at altering properties that influence the concentration of the compound (i.e., its solubility), the duration of action (which is usually related to its stability in tissue), or the ability of drug molecules to move between compartments in tissues (which is often related to its permeability in membranes). A chemical modification can effect multiple properties, so these divisions are frequently not as distinct as the section headings suggest.

8.1 ENHANCING AGENT SOLUBILITY

8.1.1 Enhancing Solubility by Non-covalent Modification

Many agents are protected from degradation within tissues by binding. Binding provides a mechanism for sequestering an unstable or potent compound within a region of a tissue. Protective binding occurs frequently within the plasma and extracellular matrix (ECM); the complex molecular composition of these tissues provides many potential binding sites. In plasma, many drugs bind to albumin. Binding to plasma proteins influences the volume of distribution of a drug, because bound drug molecules cannot permeate through membranes and, therefore, are confined to the plasma space. For this reason, drugs are often classified according to their extent of binding to plasma proteins (Table 8.1).

In the ECM, agents can bind to proteins and glycosaminoglycans. This binding in the ECM has important physiological functions. For example, basic fibroblast growth factor (bFGF) binds to heparin, which provides a stable tissue reservoir for the growth factor. Binding to heparin can be used to stabilize intravenously administered bFGF [3]. When administered by itself, the

Table 8.1   Classification of drugs according to total percentage binding to plasma proteins

Extensively bound drugs					
Erythromycin	93	Rifampicin	89	Warfarin	99
Dicoumarol	99	Phenytoin	93	Indomethacin	97
Salicylic acid	81	Digitoxin	91	Propranolol	93
Chlorpromazine	90	Clofibrate	97	Nortriptyline	95
Sertraline hydrochloride	98	Paclitaxel	89–98	Fluoxetine hydrochloride	95
Sildenafil citrate	96				
Moderately bound drugs					
Penicillin G	52	Sulphadiazine	45	Phenobarbital	50
Methotrexate	63	Cloramphenicol	70	Aspirin	61
Theophylline	59				
Poorly bound drugs					
Ampicillin	13	Cephalexin	9	Gentamicin	10
Digoxin	29	Amphetamine	22	Morphine	35
Tetracycline	24	Isoniazid	0	Levodopa	0

Drugs are grouped according to binding to plasma proteins. The values indicate percentage binding. Compiled from [1, 2]

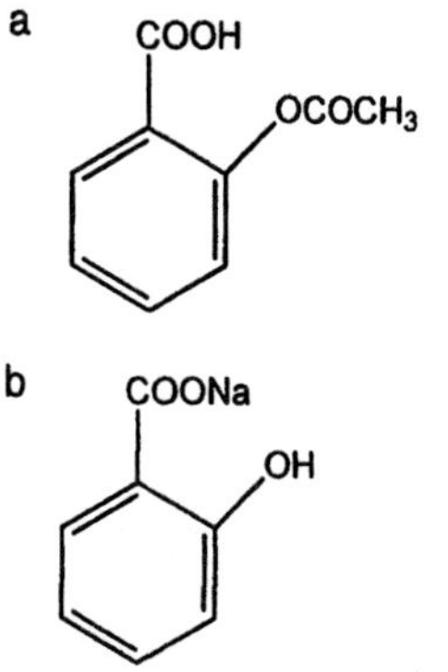

Figure 8.1
Structures of aspirin
(a) and sodium sali-
cylate (b).

half-life of bFGF administered intravenously in the rat is 1.5 min. This half-life is increased by a factor of 3 when the bFGF is injected with heparin.

Lipophilic drugs can accumulate in fat, which represents a significant fraction of the total body weight. For some very lipophilic drugs, fat reservoirs can serve as a depot from which active drug molecules are slowly released.

8.1.2 Enhancing Solubility by Chemical Modification

Simple changes in the chemical structure of an agent can have dramatic effects on solubility. Aspirin (acetylsalicylic acid) is sparingly soluble in water, but sodium salicylate is water soluble (Figure 8.1). The uptake and activity of both agents are similar, so the drug can be administered in either form. Modification of side groups on an active agent to create analogs with different solubility is a common approach to the development of new therapeutics.

Conjugation to poly(ethylene glycol) (PEG), or PEGylation, is a well-established approach in protein stabilization (which is described in Section 8.2.3). PEGylation can also be used to increase the water solubility of drugs. For example, conjugation to PEG increases the aqueous solubility of paclitaxel and its derivatives [4] (Table 8.2 and Figure 8.2). The increase in solubility is

Table 8.2 Water solubility of taxol 2'-PEG esters

M.W. of PEG	Solubility (mg/mL)	wt % paclitaxel	Paclitaxel content (mg/mL)
No PEG	~ 0	100	~ 0
5,000	660	15	100
20,000	200	8	16
40,000	125	4	5

Data from [4]. The half-life for hydrolysis of the ester (i.e., the time for 50% of the paclitaxel to be released from the PEG conjugate) was > 72 h in distilled water and 2 h in human blood at 37 °C. Solubility can be optimized by adjusting the molecular weight of the PEG.

particularly useful for delivery of paclitaxel, which has such a low solubility in water that it must be administered in solvents or by extended infusion.

Conjugation also increases the solubility of proteins in water, and makes molecules soluble over a wider range of pH values; for example, bovine catalase, which is normally soluble near neutral conditions, becomes soluble at pH 1 when conjugated to PEG.

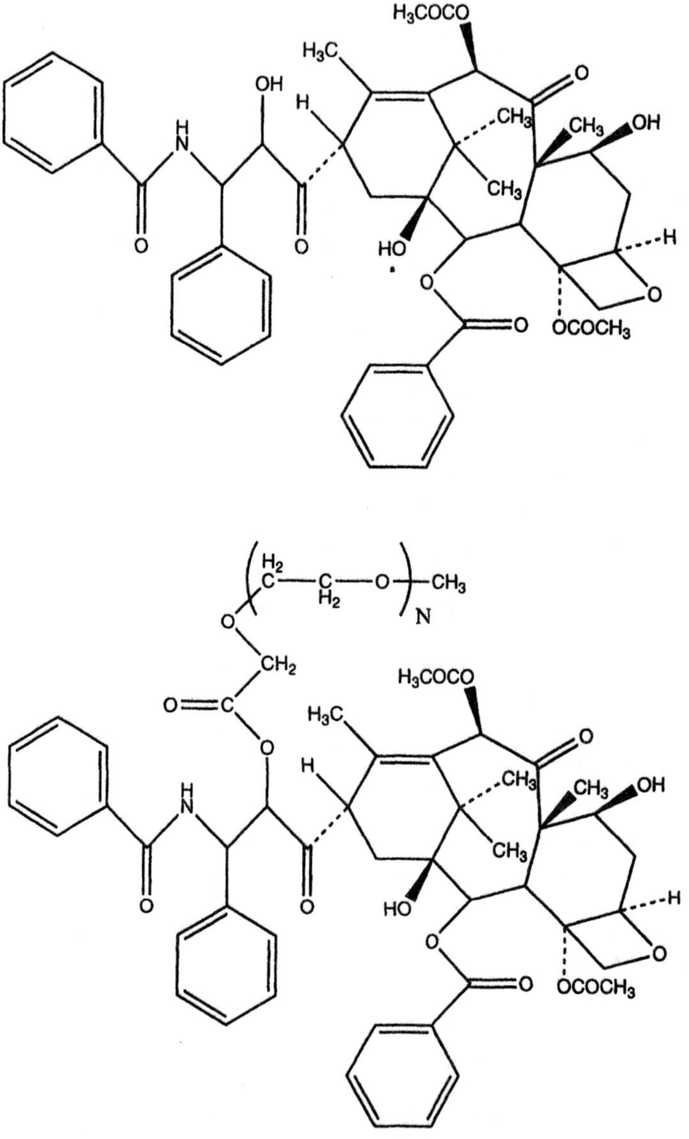

Figure 8.2 Conjugation of paclitaxel to PEG enhances solubility in water. Chemical structures of paclitaxel (Taxol®, (a)) and Taxol 2′-PEG ester (b).

8.1.3 Enhancing Solubility by Incorporation in Liposomes

Amphiphilic molecules, such as phospholipids, can be used to form hydrophobic and hydrophilic compartments within an aqueous environment (Figure 8.3). Phospholipids placed in an aqueous solution can assume three possible forms: (i) micelles, which are small spherical structures with the hydrophilic heads of the molecules on the surface and the hydrophobic tails packed into the

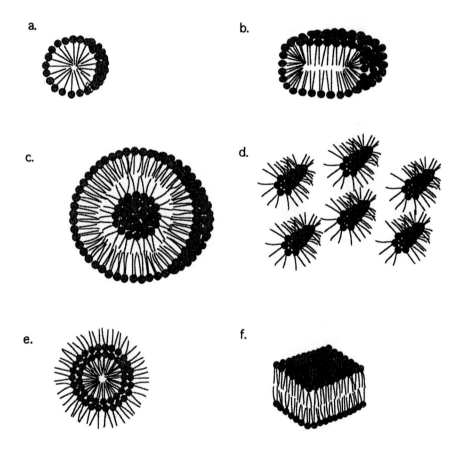

Figure 8.3 Ordered structures formed from lipids. Adapted from [5]. Lipids with hydrophilic and hydrophobic regions form structures in aqueous solution; the architecture of the structure is determined by the relative size of the hydrophobic and hydrophilic regions. When the lipid has a large polar headgroup (indicated by the gray circles), the radius of curvature of the lipid structure is large (a, b, c). Lipids with small headgroups and large tails (indicated by the lines) form inverted structures (d, e). Lipid mixtures can also form vesicles (c). Bilayers are formed when the regions are of equal size (f); the bilayer can fold into a liposome (see Figure 8.4) in which the diameter is sufficiently large that the lipids experience little curvature.

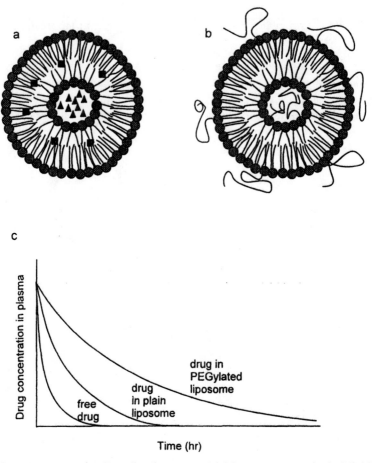

Figure 8.4 Structure of unilamellar liposomes. (a) Liposomes are spherical lipid
bilayers surrounding an aqueous core. Water-soluble drugs (triangles) can be
entrapped in the core and lipid-soluble drugs can be dissolved in the bilayer. (b)
Stealth liposomes are produced by adding lipids that are conjugated to a water-soluble polymer, most frequently PEG. (c) Liposomes can increase the circulation time
of lipid-soluble drugs; stealth liposomes circulate for longer periods than conventional
liposomes.

center; (ii) lipid bilayers, which are sheets of phospholipids two molecules thick
with the hydrophobic tails packed in the center; or (iii) liposomes, which are
spherical lipid bilayers [5]. The term "liposome" was introduced by Bangham
et al. [6] to describe one or more concentric lipid bilayers enclosing an equal
number of aqueous compartments (Figure 8.4). Liposomes form spontaneously in aqueous media; aqueous material can be entrapped in the core
during formation. The size and shape of liposomes can be varied by changing
the mixture of phospholipids, the degree of saturation of the fatty acid side
chains, and the conditions of formation [7].

Since molecules can segregate into different regions of the liposome, these multiphase systems have great potential for drug delivery. Hydrophobic drugs, which are insoluble in water, can be loaded into liposome membranes before delivery. Alternatively, water-soluble drugs can be loaded into the aqueous core region.

Liposomes disappear rapidly from the blood, probably due to phagocytosis by cells of the reticuloendothelial system, which ingests objects that are of the size of most liposomes (Figure 8.4). The circulating half-life can be increased by coupling water-soluble polymers (PEG is used most commonly) to the water-soluble region of the phospholipid. Modified liposomes, sometimes referred to as "Stealth" liposomes, have longer circulation times, presumably due to changes in the rate of phagocytosis produced by the polymer coating. A recent book provides a detailed account of the history, production, and medical applications of liposomes [8].

8.2 ENHANCING AGENT STABILITY

8.2.1 A Brief Review of Mechanisms of Elimination

In previous chapters, drug elimination is represented as a rate process. In Chapter 3, for example, a first-order rate constant (k) was used to represent the elimination of drug molecules during diffusion through a local region of tissue. Similarly, in Chapter 7, rate constants were used to indicate clearance of agents from a tissue compartment. On the microscopic scale, in which attention is focused on molecules moving through homogeneous regions of tissues, elimination can occur within a tissue by any process that removes active drug molecules from the transport stream; often this process is enzymatic conversion of the agent. On the scale of compartments, however, other elimination processes involving multistep mechanisms are possible.

Filtration and Renal Excretion. At the level of the whole body, the action of organs specialized for elimination, such as the kidney, liver, and lung, must also be considered. A few agents—such as carbon dioxide and gaseous anesthetics—are eliminated by the lung in exhaled gas; these agents are small, volatile, and able to permeate through membranes rapidly. Most water-soluble molecules are eliminated from the body in urine, which is produced by the kidneys. The first step in the formation of urine is filtration in the glomerulus; filtered plasma carries molecules that are sufficiently small through the glomerular membrane into Bowman's space (Table 8.3). As this ultrafiltrate moves through the kidney tubules, some molecules—such as water, sodium, and glucose—are reabsorbed into the blood. Other molecules—such as penicillin—are actively secreted into the tubular fluid. By this combination of processes, the kidney prepares a concentrated fluid for elimination of water-soluble wastes (Figure 8.5).

Biotransformation and Biliary Excretion. The liver is a major site of drug metabolism, but it is also an important excretory organ. Some of the

Table 8.3   Filterability of substances through the glomerular capillary

Substance	Molecular weight	Filterability
Water	18	1.0
Sodium	23	1.0
Glucose	180	1.0
Inulin	5,500	1.0
Myoglobin	17,000	0.75
Albumin	69,000	0.005

Filterability is defined as the flux of the solute relative to the flux of water under identical conditions. The glomerular capillary is more porous than most capillaries; filtration through the glomerular capillary wall depends on molecular size and charge (negatively charged molecules are less filterable than positively charged molecules, because of the negative charge of the glomerular basement membrane)

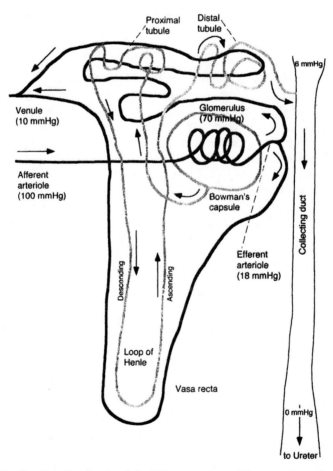

Figure 8.5   The functional unit of the kidney, nephron. The structure of a typical nephron is illustrated.

excretory action is direct. Organic anions, such as glucuronides, organic cations, and steroids are transported from hepatocytes into the liver biliary system, a branching network of vessels and reservoirs that collects bile and empties it into the small intestine. Once in the intestine, molecules in the bile can be excreted in feces, reabsorbed into the blood, or transformed through the action of intestinal enzymes.

Biliary excretion is important, but the liver's role in preparing compounds for excretion by the kidney is quantitatively more important. Many agents, including drugs and toxic compounds encountered in daily life, enter the body because of their lipid solubility; lipid-soluble compounds permeate through skin and mucosal surfaces more readily than water-soluble compounds (see Chapter 5). The kidney is inefficient at excretion of lipid-soluble molecules, which are reabsorbed in the tubules after filtration. The liver aids in elimination by converting lipid-soluble molecules into more polar compounds, which are more easily excreted by the kidney. The range of chemical transformations that can be accomplished in the liver is impressive (Table 8.4); this diversity of biochemical function partially accounts for the diverse symptoms—and high drug sensitivity—experienced by patients with advanced liver disease.

Chemical transformations in the liver are greatly facilitated by the action of enzymes within the smooth endoplasmic reticulum of hepatocytes (often called the microsomal fraction). Phase I reactions (oxidation, reduction, and hydrolysis reactions) convert molecules into more polar forms that usually differ in biological activity from the parent form.[1] Oxidation reactions occur primarily under the direction of a family of enzymes called P_{450}. Non-specific esterases in the liver, plasma, and other sites catalyze the hydrolysis of esters, whereas amides are hydrolyzed in the liver. Protein drugs are often degraded by proteases and peptidases, which are abundant in many tissues, including the intestinal tract and plasma. Phase II reactions are conjugation reactions.

8.2.2 Enhancing Stability by Chemical Modification

The susceptibility of agents to the action of enzymes is a major determinant of stability in tissues. Since enzyme activity is highly selective, subtle changes in the chemical structure can often lead to profound reductions in enzyme-mediated degradation. This approach is illustrated with examples involving analogs to acetylcholine and oligonucleotides.

---

1. For drug molecules, biotransformation reactions usually produce metabolites that are less active than the parent drug, although occasionally inactive molecules are administered with the understanding that they will be converted into an active form by the action of liver enzymes. For example, the anticancer drug cyclophosphamide is not cytotoxic but is converted to powerful alkylating agents (including phosphoramide mustard) by a series of reactions initiated by hepatic P_{450} oxidation of cyclophosphamide to 4-hydroxycyclophosphamide [10].

Table 8.4 Typical biotransformation reactions that occur in the liver

Oxidation reactions	*Examples*
1. N- and O-Dealkylation	

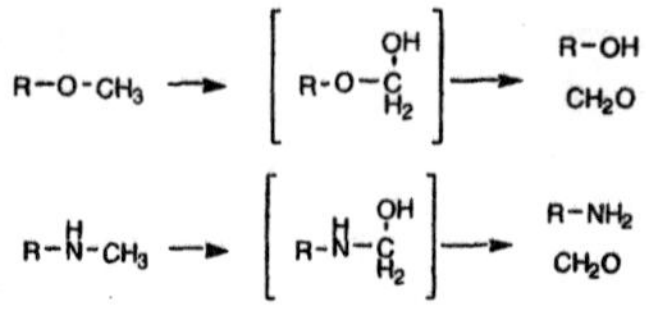

Desipramine and phenacetin

| 2. Side chain and aromatic hydroxylation | |

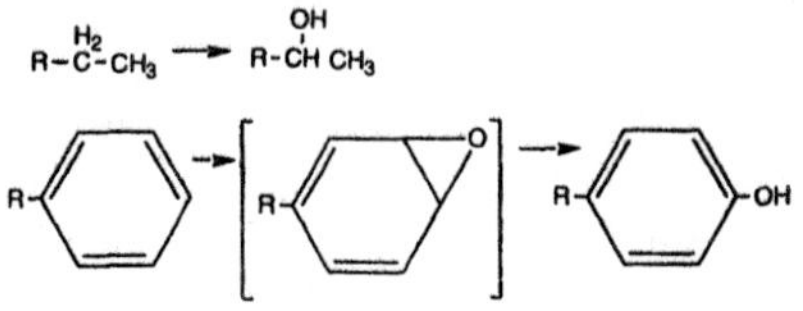

Phenobarbital and phenytoin

3. N-oxidation	Guanethidine
4. Sulfoxide formation	Chlorpromazine
5. Deamination of amines	Amphetamine
6. Desulphuration	Thiobarbital

Hydrolysis of esters and amides

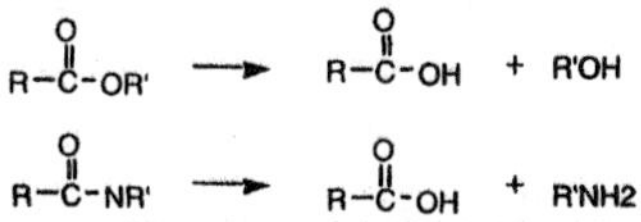

Procaine and Lidocaine

Reduction
Azo and Nitro reductions

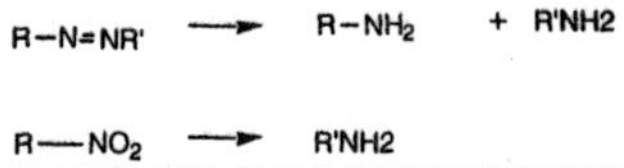

Prontosil and Chloramphenicol

Conjugation reactions
1. Glucuronidation of Ether or Ester (here shown via UDP-Glucuronic acid)

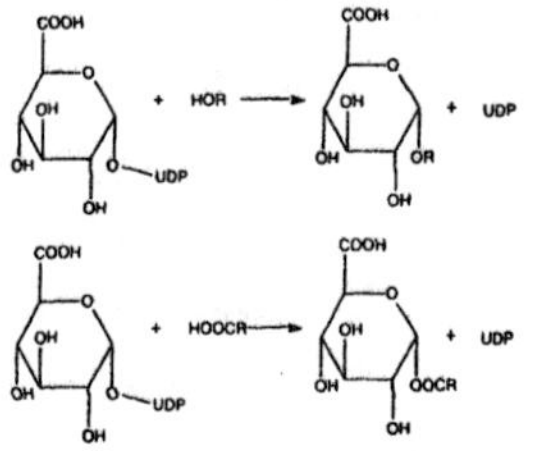

Acetaminophen and Naproxen

2. Acetylation (here via Acetyl CoA)

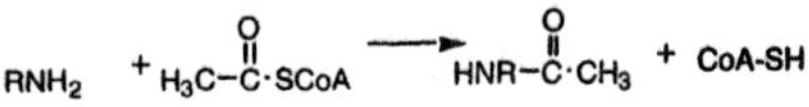

Isoniazid

3. Conjugation with glycine

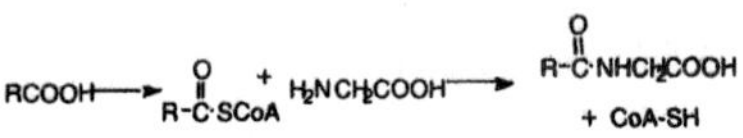

Salicylic acid

4. Conjugation with sulfate   Steroids

5. O-, S-, and N-methylation   Norepinephrine

Adapted from [9], particularly Chapter 1. These reactions can also occur in the kidney, intestinal tract, and plasma. Typical mechanisms are illustrated for some of the reactions. Reactive intermediates are enclosed in brackets.

The mechanism of action for many naturally occurring chemicals requires a short half-life (e.g., see Figure 5.21), which is often provided by the presence of degrading enzymes. Neurotransmitters, which are released by the presynaptic neuron for signal transmission across the synapse, must be rapidly removed from the synaptic space; removal is essential to restoring susceptibility to subsequent release from the presynaptic cell. One neurotransmitter, acetylcholine, is converted into an inactive form by acetylcholinesterase and other esterases that are abundant in nervous tissue and blood. Therefore, acetylcholine is not useful as a therapeutic agent, because it is degraded by enzymes before reaching the target. Cholinergic agents with sustained activity can be produced by chemical modifications that reduce susceptibility to cholinesterases; one such agent is bethanecol chloride, which has cholinergic activity but is not degraded by cholinesterases (Figure 8.6).

Oligomers of nucleotides are emerging as an important class of therapeutic agents (Figure 8.7). Antisense oligonucleotides, for example, bind to messenger RNA in the cytoplasm and, by interfering with translation, reduce the expression of specific proteins. However, both phosphodiester and phosphorothioate oligonucleotides are rapidly cleared from the blood after intravenous injection in the mouse or human. Phosphodiester oligonucleotides are also extensively metabolized, presumably by exonucleases in the serum. However, phosphorothioate oligonucleotides, although quickly cleared from the blood, persist intact within the tissues for prolonged periods. A relatively straightforward change in the backbone structure of the oligonucleotide greatly enhances the clinical value of these agents, presumably by reducing their susceptibility to degradation in plasma and in tissue (Table 8.5). Other modifications enhance oligonucleotide penetration through the plasma membrane [11].

Peptide nucleic acids (PNAs) are a more recent innovation in the design of analogs that recognize complementary nucleic acid sequences (Figure 8.7b). Because of the unusual chemical structure of the PNA backbone, these agents are resistant to protease and nuclease degradation. Chemical modification is a versatile method for enhancing the biological stability of oligonucleotides, without compromising the biological activity (i.e., specificity and affinity of binding to the complement).

8.2.3 Enhancing Stability by Conjugation to Polymers

Soluble, synthetic polymers can be used as carriers for bioactive agents, an idea first proposed by Ringsdorf [14]. A wide variety of polymer chemistries and structures have been used as carriers for drugs and bioactive macromo-

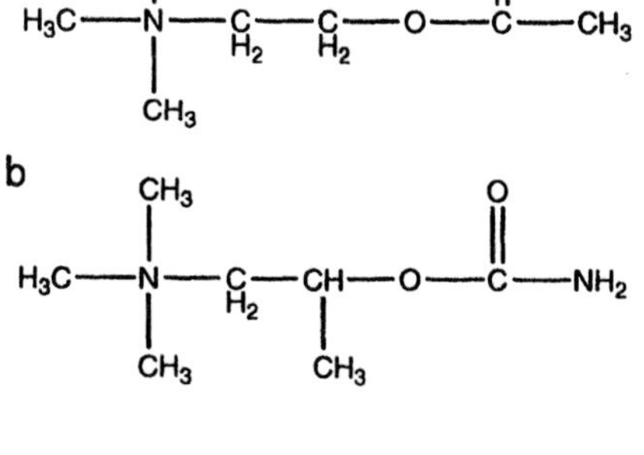

Figure 8.6
Chemical structures of acetylcholine (a) and bethanecol (b).

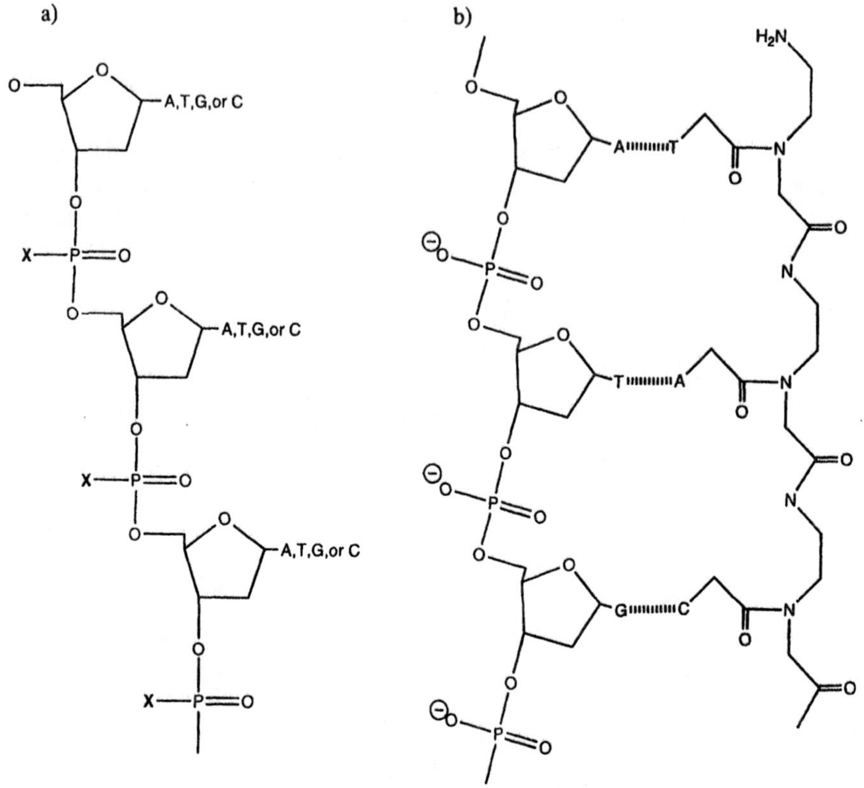

Figure 8.7 Modification of oligonucleotides to increase stability. (a) Oligonucleotides (here shown as DNA) with a phosphodiester backbone ($X=O$) are
rapidly degraded by nucleases. Modification to create phosphorothioate analogs
($X=S^-$) greatly increases half-life. (b) Peptide nucleic acids represent another DNA
analog that can be used to bind with complementary sequences of oligonucleotides.
Dashed lines represent hydrogen bonding which follows Watson–Crick base pairs.

Table 8.5   Stability of oligonucleotides in biological fluids at 37 °C

	Cytoplasmic extract	Nuclear extract	Normal human sera	Calf sera
D-oligo	+	++	++	+++
MP-oligo	−	−	+	++
S-oligo	−	+	+	+
Alt-MP-oligo	−	−	+	++

From paper by [12]. Degradation measured by polyacrylamide gel electrophoresis from radiolabeled
oligos. Complete degradation within 10 min (+++); complete degradation within 60 min (++); some
degradation observed but not complete in 60 min (+); no degradation products over 60 min (−). This is
consistent with other studies [13].

lecules (Figures 8.8 and 8.9). Carrier polymers are conjugated to drugs and proteins to stabilize the agent. Usually stabilization means extending the half-life of the agent as measured by changes in plasma concentration, but con-jugation also alters the biodistribution, as described below. Many classes of polymers have been used as carriers including PEG [15–17], *N*-(2-hydroxy-propyl)methacrylamide copolymers [18, 19], dextran [20, 21] and other poly-saccharides [22], polylysine [23], poly(iminoethylene), poly(vinyl alcohol), poly(divinyl ether-*co*-maleic anhydride), and polyurethanes [24]. Techniques for conjugation of biological molecules are summarized in the encyclopedic text by Hermanson [25].

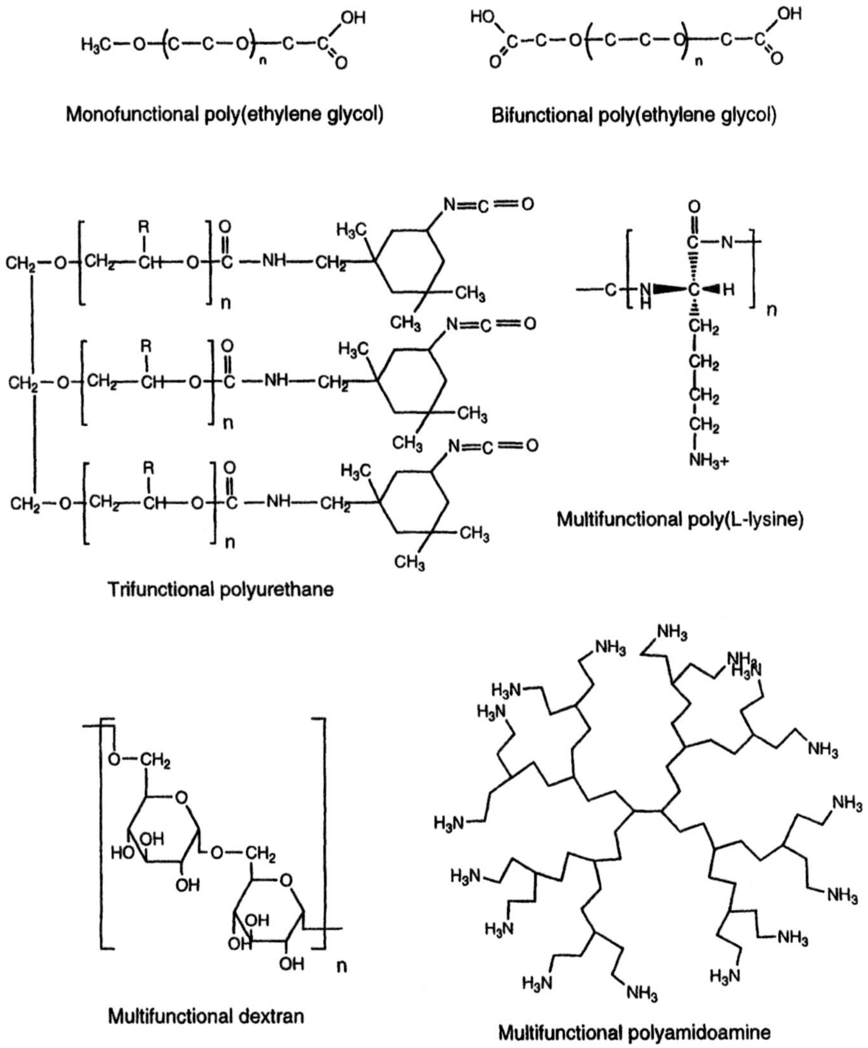

Monofunctional poly(ethylene glycol)　　Bifunctional poly(ethylene glycol)

Trifunctional polyurethane

Multifunctional poly(L-lysine)

Multifunctional dextran

Multifunctional polyamidoamine

Figure 8.8 Chemistry of drug conjugates. Examples of some of the polymeric sys-tems that have been used as drug carriers.

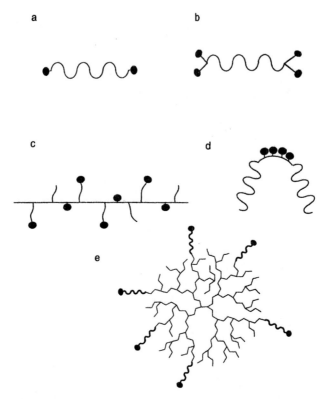

Figure 8.9 Chemical structures for some typical conjugates. Examples of some of the potential architectures that can be produced in drug conjugates. The structures can be linear (a), (b), and (d) or branched as in (c) and (e).

Cancer Therapy. Cancer chemotherapy is one of the most important applications for drug–polymer conjugates. Toxic side effects limit the use of almost every known anticancer compound (see [10] for a review of activity and toxicity of anticancer agents). Conjugation of anticancer agents to water-soluble polymers can reduce the side effects by stabilizing the agent in the blood and directing a greater fraction of the cytotoxic dose to tumor cells. The anticancer drug 5-fluorouracil (5-FU) is highly effective against a variety of tumors, but it also causes severe side effects, which are probably related to the high permeability of 5-FU through membranes. A 5-FU derivative has been coupled with a PEG carrier, creating conjugates that break down by hydrolysis to liberate 5-FU. The rate of hydrolysis of the conjugate decreases as the molecular weight of the PEG increases. These compounds possess antitumor activity, which increases with the degree of polymerization of the PEG [26]. Branched carriers with poly(L-lysine) backbones and adjustable side chains have been used as carriers for methotrexate (MTX). The degree of MTX incorporation, *in vitro* cytotoxicity, half-life in the plasma, and biodistribution depend on the molecular weight of the carrier and the terminal amino acid on the side chain [23]. Similar results have been obtained with other conjugates [18, 27, 28].

The first polymer–drug conjugate to be tested in humans for anticancer therapy is doxorubicin coupled with *N*-(2-hydroxypropyl)methacrylamide copolymer via a linker that is degraded in lysosomes, thereby releasing the drug [29]. In contrast to the free drug doxorubicin, which has a distribution half-life of 0.08 h and an elimination half-life of 30 h (see Table 8.6), the doxorubicin conjugate has a distribution half-life of 1.8 h and an elimination half-life of ~ 90 h.

PEGylation of Peptides and Proteins. Synthetic peptides can mimic aspects of protein function and are more easily administered than whole proteins, but they have short half-lives in the plasma. For example, Arg-Gly-Asp (RGD), originally derived from the fibronectin sequence, is common to many ECM proteins. RGD peptides can block tumor cell binding to ECM; therefore, they may be useful for preventing tumor metastasis, which is due to circulating tumor cells that attach to sites distant from the initial tumor. However, this activity is hindered by the short circulation time of the unmodified peptides. Intravenous administration of a PEG–RGD conjugate reduces the number of lung metastases in an experimental metastasis model [30]. In a similar approach, cell-binding peptides were conjugated to

Table 8.6 Pharmacokinetic characteristics of liposomal anticancer drug formulations

	Daunorubicin/liposome (DaunoXome)	Daunorubicin
Plasma clearance (mL/min)	17.3	236
Volume of distribution (L)	6.4	1006
Distribution half-life (h)	4.41	0.77
Elimination half-life (h)	—	55.4

	Doxil ($10\,\text{mg/m}^2$)	Doxil ($20\,\text{mg/m}^2$)	Doxorubicin ($75\,\text{mg/m}^2$)
Peak concentration	4.12	8.34	
Plasma clearance (L/h)	0.1008	0.0738	50.4
Volume of distribution (L)	5.094	4.896	1750
AUC ($\mu g/mL \cdot h$)	277	590	
First-phase half-life (h)	4.7	5.2	0.083
Second-phase half-life (h)	52.3	55	30

From the manufacturer's published information.

polyurethane prepolymers formed by end-capping a polyether polyol with diisocyanate (Figure 8.8). The circulatory half-life of the pentapeptide Tyr-Arg-Gly-Asp-Ser (YRGDS) increases from 2.7 to 11 h by conjugation [24].

As with peptides, the short circulating half-life of proteins is a major difficulty in delivery. The half-life of proteins can often be increased by con-jugation with polymers, including natural polymers such as albumin, gelatin, and dextran, and synthetic polymers such as PEG (reviewed in [28, 31]). The mechanisms by which conjugation extends circulation time are not completely understood, but conjugation appears to decrease the rate of degradation by enzymes, attenuate the immunological response, and reduce the rate of renal clearance. Because of these attributes, PEGylation is widely used to stabilize injected proteins.

Protein PEGylation was initially accomplished by reaction of the protein with activated monomethyoxy-PEG, but now a wide range of conjugation reagents are available, allowing the chemist additional control over the posi-tion and number of PEG chains added. In some cases, it is possible to produce conjugates with increased stability and nearly complete retention of biological activity. For example, conjugation of PEG to L-glutaminase-L-asparaginase [15] increases the circulating half-life of the enzyme from 2 to 24 h. When asparaginase-PEG is infused intravenously into patients, the half-life of the enzyme is 360 h, a dramatic improvement over the half-life of the unmodified protein [32].

An increase in circulation time is a tremendous potential advantage for protein drugs. In practice, this advantage is tempered by the potential for conjugation to reduce the biological activity of the proteins. For example, conjugation of PEG to urate oxidase produces a substantial increase in blood circulating life, but conjugation can harm urate oxidase activity; activity decreased with increasing PEGylation over the range 0.5 to 3.0 mol PEG/mol protein amino groups [33].

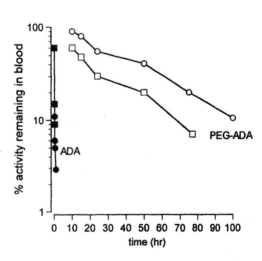

Figure 8.10
Polyethylene glycol modification prolongs the circulating half-life of proteins. Adapted from [34].

PEGylated proteins are used clinically. Conjugation of PEG to bovine adenosine deaminase (ADA) reduces the immunogenicity of the protein and increases circulation half-life from a few minutes to 24 h (Figure 8.10). Repeated injection of unmodified foreign proteins often produces an immune response that neutralizes the effect of the protein. However, when the PEG-modified ADA was repeatedly injected into mice, the extended half-life was unchanged after 13 serial injections. Weekly intramuscular injection of this preparation produced elevated ADA activity in the blood of children [35]. With the commercially available product (Adagen® from Enzon, Inc.), peak plasma levels are achieved 2–3 days following intramuscular injection. In patients the plasma half-life is variable, ranging from 3 to > 6 days, even when measured in the same child. Similarly, recombinant non-glycosylated interleukin 2 (rIL-2) was modified by covalent coupling with PEG [16]; the plasma half-life of PEG-rIL-2 in mice is 1–4 h, substantially longer than the half-life for unmodified rIL-2 ($\sim$ 30 min). This half-life depends on the density of PEG modification. Modified rIL-2 has been tested in patients with HIV infection [17].

Some proteins are used for local treatments; collagen injections, for example, are used to treat urinary incontinence and vesicoureteral reflux. Collagen has also been modified by covalent attachment of PEG [36], with the hope that the implanted collagen would be less immunogenic and would be retained at the injection site for a longer time.

Other proteins are administered locally to produce high local concentrations of a diffusable protein. For locally administered proteins, conjugation can reduce the rate of degradation within the extracellular space, the rate of permeation from the extracellular space to circulation, and the rate of uptake by receptor-mediated processes. When an agent is retained within the extracellular space, but still diffusable, the agent can penetrate a substantially greater distance from the site of administration [37]. For example, nerve growth factor (NGF) has a short half-life after direct administration into the brain (30 min) [38]. During this short period in the tissue, NGF molecules are able to penetrate a limited distance from the delivery site: $\sim$1 mm [39]. NGF covalently coupled with 70,000 M_w dextran, however, penetrates eight times further into the brain interstitium than unconjugated NGF [40].

Immobilization of Proteins on Solid Supports. Some drugs do not need to be dissolved to initiate biological activity: i.e., these agents can initiate biological activity without entering the cell. Both insulin and NGF are effective when immobilized on a surface [41, 42]. Sometimes immobilized proteins possess higher levels of activity (and greater stability) than similar concentrations of soluble factors: insulin and transferrin on poly(methyl methacrylate) films stimulate the growth of fibroblasts, even more than do comparable concentrations of soluble or physically adsorbed proteins [43]. Immobilization also permits the production of spatial patterns of protein: spatially heterogeneous immobilization of insulin on poly(ethylene terephthalate) increased Chinese hamster ovary (CHO) cell density and intracellular signaling in the treated regions [44]. The effect of growth factor binding can be attenuated by endocytosis and subsequent degradation of the receptor–ligand complex, so by presenting immobilized ligands—in a conformation that maintains biological activity but inhibits receptor internalization—the intracellular signaling cascade may be sustained for longer time periods.

The conformation of the immobilized factor is important for the induction of cellular mechanisms. In one approach for enhancement of growth factor activity, poly(ethylene oxide) (PEO) molecules were used as spacers between the surface and the factor; the spacer molecule, which is covalently bound between molecules on the surface and the growth factor, creates physical space between the protein and surface. This space allows the protein to assume a more normal conformation and, hence, improves its function. Cell growth in response to immobilized insulin was enhanced by inclusion of a short spacer between insulin and the surface [45]. Tethered epidermal growth factor (EGF), linked to the surface through PEO molecules, also had substantially more biological activity—that is, greater stimulation of either DNA synthesis or cell rounding (an indicator of cells in the growth phase)—than physically adsorbed EGF [46]. Cells expressing internalization-deficient growth factor receptors exhibit similar enhancement, compared to cells expressing the wild-type receptor. This result is consistent with the hypothesis that reduced receptor internalization, which can be achieved by either mutation or physical restraint of the ligand, leads to enhanced signaling [47].

8.3 REGULATING AGENT PERMEABILITY

8.3.1 Brief Review of Transport Mechanisms

Cell membranes act as barriers that segregate functional components of cells and tissues. As discussed in Chapter 5, molecules can move through membrane-associated barriers by dissolution and diffusion through the phospholipid bilayer (recall Equation 5-2); facilitated or active transport across the membrane; or transport through gaps or pores formed by incomplete union of adjacent cell membranes. Therefore, drug activity can be changed by altering the relative solubility of molecules in membranes; creating drug analogs that engage active or facilitated transport mechanisms; or altering the size of the drug (or drug carrier) to limit its permeation through certain barriers.

The barrier properties of membranes are essential for the function of every cell, tissue, and organ. However, membrane barriers differ from cell to cell and organ to organ. Consider, for example, the special situation of blood flow through the brain. The brain has high metabolic requirements and, therefore, requires a substantial volumetric flow rate of blood, but the composition of the extracellular fluid in the brain must not fluctuate; this fluid surrounds neurons and changes in the fluid composition could adversely effect neuron function. The blood–brain barrier refers to the regulated permeability of the capillaries in the brain, which restrict a higher number of potential permeants than do systemic capillaries (recall Figures 5.27 and 5.28). Because permeability of capillaries is a central issue in drug delivery to the brain, the discussion below includes several examples of selective permeability at the blood–brain barrier; many of these concepts also apply to systemic capillaries, although the relative importance of each pathway may differ.

8.3.2 Regulating Permeability due to Passive Diffusion

Chemical Modification. The rate of passive diffusion of an agent through a lipid bilayer correlates with relative solubility of the agent in oil and molecular size (recall Figures 5.4 and 5.5). Therefore, changes in chemical structure that increase lipid solubility (without adding substantially to the size) can often increase permeability in membranes; such a modification often results in enhanced oral uptake (because of increased permeation through the intestinal mucosa) and altered biodistribution. This effect is illustrated for barbiturate drugs (Figure 8.11 and Table 8.7). Modifications in the side groups R_1 and R_2 and X influence the solubility of the compound, as indicated by the heptane/water (pH 1) partition coefficient (Table 8.7). Bioavailability (i.e., gastric uptake) and protein binding correlate with this partition coefficient. Of course, these differences in physical properties must be considered together with the changes in biological activity that also result from chemical modification.

The endogenous opioid peptides are of pharmacological interest because they are highly potent analgesics. The enkephalin class contains small peptides

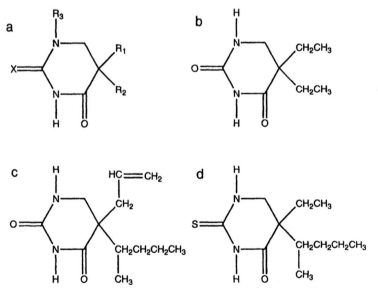

Figure 8.11 Chemical structure of barbiturates. (a) General chemical structure of barbiturate drugs. Specific chemical structures for (b) barbital; (c) secobarbital; (d) thiopental.

Table 8.7   Pharmacological properties of barbiturates

	pKa	K (heptane–0.1 N HCl)	Gastric absorption (%)	Plasma protein binding (%)	Half-life (h)	Oral dose for hypnotic effect (mg)
Barbital	7.6	< 0.001	4	5		
Secobarbital	7.8	0.10	30	65	19–34	100
Thiopental	7.9	3.30	46	80		

Properties of the compounds illustrated in Figure 8.11. These pharmacokinetic effects occur in addition to the functional effects of modifications in barbiturate structure. In particular, substitutions at positions R_1 and R_2 produce agents with increased hypnotic potency and anticonvulsant activity.

with similar structure that are derived from a precursor protein called proenkephalin; fragments of this precursor, a fraction of which is indicated in Figure 8.12, are biologically active. Physiologically, these peptides are produced within neurons and act locally after secretion. Although they are potent, they permeate poorly through the blood–brain barrier and are difficult to use as pharmaceuticals.

Various peptides from the enkephalin family, or analogs of these molecules, permeate through the blood–brain barrier at different rates, often by different mechanisms [49]. For example, D-Phe-Cys-Tyr-D-Trp-Arg-Thr-Pen-Thr-NH₂ (a cyclic peptide that binds to μ-opioid receptors, in which Pen is penicillamine) permeates by diffusion only; Tyr-D-Pen-Gly-Phe-D-Pen ([D-Pen]²,⁵-enkephalin, a cyclic Met-enkephalin analog that binds to δ-opioid receptors) permeates by diffusion and a saturable transport mechanism; (Tyr-D-Ala-Gly-Phe-NH)₂ (biphalin, a dimeric analog that is active at μ-

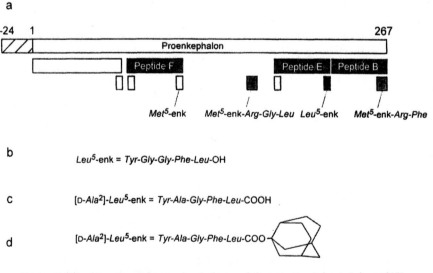

Figure 8.12   Structure of proenkephalon and fragments. Adapted from [48].

and δ-opioid receptors) permeates by diffusion and the large neutral amino acid transporter.

Chemical modification can be used to manipulate lipid solubility and permeation through membranes. Leu^5-enkephalin is a pentapeptide Tyr-Gly-Gly-Phe-Leu-OH of M_w 627; Met^5-enkephalin is identical except for the replacement of methionine for leucine. Leu-enkephalin (Figure 8.12b) has an octanol/water partition coefficient of 0.6; addition of an 1-adamantane moiety (Figure 8.12c) increases the partition coefficient to 132 and enhances permeation into the brain [50].

Other modifications enhance peptide stability. For example, [D-Ala^2]-Leu^5-enkephalin resists degradation by proteases found in the intestine; the half-life for degradation by enzymes in the jejunum was ~ 2 h for [D-Ala^2]-Leu^5-enkephalin and ~ 1 min for Leu-enkephalin [51]. Permeation of Met-enkephalin is enhanced by the addition of compounds that inhibit either aminopeptidase (which cleaves Met-enkephalin at the Tyr^1-Gly^2 bond) or neutral endopeptidase/angiotensin-converting enzyme (which both cleave at the Gly^3-Phe^4 bond) [52]. These observations suggest another target for enhancing neuropeptide permeation through epithelial or endothelial barriers: reduction of enzymatic degradation, which can be achieved by identification of enzyme-resistant analogs or co-administration of enzyme inhibitors.

Cationic Lipids for Gene Transfer. Gene therapy requires techniques for efficient introduction of DNA into cells. Since DNA molecules are large and water soluble, they do not permeate through lipid bilayers. One approach for enhancing DNA delivery into cells is by adding agents that will form complexes with DNA and facilitate cell entry. The ideal agent should form spontaneous complexes with DNA and facilitate fusion/uptake with the cell surface. Cationic lipids have both properties. The cationic lipid, N-[1-(2,3-dioleoyloxy)propyl]-N,N,N-trimethylammonium chloride (DOTMA), forms multilamellar liposomes in water; small unilamellar vesicles can be produced by sonication [53]. DOTMA can be combined with neutral phospholipids, such as dioleoyl phosphatidylethanolamine (DOPE) and dioleoyl phosphatidylcholine (DOPC), to produce liposomes with desired properties. Liposomes containing DOTMA interact spontaneously with DNA to form lipid–DNA complexes in which nearly 100% of the DNA is entrapped. This cationic lipid reagent (also called Lipofectin®) promotes gene delivery and expression in many cell lines. The DOTMA reagent was the first in a still growing category of lipid agents for gene transfer, which also includes dioleoyl trimethylammoniumpropane (DOTAP).

The mechanism of transfection by cationic lipids is not known, but cationic lipids probably form stable complexes with anionic DNA molecules to facilitate entry into the cell. Physical studies of DNA/DOTAP/DOPE complexes suggest that DNA molecules are arranged in a regular pattern between layers of lipid bilayer [54]. This multilamellar arrangement permits the charged lipid-head groups to interact with the charged DNA molecules. When mixtures of cationic and neutral lipids are used, the cationic lipids concentrate in regions nearest the DNA.

Gaining access to the cytoplasm is only the first step in DNA delivery. The foreign DNA molecules must also traverse the cytoplasm and gain entry into the nucleus. The development of transfection agents that promote overall DNA delivery efficiency is a challenge (see recent review [55]), because properties that enhance DNA migration across one barrier may impede its progress across another. Cationic lipids, for example, form complexes with DNA that facilitate transport across the plasma membrane; these same lipids interfere with DNA entry from cytoplasm into the nucleus.

8.3.3 Regulating Permeability by Active or Facilitated Transport Systems

Capillary endothelial cells possess specialized transport proteins which permit the passage of specific molecules. The endothelial cells of the blood–brain barrier have a number of specialized transport systems that are essential for brain function (see Chapter 5 and [56, 57]). Sometimes, these transport systems can be deceived by creating or finding agents that resemble the native substrate for the transport protein, but are modified to possess a desired biological activity.

The classic example of this approach involves the use of levodopa (L-3,4-dihydroxyphenylalanine, Figure 8.13) to treat Parkinson's disease [58]. Parkinson's disease is distinguished by the marked depletion of dopamine—an essential neurotransmitter—in the basal ganglia. Direct dopamine replacement is not possible, because dopamine does not permeate through the blood–brain barrier. However, the metabolic precursor of dopamine, levodopa, is transported across brain capillaries by the neutral amino acid transporter (see Table 5.5 and the related discussion). Peripheral administration of levodopa, therefore, produces an increase in levodopa concentration within the central nervous system; some of these molecules are converted into dopamine due to the presence of decarboxylate enzymes in the brain tissue, but decarboxylate activity is also present in the intestines and blood. To prevent conversion of levodopa into dopamine before entry to the brain, levodopa is usually administered with decarboxylase inhibitors.

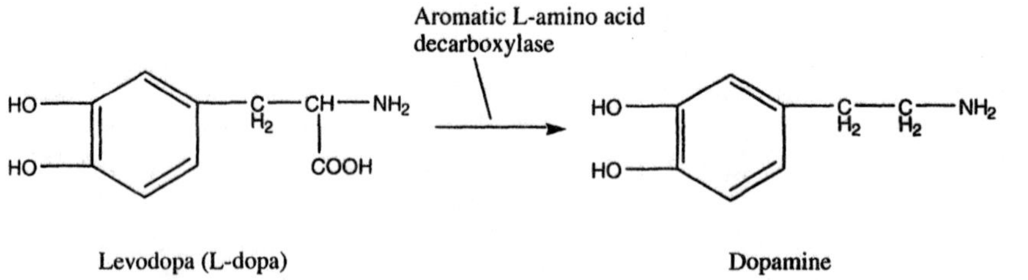

Figure 8.13   Chemical structure of dopamine and levodopa.

There is considerable interest in developing pharmaceuticals that use specific carriers to gain entry to the brain. For example, enhanced transport has been observed for agents coupled with antibodies that bind the transferrin receptor as well as other vectors [59, 60]. Increased uptake of proteins into the brain has been reported after cationization [61], which appears to enhance endocytosis, perhaps due to improved interactions between the cationic protein and the negatively charged cell surface .

Peptides Derived from Homeodomain Proteins Can Translocate across a Cell Membrane. Homeodomain proteins are transcription factors that are expressed during development and, in some tissues, throughout adult life. These factors, which act by binding to DNA within the nucleus, must translocate across membranes to function. Recent studies suggest that the ability of peptides derived from the third helix of *Drosophila* Antennapedia homeobox (Antp-HD) to cross membranes is conferred by domains rich in basic amino acids (Table 8.8). A number of other proteins—such as HIV-1 Tat protein [62], transportan [63], and basic amino acid-rich model proteins [64]—have domains that provide a similar enhanced permeation function.

The mechanism of peptide transport into the cell is still uncertain. Peptides derived from Antp-HD appear to be internalized by a receptor-independent mechanism that depends on direct interaction with membrane lipids; peptide association with lipids and internalization via an inverted micelle that crosses the membrane, has been suggested [65]. In the case of transportan (Table 8.8), binding to the cell surface and internalization occur quickly (in ~ 1 min with maximum concentration at ~ 20 min) and efficiently (10–16% uptake) [63]. As in Antp-HD, uptake does not appear to require specific receptors, as internalization is not saturable. Once in the cell, most of the peptide is associated with membranes, including the nuclear membrane. Similar observations have been reported in Tat peptide sequence derived from HIV-1 [62].

Peptides with this transport function can be used as vectors to carry passengers into cells or cell compartments. For example, homeodomain peptides can be conjugated to agents that would otherwise not permeate into the cell cytoplasm. Conjugation facilitates the entry of the agent into the cytoplasm and, sometimes, into the nucleus (Table 8.8). For example, Tat peptides were conjugated to the surface of 40-nm particles; this modification permitted cellular and nuclear uptake of the particles [66]. A PNA complementary to mRNA of a human receptor was conjugated to transportan and an Antp peptide [63]. This conjugate was efficiently internalized by cultured cells; after uptake, the PNA inhibited translation of the specific RNA sequence. When administered to rats, the PNA–peptide complex caused a decrease in receptor expression and, subsequently, inhibition of the function of that receptor.

Table 8.8 Internalization of peptides derived from the third helix of Antp-HD and other proteins

Synthesized peptide	Sequence	Entry into cytoplasm	Entry into nucleus
43 → 58	RQIKIWFQNRRMKWKK	+++	+++
58 → 43[a]	KKWKMRRNQFWIKIQR	+++	+++
Da43 → 58[b]	RQIKIWFQNRRMKWKK	+++	+++
Pro50[b]	RQIKIWFPNRRMKWKK	+++	+
3Pro[b]	RQPKIWFPNRRMPWKK	+++	±
2Phe[c]	RQIKIWFQNRRMKWKK	±	±
41–55	TERQIKIWFQNRRMK	−	−
46–60	KIWFQNRRMKWKKEN	−	−
Transportan	GWTLNSAGYLLGKINLKA LAALLAKKIL-amide	Yes	Yes
Tat protein basic domain	FITKALGISYGRKKRRQRR RPPQC	Yes	Yes
Model peptide	FLUOS- KLALKLALKALKAALKLA	Yes	No[d]
Colicin		Yes	?

These Antp-HD sequences were used to show that internalization does not require a chiral receptor. [a] does not require a helical structure; [b] does not depend solely on amphiphilicity. [c] Adapted from references cited in text. The symbols indicate the relative extent of entry into the cytoplasm and nucleus: +++ > + > ± > −. [d] Unless coupled with nuclear-localization peptide: PKKKRKV.

8.3.4 Regulating Permeability and Biodistribution by Size

After Systemic Administration.   Liposomes were introduced earlier in this chapter as a versatile method for altering the solubility of agents (Figure 8.4). Liposomes have other effects as well. For example, "Stealth" liposomes can extend the circulation half-life of drugs by decreasing the rate of phagocytosis (Figure 8.4). However, incorporation into liposomes also changes the biodistribution of the agent, because the agent is now associated with a particle that is restricted to the plasma space. A hydrophobic agent would ordinarily permeate easily through membranes, achieving a large volume of distribution, which is associated with a low peak concentration in the plasma (Table 8.6). Since the liposome remains in the plasma volume, the volume of distribution is markedly reduced, and the effective concentration (as measured in the plasma) is increased.

The consequences of liposome incorporation must be balanced. For example, loading of daunorubicin into conventional liposomes decreases the volume of distribution, increases peak concentration, and prolongs the initial rate of elimination from the plasma (Figure 8.14). The conventional liposomes are cleared from the circulation, however, producing a substantial drop in the plasma concentration of daunorubicin during the first 24 h after injection. In contrast, incorporation of doxorubicin into "Stealth" liposomes provides the same benefits with respect to distribution and concentration, but also extends the circulating half-life of the agent (Figure 8.14). It is important to remember that the goal of drug therapy is not merely to maintain high concentrations of the agent in the plasma volume; the goal is delivering active agent to the site of action, which is usually outside of the plasma volume. Therefore, while the "Stealth" liposomes are effective at increasing and maintaining concentrations, the usefulness of this approach depends on subsequent release and localization of the incorporated doxorubicin. This same reasoning—that is, balancing stabilization within the plasma with local delivery of the active compound—must be applied to polymer-conjugated drugs as well.

Liposomes, or any drug modification that increases the size of the administered molecule, can produce another effect related to permeability. The size dependence of permeation through biological barriers was discussed in Chapter 5. As the physical size of an agent increases, it becomes less permeable in most capillaries. This effect explains the reduced volume of distribution observed for liposome-entrapped drugs (Figure 8.14). This permeability effect will be most pronounced in tissues with the most restrictive barriers. If the size of an agent were to be increased incrementally, it would first be excluded from permeation through the capillaries with the lowest molecular weight cut-off;

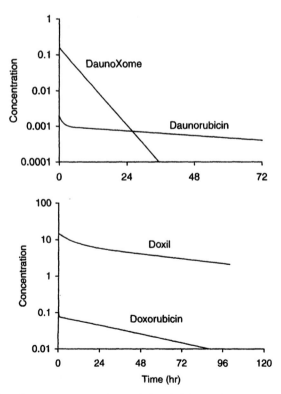

Figure 8.14 Plasma concentrations of chemotherapy drugs encapsulated in liposomes. Curves were calculated from pharmacokinetic parameters in [2], see Table 8.6.

vessels with very "leaky" capillaries (i.e., capillaries with a high molecular weight cut-off that permit the permeation of very large agents) would be affected last, only when the agent has become quite large.

It is known that tumor vessels are frequently "leaky" and, therefore, it is reasonable to speculate that this size-dependent permeation could be used to deliver agents preferentially to tumors. Delivery of large molecules with long-circulating half-lives should encourage drug accumulation within regions of increased permeability, such as tumors. This effect is sometimes called the "enhanced permeability and retention (EPR) effect," and occurs after the intravenous injection of drug–polymer conjugates. The precise mechanisms are unknown, and may differ among solid tumors, but appear to be due to both rates of permeation into the tumor and rates of clearance from the tumor site [28].

After Local Adminstration. Local administration is sometimes used to produce high drug concentrations at a particular tissue site, while sparing other tissues from drug exposure. One would expect locally injected agents to diffuse slowly from the site of injection, eventually dispersing throughout a region of the tissue. This effect has been well studied in the brain; previous investigators have measured the penetration of neuroactive compounds in brain tissue (Table 8.9). Although a variety of different methods were used to administer agents with a correspondingly wide range of physicochemical and biological properties, the extent of penetration into the brain tissue is limited in every case. This observation may have important implications for the development of drug delivery strategies for the human central nervous system. This limitation on penetration is predicted by simple models of diffusion and elimination in tissue (Figure 8.15).

8.4 DRUG TARGETING AND DRUG VALENCY

8.4.1 Targeting Agents to Specific Cells or Tissues

Drug targeting is one of the most intellectually attractive (and vigorously pursued) concepts in all of drug delivery. Drug targeting adds chemical speci-ficity to an agent, making it "targeted" in the sense that it can differentiate cells (or other anatomical sites) that are appropriate for treatment from other cells (Figure 8.16). The chemical specificity is often achieved by isolating protein receptors that distinguish a particular cell from other cells; for example, tumor cell antigens that are not found on normal cells are attractive targets for chemotherapy agents. Unfortunately, despite intensive effort and creative approaches, drug targeting has proven to be a difficult concept to achieve in humans.

In most drug targeting scenarios, a targeting moiety is attached to an active drug compound; the variations on this theme are manifold, as carriers and spacers can be added to facilitate the targeting effect (Figure 8.16). Current thinking about targeting derives largely from observations on the specificity of

Table 8.9 Penetration distances for agents administered to the brain

Mode of administration	Agent	Penetration distance (mm)	Period for diffusion (h)	Reference
Ventriculocisternal perfusion	Hydroxyurea	2	1	[81]
	Methotrexate	1.5	1	[81]
	Nerve growth factor	1.2	< 1	[82]
	Cytosine arabinoside	1	1	[81]
	Sucrose	~ 1	4	[83]
	Inulin	1	2–3	[84]
	Thiotepa	0.7	1,4	[81]
	Albumin	~ 0.5	2–3	[84]
	BCNU	0.5	1	[81]
	Brain-derived neurotrophic factor	0.3	< 1	[82]
Intracerebral microinfusion	Cisplatin	0.8	160	[85]
	Dopamine, methotrexate, and antipyrine	< 1	144	[86]
	6-Hydroxydopamine and norepinephrine	< 1	—	[87]
Intracerebral microdialysis	Sucrose	1.3	1	[88]
Intracerebral polymer implant	IgG	> 2	48	[89]
	Nerve growth factor	~ 2	48	[38]
	Dextran	2–7	24	[90]
	BCNU	1.2	24	[90]
	Iodoantipyrine	0.8	24	[90]
	Taxol	~ 1	72	[91]

Penetration distance equals the distance measured from site of administration to the point where concentration drops to 10% of maximum. In ventriculocisternal perfusion, drug solution is infused into the ventricular fluid; in intracerebral micro-infusion or microdialysis, drug solution is either infused slowly or equilibrated through a membrane into the brain parenchyma.

action within the immune system. Antibodies, which can be generated to bind to virtually any target antigen, are logical targeting moieties. In fact, the concept of targeting can be traced back to Ehrlich, who predicted that the availability of antisera would lead to the development of agents that could provide selective protection from toxic compounds [67]. Antibodies have several potential advantages as targeting agents:

1. Antibodies can be produced with virtually any desired specificity;
2. Antibodies are large molecules, so drugs can be conjugated without altering the binding properties;

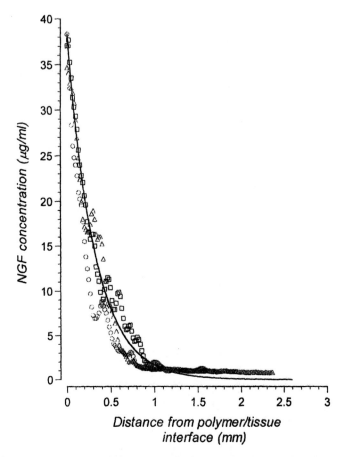

Figure 8.15 Penetration of NGF into the brain after local administration. Concentrations were measured (by quantitative autoradiography) as a function of distance from the administration site. The drop in concentration as a function of distance matches the concentration profile predicted by models of diffusion and elimination (Equation 3-58).

3. Antibodies—or antibody fragments—can now be produced by a number of mechanisms including immunized animals [68], mammalian cells [69], bacteria [70], and plants [71].

Recombinant DNA technology—and technologies for creating combinatorial libraries of antibody binding sites—have provided opportunities to modify and improve antibody properties. In cancer therapy, the targeting agent is used to direct cytotoxic compounds to tumor cells. Drugs, toxins, and radioactive isotopes have been targeted to tumors. The most successful applications of this concept have been in the treatment of lymphomas or leukemias, in which the tumor cells circulate through the blood. Solid tumors have proven more difficult to treat, presumably due to limited penetration of the targeted agent (which is usually a large molecule) into the solid tumor.

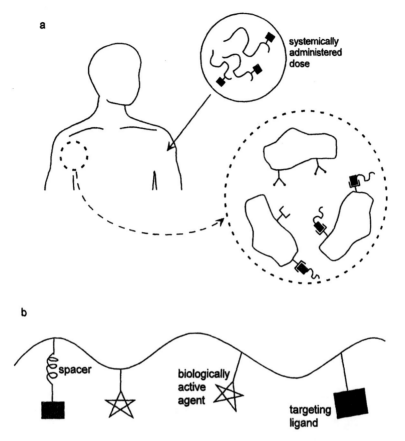

Figure 8.16 Illustration of drug targeting. The addition of ligand moieties to a drug molecule can make it selective for cells that express a complementary receptor.

Imagine a tumor that is bathed on its external surface with a cytotoxic agent that binds to a tumor antigen. To kill cells in the center of the tumor, the agent must penetrate through outer cell layers. For small tumor nodules (radius $\sim 150\,\mu$m), without any vascular system, penetration must occur by diffusion through the tumor interstitial space, which is slow (recall the discussion in Chapter 4 on diffusion through solid tissues). The targeted agent binds to tumor cells, enhancing retention at the tumor site, but further slowing the rate of penetration of agent into the tumor. In fact, under many circumstances, no targeted molecules will reach the center of the tumor nodule because they are bound to tumor cells at the periphery. A variety of models—formulated using the principles described in Chapter 3 and 4—have been developed to evaluate this phenomenon, which is sometimes called the "binding-site barrier" to drug penetration [72–74]. Recently these models have been extended to account for internalization and intracellular processing of the targeted agent [75–77].

8.4.2 Enhancing Interactions by Increasing Valence

Receptor clustering after binding of multivalent ligands occurs frequently in cultured cells and is probably an important mechanism for cell activation and signaling, particularly in the immune system. In addition, multivalent ligands have a greatly enhanced avidity for binding, due to the presence of multiple binding partners, which decreases the probability of ligand release. Therefore, it may be useful to create pharmaceuticals that are multivalent in order to enhance potency or binding properties.

Dextran has been used as a multivalent carrier by covalent coupling with drugs and proteins (Figure 8.8). Conjugation of weakly immunogenic molecules, such as BSA and other T-cell dependent antigens, to high M_w dextran (a T-cell independent carrier) stimulates high and persistent serum antibody response following intravenous injection [78]. Dose-dependent antibody titers were detected as early as 7 days, peaking at ~ 14 days, for 5×10^5 or 2×10^6 M_w dextran (70,000 M_w was not effective). Anti-hapten responses were also elicited by conjugation of haptens to the protein on the dextran–protein conjugate. A similar approach has been used to produce vaccines from poorly immunogenic antigens (such as meningococcal C polysaccharide) that are effective in stimulating immune responses in children [79].

The creation of multivalent ligands can also lead to new biological activity. As an example, consider the adhesion of neutrophils to the vasculature at sites of inflammation. Selectins, a class of protein receptors on the surface of neutrophils, mediate the binding of cells to counterligands that are transiently expressed on endothelial cells. Selectin-mediated binding is an important initial step in the recruitment of neutrophils to sites of inflammation, so this specific binding activity is a potential target for anti-inflammatory drugs. The ligands of L-selectin, for example, are glycosylated, mucin-like molecules, which possess repeated saccharide units, such as sialyl Le^x.

Multivalent ligands for L-selectin were produced by polymerization of saccharide analogs [80]. These ligands bind to the cell surface and cause the cells to lose a fraction of their L-selectin surface proteins (Figure 8.17); this loss of receptor does not occur after binding of monovalent ligands. Presumably, the multivalent binding induces receptor clustering, which activates some natural proteolytic mechanisms for L-selectin release from the cell surface. Because the multivalent ligand binds and produces soluble receptor fragments, these agents can potentially block cell adhesion in several ways: (1) by competitively binding cell adhesion receptors; (2) by producing soluble receptors that competitively bind the adhesive counterligand; and (3) by reducing the number of viable receptors on the cell surface (Figure 8.17). Many cell-surface proteins are released from the cell surface during biological activity, so this mechanism for enhancing the activity of a receptor-binding drug may have multiple clinical applications.

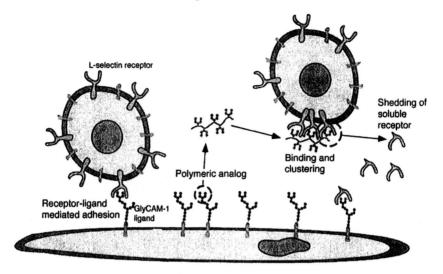

Figure 8.17 Ligands bound to polymer carriers can induce ligand shedding. Based on the work reported in [80].

SUMMARY

- Modification is a versatile means for creating novel agents with enhanced biological activity by virtue of their increased concentration, improved stability, or enhanced ability to enter cellular or tissue compartments.

- Solubility of agents in water and lipid can be manipulated by formulation (e.g., adding lipids) and by chemical modification.

- Stability of agents can be enhanced by making changes that reduce elimination or restrict the agent to certain compartments of the body.

- The ability of agents to permeate through membranes can be varied by modifications that effect diffusion in membranes, engage protein carriers in the membrane, or regulate overall size.

REFERENCES

1. Laubane, J.-P., in M. Rubinstein, *Handbook of Pharmacokinetics. Pharmaceutical Technology*, Chichester, England: Ellis Horwood, 1989.
2. Various, *Physicians' Desk Reference*. Montvale, NJ: Medical Economics Co, 1998.
3. Whalen, G.F., Y. Shing, and J. Folkman, The fate of intravenously administered bFGF and the effect of heparin. *Growth Factors*, 1989, **1**, 157–164.
4. Greenwald, R.B., *et al.*, Drug delivery systems: water soluble taxol 2'-poly(ethylene glycol) ester prodrugs-design and *in vivo* effectiveness. *Journal of Medical Chemistry*, 1996, **39**, 424–431.
5. Lasic, D., Liposomes. *American Scientist*, 1992, **80**, 20–31.

6. Bangham, A., M. Standish, and J. Watkins, Diffusion of univalent ions across the lamellae of swollen phospholipids. *Journal of Molecular Biology*, 1965, **13**, 238–252.

7. Szoka F. Jr. and D. Papahadjopoulos, Procedure for preparation of liposomes with large internal aqueous space and high capture by reverse-phase evaporation. *Proceedings of the National Academy of Sciences USA*, 1978, **75**, 4194–4198.

8. Lasic, D.D. and D. Papahadjopoulos, eds., *Medical Applications of Liposomes*, Amsterdam: Elsevier Science, 1998.

9. Goodman Gilman, A., *et al.*, eds., *The Pharmacological Basis of Therapeutics*. 8th ed. New York: Pergamon Press, 1990, 1811 pp.

10. Pratt, W.B., *et al.*, *The Anticancer Drugs*. 2nd ed. New York: Oxford University Press, 1994.

11. Flanagan, W., *et al.*, Cellular penetration and antisense activity by a phenoxazine-substituted heptanucleotide. *Nature Biotechnology*, 1999, **17**, 48–52.

12. Akhtar, S., R. Kole, and R.L. Juliano, Stability of antisense DNA oligodeoxynucleotide analogs in cellular extracts and sera. *Life Sciences*, 1991. **49**, 1793–1801.

13. Wickstrom, E., Oligodeoxynucleotide stability in subcellular extracts and culture media. *Journal of Biochemical and Biophysical Methods*, 1986, **13**, 97–102.

14. Ringsdorf, H., Structure and properties of pharmacologically active polymers. *Journal of Polymer Science, Polymer Symp.*, 1975, **51**, 135–153.

15. Abuchowski, A., *et al.*, Treatment of L5178Y tumor-bearing BDF1 mice with a nonimmunogenic L-glutaminase-L-asparaginase. *Cancer Treatment Reports*, 1979, **63**(6), 1127–1132.

16. Katre, N.V., M.J. Knauf, and W.J. Laird, Chemical modification of recombinant interleukin 2 by polyethylene glycol increases its potency in the murine Meth A sarcoma model. *Proceedings of the National Academy of Sciences USA*, 1987, **84**, 1487–1491.

17. Teppler, H., *et al.*, Prolonged immunostimulatory effect of low-dose polyethylene glycol interleukin 2 in patients with human immunodeficiency virus type 1 infection. *Journal of Experimental Medicine*, 1993, **177**, 483–492.

18. Duncan, R., Drug–polymer conjugates: potential for improved chemotherapy. *Anti-Cancer Drugs*, 1992, **3**, 175–210.

19. Duncan, R., *et al.*, Macromolecular prodrugs for use in targeted cancer chemotherapy: melphalan covalently coupled to *N*-(2-hydroxypropyl) methacrylamide copolymers. *Journal of Controlled Release*, 1991, **16**, 121–136.

20. Molteni, L., Dextrans as carriers, in G. Gregoriadis, *Drug Carriers in Biology and Medicine*, New York: Academic Press, 1979.

21. Dang, W.B., *et al.*, Covalent coupling of methotrexate to dextran enhances the penetration of cytotoxicity into a tissue-like matrix. *Cancer Research*, 1994, **54** 1729–1735.

22. Schacht, E., Polysaccharide macromolecules as drug carriers, in L. Illum and S.S. Davis, *Polymers in controlled drug delivery*, Bristol: Wright, 1987, pp. 131–187.

23. Hudecz, F., *et al.*, Influence of carrier on biodistribution and *in vitro* cytotoxicity of methotrexate-branched polypeptide conjugates. *Bioconjugate Chemistry*, 1993, **4**, 25–33.

24. Braatz, J.A., *et al.*, Functional peptide–polyurethane conjugates with extended circulatory half-lives. *Bioconjugate Chemistry*, 1993, 4, 262–267.

25. Hermanson, G.T., *Bioconjugate Techniques*. San Diego, CA: Academic Press, 1996.

26. Ouchi, T., H. Yuyama, and O. Vogl, Synthesis of poly(ethylene glycol)-bound 3-(5-fluorouracil-1-yl)propanoic acid, its hydrolysis reactivity and antitumor activity. *Makromol. Chem., Rapid Commun.*, 1985, **6**, 815–819.

27. Duncan, R., Polymer therapeutics for tumour specific delivery. *Chemistry and Industry*, 1997, 262–264.
28. Duncan, R., S. Dimitrijevic, and E.G. Evagorou, The role of polymer conjugates in the diagnosis and treatment of cancer. *S.T.P. Pharma Sciences*, 1996, **6**(4), 237–263.
29. Vasey, P., *et al.*, Phase I clinical and pharmacokinetic study of PK1: first member of a new class of chemotherapeutic agents—Drug-polymer conjugates. *Clinical Cancer Research*, 1999, **5**(1), 83–94.
30. Saiki, I., *et al.*, Antimetastatic activity of polymeric RGDT peptide conjugates with poly(ethylene glycol). *Japanese Journal of Cancer Research*, 1993, **84**, 558–565.
31. Gombotz, W.R., Pettit, D.K., Biodegradable polymers for protein and peptide drug delivery. *Bioconjugate Chemistry*, 1995, **6**, 332–351.
32. Ho, D.H., *et al.*, Clinical pharmacology of polyethylene glycol-L-asparaginase. *Drug Metabolism and Disposition*, 1986, **14**, 349–352.
33. Chen, R.H.-L., *et al.*, Properties of two urate oxidases modified by the covalent attachment of poly(ethylene glycol). *Biochimica et Biophysica Acta*, 1981, **660**, 293–298.
34. Davis, S., *et al.*, Alteration of the circulating life and antigenic properties of bovine adenosine deaminase in mice by attachment of polyethylene glycol. *Clin. Exp. Immunol.*, 1981, **46**, 649–652.
35. Hershfield, M.S., *et al.*, Treatment of adenosine deaminase deficiency with poly-ethylene glycol-modified adenosine deaminase. *New England Journal of Medicine*, 1987, **316**(10), 589–596.
36. Rhee, W., *et al.*, Bovine collagen modified by PEG, in J. M. Harris, *Poly(ethylene glycol) Chemistry: Biotechnical and Biomedical Applications*, New York: Plenum Press, 1992
37. Saltzman, W.M. and M.L. Radomsky, Drugs released from polymers: diffusion and elimination in brain tissue. *Chemical Engineering Science*, 1991, **46**, 2429–2444.
38. Krewson, C.E., M. Klarman, and W.M. Saltzman, Distribution of nerve growth factor following direct delivery to brain interstitium. *Brain Research*, 1995, **680**, 196–206.
39. Mahoney, M.J. and W.M. Saltzman, Millimeter-scale positioning of a nerve-growth-factor source and biological activity in the brain. *Proceedings of the National Academy of Sciences USA*, 1999, **96**, 4536–4539.
40. Krewson, C.E., *et al.*, Stabilization of nerve growth factor in polymers and in tissues. *Journal of Biomaterials Science*, 1996, **8**, 103–117.
41. Cuatercasas, P., Interaction of insulin with the cell membrane: the primary action of insulin. *Proceedings of the National Academy of Science USA*, 1969, **63**, 450–457.
42. Frazier, W.A., L.F. Boyd, and R.A. Bradshaw, Interaction of nerve growth factor with surface membranes: Biological competence of insoluble nerve growth factor. *Proceedings of the National Academy of Science USA*, 1973, **79**(10), 2931–2935.
43. Ito, Y., S.Q. Liu, and Y. Imanishi, Enhancement of cell growth on growth factor-immobilized polymer film. *Biomaterials*, 1991, **12**, 449–453.
44. Ito, Y., *et al.*, Patterned artificial juxtacrine stimulation of the cells by covalently immobilized insulin. *FEBS Letters*, 1997, **403**, 159–162.
45. Ito, Y., *et al.*, Cell growth on immobilized cell growth factor. 6. Enhancement of fibroblast cell growth by immobilized insulin and/or fibronectin. *Journal of Biomedical Materials Research*, 1993, **27**, 901–907.
46. Kuhl, P.R. and L.G. Griffith-Cima, Tethered epidermal growth factor as a paradigm for growth factor-induced stimulation from the solid phase. *Nature Medicine*, 1996, **2**(9), 1022–1027.

47. Reddy, C.C., A. Wells, and D.A. Lauffenburger, Proliferative response of fibro-blasts expressing internalization-deficient epidermal growth factor (EGF) receptor is altered via differential EGF depletion effect. *Biotechnological Progress*, 1994, **10**, 377–384.

48. Kandell, E.R., J.H. Schwartz, and T. Jessell, M., eds., *Principles of Neural Science*. 3rd ed. Norwalk, CT: Appleton & Lange, 1991.

49. Egleton, R.D., *et al.*, Transport of opioid peptides into the central nervous system. *Journal of Pharmaceutical Sciences*, 1998, **87**(11), 1433–1439.

50. Tsuji, A., Peptide delivery across the blood–brain barrier, in V.H.L. Lee, M. Hashida, and Y. Mizushima, *Trends and Future Perspectives in Peptide and Protein Drug Delivery*, Amsterdam: Harwood Academic, 1995, pp. 153–171.

51. Uchiyama, T., *et al.*, Effects of various protease inhibitors on the stability and permeability of [D-Ala, D-Leu]enkephalin in the rat intestine: comparison with leu-cine enkephalin. *Journal of Pharmaceutical Sciences*, 1998, **87**(4), 448–452.

52. Brownson, E.A., *et al.*, Effect of peptidases at the blood–brain barrier on the per-meability of enkephalin. *Journal of Pharmacology and Experimental Therapeutics*, 1994, **270**(2), 675–680.

53. Felgner, P., *et al.*, Lipofection: a highly efficient, lipid-mediated DNA-transfection procedure. *Proceedings of the National Academy of Sciences*, 1987, **84**, 7413–7417.

54. Radler, J.O., *et al.*, Structure of DNA-cationic liposome complexes: DNA interca-lation in multilamellar membranes in distinct interhelical packing regimes. *Science*, 1997, **275**, 810–814.

55. Luo, D. and W. Saltzman, Synthetic DNA delivery systems. *Nature Biotechnology*, 2000, **18**, 33–37.

56. Pardridge, W.M., ed., *The Blood–Brain Barrier: Cellular and Molecular Biology*. New York: Raven Press, 1993.

57. Davson, H. and M.B. Segal, eds., *Physiology of the CSF and Blood-Brain Barrier*. Boca Raton, FL: CRC Press, 1996.

58. Nutt, J.G. and J.H. Fellman, Pharmacokinetics of levodopa. *Clinical Neuropharmacology*, 1984, **7**, 35–49.

59. Friden, P., *et al.*, Anti-transferrin receptor antibody and antibody–drug conjugates cross the blood–brain barrier. *Proceedings of the National Academy of Sciences USA*, 1991, **88**, 4771–4775.

60. Pardridge, W.M., Y.-S. Kang, and J.L. Buciak, Transport of human recombinant brain-derived neurotrophic factor (BDNF) through the rat blood–brain barrier *in vivo* using vector-mediated peptide drug delivery. *Pharmaceutical Research*, 1994, **11**(5), 738–746.

61. Triguero, D., *et al.*, Blood–brain barrier transport of cationized immunoglobulin G: enhanced delivery compared to native protein. *Proceedings of the National Academy of Sciences USA*, 1989, **86**, 4761–4765.

62. Vives, E., P. Brodin, and B. Lebleu, A truncated HIV-1 Tat protein basic domain rapidly translocates through the plasma membrane and accumulates in the cell nucleus. *Journal of Biological Chemistry*, 1997, **272**(25), 16010–16017.

63. Pooga, M., *et al.*, Cell penetrating PNA constructs regulate galanin receptor levels and modify pain transmissions *in vivo*. *Nature Biotechnology*, 1998, **16**, 857–861.

64. Oehlke, J., *et al.*, Cellular uptake of an alpha-helical amphipathic model peptide with the potential to deliver polar compounds into the cell interior non-endocyti-cally. *Biochimica et Biophysica Acta*, 1998, **1414**, 127–139.

65. Derossi, D., *et al.*, Cell internalization of the third helix of the antennapedia home-odomain is receptor-independent. *Journal of Biological Chemistry*, 1996, **271**(30), 18188–18193.

66. Josephson, L., *et al.*, High-efficiency intracellular magnetic labeling with novel superparamagnetic-Tat peptide conjugates. *Bioconjugate Chemistry*, 1999, **10**, 186–191.

67. Ehrlich, P., On immunity with special reference to cell life. *Proceedings of the Royal Society (London)*, 1900, **66**, 424–448.

68. von Behring, E. and S. Kitasato, On the acquisition of immunity against diphtheria and tetanus in animals. *Deutsch. Med. Wochenschr.*, 1890, **16**, 1113.

69. Kohler, G. and C. Milstein, Continuous cultures of fused cells secreting antibody of predefined specificity. *Nature*, 1975, **256**, 495–497.

70. Huse, W.D., *et al.*, Generation of a large combinatorial library of the immunoglobulin repertoire in phage lambda. *Science*, 1989, **246**, 1275–1281.

71. Ma, J.K.-C., *et al.*, Generation and assembly of secretory antibodies in plants. *Science*, 1995, **268**, 716–719.

72. van Osdol, W., K. Fujimori, and J. Weinstein, An analysis of monoclonal antibody distribution in microscopic tumor nodules: consequences of a "binding site barrier". *Cancer Research*, 1991, **51**, 4776–4784.

73. Sung, C., *et al.*, The spatial distribution of immunotoxins in solid tumors: assessment by quantitative autoradiography. *Cancer Research*, 1993, **53**, 2092–2099.

74. Sung, C., *et al.*, Streptavidin distribution in metastatic tumors pretargeted with a biotinylated monoclonal antibody: theoretical and experimental pharmacokinetics. *Cancer Research*, 1994, **54**, 2166–2175.

75. Chu, L., H.S. Wiley, and D.A. Lauffenburger, Endocytic relay as a potential means for enhancing ligand transport. *Tissue Engineering*, 1996, **2**, 17–38.

76. Lauffenburger, D.A., *et al.*, Engineering dynamics of growth factors and other therapeutic ligands. *Biotechnology and Bioengineering*, 1996, **52**, 61–80.

77. Rippley, R.K., R.C. Wilson, and C.L. Stokes, Effect of cellular pharmacology on drug distribution in tissues, *Biophysical Journal*, 1995, **69**, 825–839.

78. Lees, A., *et al.*, Enhanced immunogenicity of protein-dextran conjugates: I. Rapid stimulation of enhanced antibody responses to poorly immunogenic molecules. *Vaccine*, 1994, **12**, 1160–1166.

79. MacDonald, N., *et al.*, Induction of immunologic memory by conjugated vs plain meningococcal C polysaccharide vaccine in toddlers. *JAMA*, 1998. **280**(19), 1685–1689.

80. Gordon, E.J., W.J. Sanders, and L.L. Liessling, Synthetic ligands point to cell surface strategies. *Nature*, 1998, **392**, 30–31.

81. Blasberg, R., C. Patlak, and J. Fenstermacher, Intrathecal chemotherapy: brain tissue profiles after ventriculocisternal perfusion. *Journal of Pharmacology and Experimental Therapeutics*, 1975, **195**, 73–83.

82. Anderson, K.D., *et al.*, Distribution of exogenous BDNF and NGF delivered into the brain. *Society for Neuroscience Abstracts*, 1993, **19**, 662.

83. Rosenberg, G., W. Kyner, and E. Estrada, Bulk flow of brain interstitial fluid under normal and hyperosmolar conditions. *American Journal of Physiology*, 1980, **238**, F42–F49.

84. Curran, R.E., *et al.*, Cerebrospinal fluid production rates determined by simultaneous albumin and inulin perfusion. *Experimental Neurology*, 1970, **29**, 546–553.

85. Morrison, P. and R.L. Dedrick, Transport of cisplatin in rat brain following micro-infusion: an analysis. *Journal of Pharmaceutical Sciences*, 1986, **75**, 120–128.

86. Lee Sendelbeck, S. and J. Urquhart, Spatial distribution of dopamine, methotrexate and antipyrine during continuous intracerebral microperfusion. *Brain Research*, 1985, **328**, 251–258.

87. Kasamatsu, T., T. Itakura, and G. Jonsson, Intracortical spread of exogenous catecholamines: effective concentration for modifying cortical plasticity. *Journal of Pharmacology and Experimental Therapeutics*, 1981, **217**, 841–850.

88. Dykstra, K.H., *et al.*, Quantitative examination of tissue concentration profiles associated with microdialysis. *Journal of Neurochemistry*, 1992, **58**, 931–940.

89. Salehi-Had, S. and W.M. Saltzman, Controlled intracranial delivery of antibodies in the rat, in R. Langer and J. Cleland, *Protein Formulations and Delivery*, Washington, DC: ACS Symposium Series, 1994, pp. 278–291.

90. Strasser, J.F., *et al.*, Distribution of 1,3-bis(2-chloroethyl)-1-nitrosourea (BCNU) and tracers in the rabbit brain following interstitial delivery by biodegradable polymer implants. *Journal of Pharmacology and Experimental Therapeutics*, 1995, **275**(3), 1647–1655.

91. Walter, K.A., et al., Interstitial taxol delivered from a biodegradable polymer implant against experimental malignant glioma. *Cancer Research*, 1994, **54**, 2207–2212.

9

Controlled Drug Delivery Systems

O the joy of my soul leaning pois'd on itself—receiving identity through materials, and loving them.

Walt Whitman, *Poem of Joys*

In most forms of drug delivery, spatial localization and duration of drug concentration are constrained by organ physiology and metabolism. For example, drugs administered orally will distribute to tissues based on the principles of diffusion, permeation, and flow presented in Part II of this book. If the duration of therapy provided by a single administration is insufficient, the drug must be readministered. Localization of drug can be controlled by injection, but only within limited spatial constraints, and effectiveness after an injection is usually short-lived.

Controlled-delivery systems offer an alternative approach to regulating both the duration and spatial localization of therapeutic agents. In controlled delivery, the active agent is combined with other (usually synthetic) components to produce a delivery system. Unlike drug modification, which results in new agents that are single molecules (Figure 8.2), or assemblies of a limited number of molecules (Figure 8.4), drug delivery systems are usually macroscopic. Like drug modification, controlled-delivery systems frequently involve combinations of active agents with inert polymeric materials.

In this text, controlled-delivery systems are distinguished from "sustained-release" drug formulations. Sustained release is often achieved by mixing an active agent with excipients or binders that alter the agent's rate of dissolution in the intestinal tract or adsorption from a local injection site. The distinction between sustained release (often achieved by drug formulation) and controlled delivery or controlled release is somewhat arbitrary. In our definition, controlled delivery systems must (1) include a component that can be engineered to regulate an essential characteristic (e.g., duration of release, rate of release, or targeting) and (2) have a duration of action longer than a day.

9.1 RESERVOIR AND TRANSDERMAL DELIVERY SYSTEMS

Many polymeric materials are available for the development of drug delivery systems (see Appendix A). Non-degradable, hydrophobic polymers have been used the most extensively. Reservoir drug delivery devices, in which a liquid reservoir of drug is enclosed in a silicone elastomer tube, were first demonstrated to provide controlled release of small molecules several decades ago [1]. This discovery eventually led to clinically useful devices, including the Norplant® (Wyeth-Ayerst Laboratories) contraceptive delivery system, which provides reliable delivery of levonorgestrel for 5 years following subcutaneous implantation. Norplant® has been available to women in the United States since 1990, after use by millions of women in other countries, and it has been generally well received [2]. Other polymers, most notably poly[ethylene-*co*-(vinyl acetate)] (EVAc), have been used to control the delivery of contraceptive hormones to the female reproductive tract (Progestasert®, Alza Corp.) and lipophilic drugs to the eye or the skin (Ocusert®, Alza Corp., Estraderm® and Transderm Nitro®, Ciba-Geigy Corp., see Figure 9.1). The reservoir configuration offers a number of potential advantages, including the possibility of long service life (because large quantities of drug can be stored in the reservoir) and nearly constant release rates, as described below.

9.1.1 Diffusion through Planar Membranes

In most reservoir and transdermal systems (Figure 9.2), release of drug from the reservoir into the external solution occurs in three steps: (1) dissolution of the drug in the polymer, (2) diffusion of drug across the polymer membrane, and (3) dissolution of the drug into the external phase. Assuming that the rate of diffusion across the membrane is much slower than the rate of dissolution/

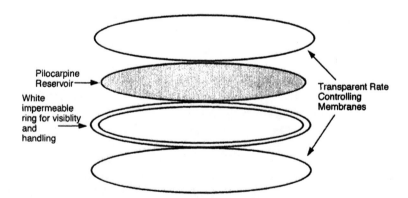

Figure 9.1   Schematic diagram of the Ocusert reservoir delivery system. The figure shows an exploded view of the different elements of the Ocusert system. Two rate-controlled membranes composed of poly[ethylene-*co*-(vinyl acetate)] enclose a drug reservoir. An opaque ring is added for ease of handling and visibility.

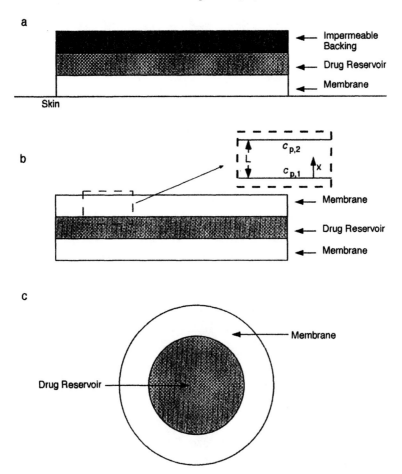

Figure 9.2 Reservoir delivery systems based on rate-limiting polymer membranes. Rate-limiting polymer membranes can be used to produce several different types of drug delivery devices including (a) transdermal delivery systems, (b) planar controlled-release systems, and (c) cylindrical controlled-release systems.

partitioning at either interface, so that partitioning can be assumed to be at equilibrium, the kinetics of release can be modeled by considering a mass balance on the drug within the polymer membrane (Figure 9.2b). For a differential control volume in the membrane, Δx, a mass balance on the diffusing drug molecule A yields (recall Equation 3-32):

$$\frac{\partial c_p}{\partial t} = D_{i:p} \frac{\partial^2 c_p}{\partial x^2} \tag{9-1}$$

where $D_{i:p}$ is the diffusion coefficient for the drug within the polymer material and c_p is the concentration of the drug (mg/mL) within the polymer. Equation 9-1 assumes that there is no bulk flow or generation/consumption of the drug within the polymer. For a drug diffusing within a polymeric material, this first assumption also implies that the density of the material is not changing with time (and therefore not changing with drug concentration). This usually means that the drug is always present at low concentrations (i.e., drug is dilute) within the material.

To calculate concentrations within the membrane as a function of time, Equation 9-1 can be solved subject to the following conditions (see Figure 9.2b):

$$
\begin{aligned}
c_p &= c_{p,0} & t = 0 \quad & 0 < x < L \\
c_p &= c_{p,1} & t > 0 \quad & x = 0 \\
c_p &= c_{p,,2} & t > 0 \quad & x = L
\end{aligned}
\tag{9-2}
$$

where $c_{p,0}$ is the initial concentration of drug within the polymer, which is assumed to be uniformly distributed throughout the membrane, and $c_{p,1}$ and $c_{p,2}$ are the concentrations within the polymer at the surfaces. Equations 9-1 and 9-2 yield:

$$
c_p = c_{p,1} + (c_{p,2} - c_{p,1})\frac{x}{L} + \frac{2}{\pi}\sum_{n=1}^{\infty}\frac{c_{p,2}\cos(n\pi) - c_{p,1}}{n}\sin\left(\frac{n\pi x}{L}\right)\exp\left(-\frac{D_{i:p}n^2\pi^2 t}{L^2}\right)
$$

$$
+ \frac{4c_{p,0}}{\pi}\sum_{m=0}^{\infty}\frac{1}{2m+1}\sin\left(\frac{(2m+1)\pi x}{L}\right)\exp\left(-\frac{D_{i:p}(2m+1)^2\pi^2 t}{L^2}\right)
$$

$$
\tag{9-3}
$$

which provides the concentration of drug in the polymer as a function of position and time, given two simple characteristics of the system (i.e., the thickness of the membranes, L, and the diffusion coefficient of the drug in the polymer, $D_{i:p}$).

The total amount of material leaving the membrane, M_t, can be determined by calculating the flux at the external surface and integrating with respect to time:

$$
M_t = A\int_0^t -D_{i:p}\left(\frac{\partial c_p}{\partial x}\right)_{x=L} dt
$$

$$
= -AD_{i:p}\left\{
\begin{array}{l}
\dfrac{(c_{p,2} - c_{p,1})}{L} + \dfrac{2L}{D_{i:p}\pi^2}\sum_{n=1}^{\infty}(c_{p,2}\cos n\pi - c_{(p,1)})\dfrac{\cos n\pi}{n^2} \\[2ex]
\left[1 - \exp\left(-\dfrac{D_{i:p}n^2\pi^2 t}{L^2}\right)\right] \\[2ex]
+ \dfrac{4c_{p,0}L}{D_{i:p}\pi^2}\sum_{m=0}^{\infty}\dfrac{\cos(2m+1)\pi}{(2m+1)^2}\left[1 - \exp\left(-\dfrac{D_{i:p}(2m+1)^2\pi^2 t}{L^2}\right)\right]
\end{array}
\right\}
$$

$$
\tag{9-4}
$$

where A is the exposed area of the external surface of the device, which is equal to the cross-sectional area for the transdermal system (Figure 9.2a) and twice the cross-sectional area for the planar reservoir system (Figure 9.2b).

When the initial concentration within the membrane, $c_{p,0}$, and the concentration at the external surface of the membrane, $c_{p,2}$, are both equal to zero, Equation 9-4 can be simplified considerably:

$$M_t = Ac_{p,1}L\left\{\frac{D_{i:p}t}{L^2} - \frac{1}{6} - \frac{2}{\pi^2}\sum_{n=1}^{\infty}\frac{(-1)^n}{n^2}\exp\left(-\frac{D_{i:p}n^2\pi^2t}{L^2}\right)\right\} \qquad (9\text{-}5)$$

Typical profiles of cumulative mass released are provided in Figure 9.3. When t is sufficiently large, Equation 9-5 becomes:

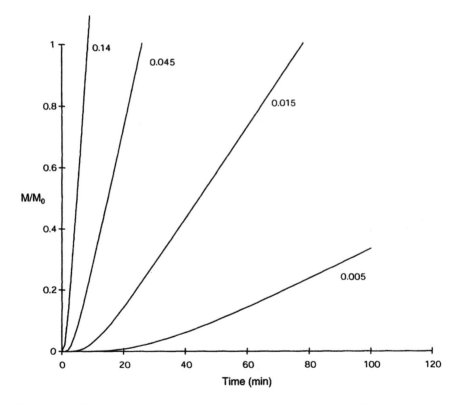

Figure 9.3   Drug release from a planar membrane-reservoir drug delivery system. The cumulative mass of drug released is plotted versus time for planar reservoir devices with a range of physical characteristics, which are determined principally by the diffusion coefficient for the drug in the polymer and the membrane thickness. The cumulative mass released on the y-axis is scaled by $M_0 = Ac_{p,1}L$ and each of the separate curves represents the normalized mass released at a particular value of $D_{i:p}/L^2$ (in min^{-1}). If $D_{i:p}$ is equal to 1×10^{-8} cm^2/s, then these curves correspond to thicknesses (L) of 20, 40, 60, and 120 μm (from left to right).

$$M_t = \frac{AD_{i:p}c_{p,1}}{L}\left(t - \frac{L^2}{6D_{i:p}}\right) \tag{9-6}$$

In situations where the concentrations at each surface of the polymer, $c_{p,1}$ and $c_{p,2}$, are maintained for a sufficient period, the drug diffusion will achieve steady state, which can be identified by examining the limit of Equation 9-3 as $t \rightarrow \infty$:

$$c_p = c_{p,1} + (c_{p,2} - c_{p,1})\frac{x}{L} \tag{9-7}$$

At steady state, the rate of drug release from the membrane is:

$$\frac{dM_t}{dt} = -AD_{i:p}\frac{(c_{p,1} - c_{p,2})}{L} \tag{9-8}$$

where M_t is the cumulative mass of drug released from the device—which is equal to J, the mass flux of drug from the surface of the polymer, times A, the external area available for transport (which must include the two identical free surfaces of a reservoir device). To use this expression for the description of real situations, it is necessary to relate the concentrations within the polymer membrane—i.e., $c_{p,1}$ and $c_{p,2}$—to concentrations in the adjacent phases. This is most commonly accomplished by assuming that the material on either side of the membrane interface is at equilibrium. For example, $K_{p:r}$ and $K_{p:w}$ are the equilibrium partition coefficients for the drug, which are defined:

$$K_{p:r} = \left[\frac{c_p}{c_{\text{reservoir fluid}}}\right]_{\text{equilibrium}} \qquad K_{p:w} = \left[\frac{c_p}{c_{\text{water}}}\right]_{\text{equilibrium}} \tag{9-9}$$

With these partition coefficients, the rate of release from the membrane can be written as:

$$\frac{dM_t}{dt} = -AD_{i:p}\frac{(K_{p:r}c_r - K_{p:w}c_w)}{L} \tag{9-10}$$

Table 9.1 lists diffusion coefficients and partition coefficients for some polymer/drug combinations.

When a transdermal device is applied to the patient's skin, the agent is released from the device, partitioning into the skin and diffusing through the stratum corneum—eventually reaching the capillaries. Drug concentration rises above the baseline level in the plasma, and is maintained at a nearly constant elevated level for the duration of release (Figure 9.4).

9.1.2 Diffusion through Cylindrical Membranes

For implantable reservoir devices, a cylindrical geometry is often more practical than a planar arrangement. Consider a cylindrical reservoir surrounded by a polymeric membrane (Figure 9.2c); the cylinder has a length L, cross-sectional radius b, and wall thickness $b - a$. In practice, the rate of drug release from this cylinder can be modified by changing the geometry of the device (by

Table 9.1 Diffusion coefficients and partition coefficients for some typical polymer/drug combinations

Solute (i)	M_w	$D_{i:p}$	$K_{p:w}$
Polyethylene			
Acetophenone	120	3.55×10^{-8}	3.16
Benzaldehyde	106	3.39×10^{-8}	3.74
Benzoic acid	122	5.29×10^{-10}	6.25
4-Methylacetophenone	134	1.79×10^{-8}	12.5
4-Methylbenzaldehyde	120	1.37×10^{-8}	25.6
Silicone rubber			
p-Aminoacetophenon	135	2.44×10^{-6}	
Androstenedione	286	1.48×10^{-6}	7.4
Chlormadione acetate	405	3.03×10^{-7}	82
Delmadinone acetate	403	3.80×10^{-8}	140
Deoxycorticosterone acetate	373	4.94×10^{-7}	
Estrone	270	2.40×10^{-7}	8
Ethyl-p-aminobenzoate	165	2.67×10^{-6}	0.97
Ethyl-p-aminobenzoate	165	1.78×10^{-6}	
Ethynodiol diacetate	385	3.79×10^{-7}	
Hydrocortisone	362	4.50×10^{-7}	0.05
17a-Hydroxyprogesterone	330	5.65×10^{-7}	0.89
Medroxyprogesterone acetate	387	4.17×10^{-7}	26.9
6a-Methyl-11b-hydroxyprogesterone	345	2.84×10^{-7}	
Norprogesterone	300	1.85×10^{-6}	22.4

Solute (i)	M_w	$D_{i:p}$	$K_{p:w}$
Poly(ethylene-co-vinyl acetate)			
Androstenedione	286	5.50×10^{-9}	
Adrenosterone	300	5.70×10^{-9}	
Corticosterone	346	5.70×10^{-9}	9.79
Cortisone	360	2.30×10^{-9}	1.71
11-Dehydrocorticosterone	344	3.60×10^{-9}	
11-Deoxycorticosterone	330	4.90×10^{-9}	
11-Deoxy-17-hydroxycorticosterone	346	4.70×10^{-9}	
11-Hydroxyandrostenedione	302	3.80×10^{-9}	
Hydrocortisone	362	2.80×10^{-9}	
L-Methadone			7,800
Testosterone			165
Poly(methyl methacrylate)			
Fluphenazine	438	1.74×10^{-17}	
Fluphenazine enanthate	551	1.12×10^{-17}	
Hydroquinone	110	5.75×10^{-15}	
4-Nitroaniline	138	3.02×10^{-15}	
Procaine	236	1.35×10^{-15}	
Promethazine	284	1.41×10^{-17}	
Salicylic acid	138	9.55×10^{-15}	
Trinitrophenol	229	3.55×10^{-16}	

continued

Table 9.1 *Continued*

Solute (*i*)	M_w	$D_{i;p}$	$K_{p;w}$	Solute (*i*)	M_w	$D_{i;p}$	$K_{p;w}$
Progesterone	314	5.78×10^{-7}	45	*Poly(vinyl acetate)*			
Progesterone	314	6.40×10^{-7}		Fluphenazine	438	1.05×10^{-12}	
Pyrimethamine	249	1.10×10^{-10}		Fluphenazine enanthate	551	1.82×10^{-12}	
Poly(ether urethane)				Hydrocortisone	362	4.31×10^{-12}	
Androstenedione	286	9.00×10^{-9}		4-Nitroaniline	138	3.02×10^{-11}	
Adrenosterone	300	8.90×10^{-9}		Procaine	236	1.45×10^{-11}	
Corticosterone	346	6.30×10^{-9}		Promethazine	284	1.45×10^{-12}	
Cortisone	360	5.80×10^{-9}		Salicylic acid	138	4.37×10^{-11}	
11-Dehydrocorticosterone	344	5.20×10^{-9}		Trinitrophenol	229	7.59×10^{-12}	
11-Deoxycorticosterone	330	5.40×10^{-9}					
11-Deoxy-17-hydrocorticosterone	346	5.60×10^{-9}					
Estriol	288	2.00×10^{-9}	133				
11-Hydroxyandrostenedione	302	8.10×10^{-9}					
Hydrocortisone	362	4.80×10^{-9}					

Adapted from [3].

a

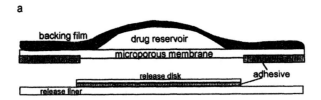

b

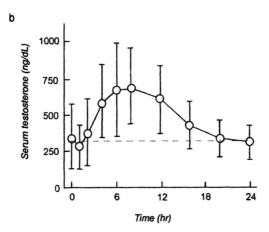

Time (hr)

Figure 9.4 Serum concentrations after placement of a transdermal device.
(a) Schematic diagram of the transdermal testosterone-releasing system. To initiate
therapy, a protective backing layer (composed of the release liner and disk) is
removed and the device is applied to the skin with the microporous membrane con-
tacting the external surface of the skin. (b) A transdermal patch that releases testosterone was applied and serum concentrations were monitored over the subsequent
24 h. Serum concentration of testosterone increases during the first few hours and
then remains nearly constant for the remainder of the 24-h measurement period.
Data from [4] for Androderm® (SmithKline Beecham).

changing b, b/a, or L) or by changing the drug/polymer combination (which
changes K and $D_{i:p}$).

As in the planar membrane, the concentration of drug in the cylinder wall
can be found by solving the diffusion equation, now written in cylindrical
coordinates:

$$\frac{\partial c}{\partial t} = D_{i:p} \frac{1}{r}\left(\frac{\partial}{\partial r} r \frac{\partial c}{\partial r}\right) \tag{9-11}$$

If the inside of the cylinder is maintained at a constant concentration of drug,
$c = c_1$ at $r = a$, and the outside of the cylinder is free of drug, $c = 0$ at $r = b$, and
the cylinder wall is initially saturated with drug, $c = c_1$ at $a < r < b$, then
Equation 9-11 can be solved analytically to obtain c as a function of position
in the cylinder wall (analogous to Equation 9-3) [5]. The total mass of drug

released at time t, M_t, is found by (1) calculating the flux from the surface of the ring via Fick's law, $J_r(b) = -D_{i:p}(\partial c/\partial r)_{r=b}$; (2) multiplying the flux by the total surface area available for release, $2\pi bL$; and (3) integrating with respect to time, t, to give:

$$\frac{M_t}{2\pi c_1 L} = \frac{D_{i:p}t}{\ln(b/a)} - 2\sum_{n=1}^{\infty}\left(\frac{J_0(a\alpha_n)}{J_0(a\alpha_n) - J_0(b\alpha_n)} + \frac{J_0(a\alpha_n)J_0(b\alpha_n)}{j_0^2(a\alpha_n) - J_0^2(b\alpha_n)}\right)$$
$$\left(e^{-D_{i:p}\alpha_n^2 t} - 1\right)$$

(9-12)

where $U_0(r\alpha_n) = J_0(r\alpha_n)\,Y_0(b\alpha_n) - J_0(b\alpha_n)\,Y_0(r\alpha_n)$, α_n are roots of $U_0(b\alpha_n) = 0$, and $J_0(\cdot)$ and $Y_0(\cdot)$ are Bessel functions of the first and second kind [6].

Since the drug must dissolve in the polymer before diffusing through the cylinder wall, c_1 is equal to the concentration of drug in the core of the cylinder times the partition coefficient. In the case of most interest, where the concentration of drug in the interior of the polymer is very high, c_1 will be the solubility of the drug in the polymer. Equation 9-12 can be used to predict the release characteristics for drug-containing cylinders with a range of shapes (by changing L, a, and b) and polymer/solute combinations (by changing $D_{i:p}$) (Figure 9.5). The rate of release can also be modified by changing c_1; those calculations are not shown because M_t varies linearly with c_1, so the effect can

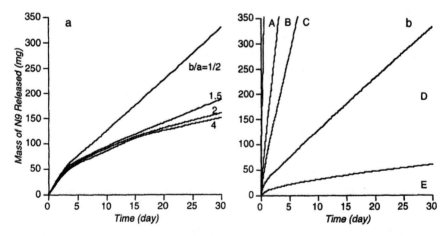

Figure 9.5   Drug release from a cylindrical-reservoir drug delivery system. The cumulative mass of drug released is plotted versus time for cylindrical-reservoir devices with a range of physical characteristics, which are determined principally by the diffusion coefficient for the drug in the polymer and the membrane thickness. In all cases, the overall length of the device, L, and the cross-sectional radius, b, were fixed at 2.7 and 0.5 cm, respectively. In each panel, one of the relevant design parameters was varied: (a) b/a was varied between 1.2 and 4 with $D_{i:p} = 1 \times 10^{-8}$ cm^2/s and $c_1 = 20$ mg/mL; (b) $D_{i:p}$ was varied (for curve A, $D_{i:p} = 5 \times 10^{-7}$ cm^2/s; B, 1×10^{-7} cm^2/s; C, 5×10^{-8} cm^2/s; D, 1×10^{-8} cm^2/s; E, 1×10^{-9} cm^2/s) with $a/b = 1.2$ and $c_1 = 20$ mg/mL. From [7].

be easily obtained from Figure 9.5. In most cases the concentration of drug will be high in the core of the device, so that c_1 is near the solubility of drug in the polymer, and a constant rate of drug release is maintained.

When observations are made after sufficient time following immersion of the cylinder in water, so that steady state can be assumed, Equation 9-12 reduces to:

$$M_t = \frac{2\pi c_1 L D_{i:p} t}{\ln(b/a)} \tag{9-13}$$

Equation 9-13 is useful for the design of drug-releasing devices and is analagous to Equation 9-6 for planar devices. Reservoirs devices are conceptually simple: drug release from the device is controlled primarily by controlling the thickness and composition of the membrane surrounding the drug reservoir. Therefore, the devices are easy to design and alternate release rates are simple to obtain.

9.2 MATRIX DELIVERY SYSTEMS

For some therapeutic agents, it is not possible to find membrane materials that provide adequate permeability to permit release from a reservoir device. Proteins, for example, do not diffuse readily through any of the hydrophobic, biocompatible[1] polymers that are commonly used for implantable reservoir systems (Table 9.1), or they diffuse very slowly ($D_{i:p} < 10^{-13}$ cm^2/s)[2]. In addition, reservoir devices are structurally complex, requiring several manufacturing steps, with resulting additional expense. For this combination of reasons, matrix systems for the delivery of agents (particularly protein or large molecular weight drugs) have been examined in considerable detail.

In a matrix system, the drug molecules are dissolved or dispersed throughout a solid polymer phase. In many cases, the polymer materials are the same as those used for the rate-limiting membrane in reservoir devices: silicone elastomers or EVAc. Alternatively, a variety of new materials have been developed for use in matrix drug delivery systems, particularly biodegradable polymers that slowly dissolve. Degradable materials disappear after implantation, an important advantage for use in patients. In addition, by careful design of the material and the device, it is sometimes possible to design delivery systems in which the rate of polymer degradation and dissolution controls the rate of drug delivery, providing a new element for controlling the rate of release of dispersed or dissolved drugs.

---

1. See footnote in Chapter 1 for the definition of biocompatible.
2. With this diffusion coefficient, assuming a 100-μm thick membrane, the characteristic diffusion time is $\sim 10^5$ days ($t = L^+2/2D$).

9.2.1 Matrix Delivery Systems with Dissolved Drugs

In a matrix drug delivery system, molecules of drug are dissolved in a biocompatible polymer, producing a homogeneous device with drug molecules uniformly dispersed throughout the material (Figure 9.6). In this case, the drug molecules are released by diffusing through the polymer to the surface of the device, from which they are released into the external environment.

Desorption of Dispersed Drug from a Slab. The release of a drug that is initially dissolved within a polymer matrix can be predicted by solving the equations for drug diffusion within the polymer slab:

$$\frac{\partial c}{\partial t} = D_{i:p} \frac{\partial^2 c}{\partial x^2} \tag{9-14}$$

where c is the concentration of the drug in the polymer and $D_{i:p}$ again represents the diffusion coefficient for molecules dissolved in the polymer. This

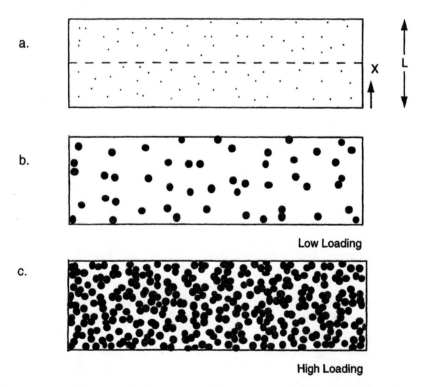

Figure 9.6 Schematic of matrix-type systems for controlled drug delivery. Matrix delivery systems can be constructed with drug dissolved in the matrix material (a) or particles of drug dispersed to form a composite material (b and c). For dispersed drug particles, the overall loading is an important determinant of the dynamics of release.

equation can be solved subject to the boundary and initial conditions (Figure 9.6):

$$c = c_0 \quad t = 0 \quad 0 < x < L$$
$$c = c_{ext} \quad t > 0 \quad x = 0, L \qquad (9\text{-}15)$$
$$\frac{\partial c}{\partial x} = 0 \quad t > 0 \quad x = L/2$$

where c_{ext} is the drug concentration in the external reservoir. The differential equation can be solved to yield:

$$\frac{c - c_{ext}}{c_0 - c_{ext}} = \frac{4}{\pi} \sum_{n=0}^{\infty} \frac{1}{2n+1} \exp\left(-\frac{D_{i:p}(2n+1)^2 \pi^2 t}{L^2}\right) \sin\left(\frac{(2n+1)\pi x}{L}\right) \qquad (9\text{-}16)$$

The total amount of drug released from the matrix can be determined by integration:

$$M_t = c_0 AL - \int_0^L c(x, t) A dx \qquad (9\text{-}17)$$

where the total amount of drug initially within the matrix is $M_\infty = c_0 AL$. Integration results in the following expression:

$$\frac{M_t}{M_\infty} = 1 - \frac{8}{\pi^2} \sum \frac{1}{(2n+1)^2} \exp\left\{-\frac{D_{i:p}(2n+1)^2 \pi^2 t}{L^2}\right\} \qquad (9\text{-}18)$$

For the early stages of release ($M_t < 0.6 M_\infty$, this expression is closely approximated by:

$$\frac{M_t}{M_\infty} \sim 4\sqrt{\frac{D_{i:p} t}{L^2 \pi}} \qquad (9\text{-}19)$$

Figure 9.7 shows the cumulative mass released as predicted by Equation 9-18 for materials with a range of diffusion coefficients ($D_{i:p}$). This simple model has been used to characterize the release of drugs from a large number of experimental systems: for example, the release of dispersed dexamethasone from an EVAc matrix follows this model very well (Figure 9.8).

This simple approach for modeling the release of dispersed or dissolved drugs from a matrix can be easily extended to cylindrical or spherical shapes, by writing and solving the analogous differential equation. The macroscopic geometry of a matrix can influence the rate and pattern of protein release. Increasing the surface area-to-volume ratio of the matrix increases the release rate by allowing more particles direct access to the matrix exterior.

The rate of release decreases with time since drug molecules near the surface are released first (they have the shortest distance to travel by diffusion). The common matrix form of a slab has a cumulative release proportional to the square root of time; therefore, the release rate decreases with the square root of time (this fact can be demonstrated by taking the first derivative of Equation 9-19). However, if the matrix is formed as a hemisphere with an

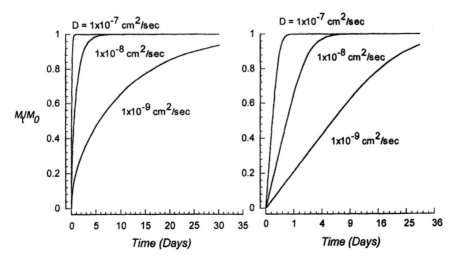

Figure 9.7 Drug release from a planar matrix drug delivery system. The cumulative mass of drug released from a planar drug delivery system as predicted by Equation 9-18 is shown as a function of the rate of diffusion of the dissolved drug in the matrix, $D_{i:p}$. The left panel shows release as a function of time and the right panel shows release as a function of the square root of time (with the expected linear dependence). The thickness of the matrix is 1 mm.

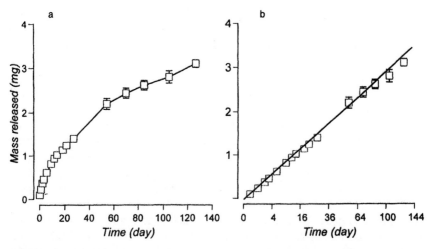

Figure 9.8 Release of dissolved dexamethasone from an EVAc matrix. The mass of dexamethasone released from a small polymer matrix (~ 10 mg) during incubation in phosphate-buffered saline is plotted versus (a) time and (b) the square root of time. Data from [8].

impermeable surface coating, which has a small defect in its planar surface, essentially zero-order kinetics can be obtained [9]. In cases with hemispherical geometry, releasing drug from the center of the planar face, the longer diffusion distance for molecules on the outside of the hemisphere is balanced by the increase in surface area.

Pseudo-steady-state Approximation for Dispersed Drugs Loaded above the Solubility. Drug molecules can be loaded into the matrix as fine solid particles. In this situation, the drug concentration within the matrix may be much higher than the drug solubility in the aqueous solution. A boundary between dissolved and dispersed drug may be present; as release proceeds, this boundary would move from the outer surface of the matrix to the center. A pseudo-steady-state model has been used to describe these systems [10]:

$$M_t = A\sqrt{D_{i:p}c_s(2c_a - c_s)t} \qquad (9\text{-}20)$$

where A is the surface area for release, c_s is the solubility of the agent in the aqueous medium, and c_a is the total amount of agent present per unit volume of matrix. In addition to the assumption of pseudo-steady state, which implies a linear concentration gradient from the solid/dissolved drug interface to the releasing surface, this model also requires that the total concentration is much higher than the drug solubility (i.e., $c_a \gg \epsilon c_s$ where ϵ is the porosity of the matrix).

9.2.2 Matrix Delivery Systems for Water-soluble Drugs and Proteins

Many drug molecules, particularly water-soluble compounds, do not dissolve appreciably in hydrophobic polymers and, therefore, can not be easily formulated into the matrix drug delivery systems presented in the previous section. For these drugs, it is possible to construct matrix drug delivery systems by dispersing small particles of the drug throughout a polymer matrix (Figure 9.6b, c). The first observations of release of water-soluble molecules from polymer films were made in paints [11] and polyolefins [12]. Variations on this approach have been developed to permit the controlled release of small, water-soluble molecules such as dopamine [13, 14] and large molecules such as proteins [15, 16] and DNA [17].

The mechanisms of drug release appear to be independent of the size of the dispersed molecule. This section will focus primarily on matrix systems for dispersed proteins, with an understanding that the results can be extended to other systems. The development of drug delivery systems that are suitable for the controlled release of proteins has been of considerable interest over the last several decades, primarily due to the rapid growth of the biotechnology industry and the realization that protein drugs, which can now be produced in large quantities with high purity, present formidable challenges for administration to humans.

Matrix Systems for Proteins.   Because macromolecules such as proteins diffuse very slowly through films of silicone and EVAc, many investigators believed that it was impossible to develop matrix delivery systems capable of releasing proteins [18]. In the late 1970s, however, a method for achieving controlled release of proteins from non-degradable polymers was described [15]. These polymers provide sustained release of biologically active molecules for extended periods, up to several years in some cases [16]. One particular hydrophobic polymer, EVAc, has been investigated extensively as a matrix system for drug delivery. Other classes of hydrophobic polymers, such as silicone elastomers and polyurethanes, may also be useful for controlled protein delivery, although there are fewer examples available in the literature. Non-degradable, hydrophilic polymers, such as poly(2-hydroxyethyl methacrylate), are also biocompatible, but usually release proteins over a relatively short period of several hours and so may be useful for protein delivery in cases where prolonged release is not needed.

In general, proteins are loaded into a polymer matrix by dispersing solid particles of protein throughout the polymer (Figure 9.6b). When protein-loaded matrices are immersed in water, the proteins are slowly released (Figure 9.9). The initial rate of release is higher for matrices with higher loading (initial mass fraction of protein particles within the matrix) (Figure 9.9a).

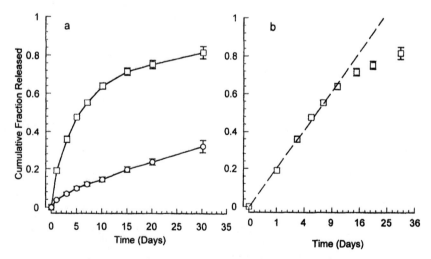

Figure 9.9   Controlled release of ferritin from an EVAc matrix. Release of ferritin (500 kDa protein) from a matrix of EVAc. (a) The cumulative fraction of mass released from matrices containing 35% (circles) or 50% (squares) ferritin by mass. (b) The same cumulative mass fraction released from the 50% loaded matrices is plotted versus the square root of time. The dashed line represents the fit to the linear model of desorption from a slab, Equation 9-19. Data points represent the mean cumulative fraction of mass of ferritin released from four EVAc matrices incubated in buffered saline at 37 °C. The error bars represent ± one standard deviation from the mean. Note that some error bars are smaller than the symbols.

This release is frequently linear with respect to the square root of time (Figure 9.9b), consistent with a diffusive release mechanism (Equation 9-14). To account for the complex structure of the composite matrix material, the diffusion equation incorporates an effective diffusion coefficient for the protein in the polymer matrix (see the section below). This effective diffusion coefficient, which is typically much lower than the diffusion coefficient of the protein in water, provides a quantitative measure of the rate of protein release, decreasing as the rate of protein release from the matrix decreases.

EVAc matrix systems have been used to release a variety of macromolecules, such as polypeptide and protein hormones [19, 20], heparin [21], growth factors [22–25], inhibitors of tumor angiogenesis [26], polyclonal antibodies [16], monoclonal antibodies (mAb) [27–31], antigens [32, 33], and DNA [17]. Macromolecules retain their biological activity after release from EVAc. For example, mAb against human chorionic gonadatropin (hCG) retained their ability to bind to hCG after release from an EVAc matrix [28, 29] and mAb that neutralize herpes virus were effective after long-term delivery in animals [34]. In addition, when released from EVAc matrices, nerve growth factor stimulated neurite outgrowth in cultured cells [22], and enhanced choline acetyltransferase activity in neurons in the brain [35], insulin altered the blood glucose levels in diabetic rats [19], and angiogenesis inhibitors blocked new blood-vessel growth [26].

EVAc matrices, usually prepared by solvent evaporation [36], consist of protein particles dispersed throughout a continuous polymer phase (as in Figures 9.6b and 9.10a). When matrices are placed in an aqueous environment, particles at the surface of the matrix can dissolve. Since water-soluble molecules diffuse very slowly through the continuous polymer phase, and since the

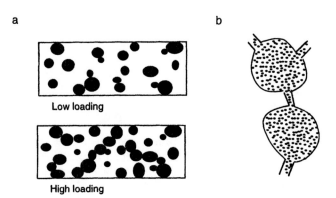

a   b

Low loading

High loading

Figure 9.10 Proteins are released from matrices by diffusion through constricted pores. Schematic view of a protein/hydrophobic polymer matrix system. (a) Cross-sectional views of matrices at two different loadings. At higher loadings connected clusters of pores are formed, providing continuous paths for diffusion to the surface. (b) Typical pores within the matrix are enlarged. Large pores (diameters of 100–400 μm) are connected by smaller channels (diameters of 1–10 μm). The constricted diffusional pathways contribute significantly to the slowness of release.

polymer is hydrophobic and does not swell appreciably in water, protein release must occur through pores in the polymer, which form as the dispersed protein particles dissolve. In fact, microscopic observations of the matrix structure reveal a network of interconnected pores in which large pores (diameters of 100–400 μm) are connected by smaller pores or channels (1–10 μm in diameter) (Figure 9.10b) [36–38]. Connected clusters of pores that contact the matrix boundary can release protein to the surrounding environment. At loadings higher than 35%, most protein particles are found in clusters that reach the matrix surface, while at lower loadings most particles are disconnected and, therefore, not releasable (Figures 9.6 and 9.10).

Modeling Release of Proteins from Matrices. Consider a protein-loaded matrix that is constructed as a thin slab. Release of the protein occurs essentially through the top and bottom faces of this slab since the thickness of the slab is small compared to the other dimensions. The desorption of protein from this slab can be described by Fick's second law of diffusion, Equation (9-14), in which an effective diffusion coefficient, D_{eff}, is used in place of $D_{i:p}$. For example:

$$\frac{\partial c}{\partial t} = D_{eff} \frac{\partial^2 c}{\partial x^2} \tag{9-21}$$

The effective diffusion coefficient accounts for the rate of diffusion of the protein through the complex, porous polymer matrix. Incorporation of an effective diffusion coefficient is necessary because protein molecules do not diffuse through the pure polymer phase, but must find a path out of the slab by diffusing through a tortuous, water-filled network of pores. D_{eff} is assumed to be independent of position in the slab; Equation 9-14 is justified in this case by local averaging over a volume that is large compared to a single pore. Characteristic desorption or protein release curves are represented by Equation 9-18, with substitution of D_{eff} for $D_{i:p}$:

$$\frac{M_t}{M_\infty} = 1 - \frac{8}{\pi^2} \sum \frac{1}{(2n+1)^2} \exp\left\{-\frac{D_{eff}(2n+1)^2\pi^2 t}{L^2}\right\} \tag{9-22}$$

Typical release curves are shown in Figure 9.7. The similarity between these curves and the experimentally determined protein-release profile is obvious (Figure 9.9). By fitting Equation 9-22 to data for protein release with the square root of time, D_{eff} can be estimated from experimental observations (see [16] for details).

The effective diffusion coefficient is related to the molecular diffusion coefficient of the protein in water, D_0, by:

$$D_{eff} = \frac{D_0}{F\tau} \tag{9-23}$$

where τ is the tortuosity and F is the shape factor. This definition of the effective diffusion coefficient was introduced in Chapter 4. Using the overall tortuosity ($F\tau$) as an adjustable parameter—and assuming that the diffusion coefficient in water, D_0, is known—Equation 9-18 or 9-19 can be fit to experimental release data for different proteins. The fits to individual release profiles, which represent one protein at fixed fabrication conditions (loading of matrix, particle size range of protein and polymer molecular weight), are reasonable; that is, this simple model appears to provide an adequate description of the release behaviour [39].

For a protein incorporated in a matrix, the rate of release (i.e., the effective diffusion coefficient and $F\tau$) depends on the molecular weight of the protein, the size of the dispersed particles, the loading, and the molecular weight of the polymer. For example, matrices of EVAc containing dispersed particles of bovine serum albumin (BSA) release protein for an extended period (Figure 9.11). The rate of release from the matrix depends on BSA loading (the number of BSA particles dispersed per unit volume of matrix) and BSA particle size. The rate of release from these matrices can be characterized by an effective diffusion coefficient or tortuosity, which must depend on the characteristics (i.e., particle size and loading) of the device. Tables 9.2 and 9.3 list the tortuosity values calculated from fits of the desorption equation to available literature data of protein release from EVAc slabs.

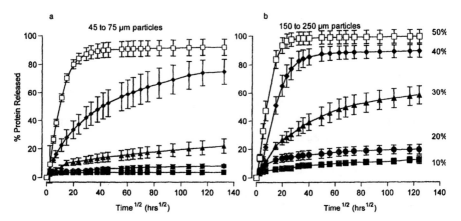

Figure 9.11   Release of bovine serum albumin from EVAc matrices. Controlled release of BSA from an EVAc matrix. Solid particles of BSA either (a) 45 to 75 μm or (b) 150 to 250 μm in diameter were dispersed within EVAc by solvent evaporation to achieve final protein loadings from 10, 20, 30, 40, or 50%. In each case five identical slabs, each ~70 mg and 1 mm thick, were incubated in cacodylate-buffered water containing 0.02% gentamicin. Periodically, the buffered water was replaced, and the amount of protein released from the matrix was determined by measuring the concentration of protein in the solution that was removed. Each symbol represents the average cumulative fraction of protein released (cumulative mass of protein released/initial mass of protein within the matrix) for the five samples; error bars indicate the standard deviation, which in some cases are smaller than the symbols. Data from [16].

Table 9.2   Tortuosity of protein release from EVAc matrices

Macromolecule (M_w)	Particle size (μm)	Loading (%)						
		10	20	25	30	35	40	50
Ferritin (5 × 10⁵)	< 178[a]					2,800		200
γ-Globulin (1.5 × 10⁵)	150–250[b]	30,000	56,000	47,000	9,300	12,000	5,900	580
BSA (68,000)	< 75[d]	37,500		28,000				220
	45–75[b]	20,000	46,000	19,000	29,000	22,000	4,500	600
	106–150[e]		11,000	2,300	690			
	150–180[e]		11,000	2,300	860			
	150–250[b]	26,000	7,400	5,700	9,400	1,500	870	160
	75–250[d]	5,700		1,300				59
	250–425[d]	1,700		660				47
	250–425[e]			560	180			
	300–425[e]		95			55	19	
β-Lactoglobulin (18K)	75–150[e]			10,000			1,600	1,100
	106–250[d]			1,500				
Lysozyme (14K)	75–250[d]			19,000				
	106–150[e]						2,100	
Heparin (14K)	< 104[f]						30,000	18,000
Insulin (6K)	< 75[g]		69			4,800		3,700
	75–250[g]		560			760		790
	250–450[g]		270			180		190

The total tortuosity ($F\tau$) is listed for protein release from EVAc matrices which were prepared as described in the following references; (a) [33], (b) [16], (c) [41], (d) [40], (e) [39], (f) unpublished data, Saltzman, (g) [36].

Increasing protein particle size increases the rate of protein release (Table 9.2 and Figure 9.12). The size of the protein particles in the matrix affects the size of the water-filled channels formed as the particles dissolve. Larger particles occupy more volume in a matrix, increasing the pore-to-pore connectivity. Increased connectivity provides simpler pathways (i.e., less tortuous and less constricted pathways) for diffusion of protein molecules [16, 36, 40]. This can be seen in Table 9.2 for a variety of proteins; as the average particle size increases, the overall tortuosity ($F\tau$) decreases.

Table 9.3   Tortuosity measured in EVAc matrices with different M_w fractions

Protein	Molecular weight (M_n/M_w in kDa)								
	167/253	76/106	71/169	56/82	33/72	32/54	23/33	14/42	13/29
γ-Globulin	44,000	490	490	74		44	22		
BSA					43			6.2	1.5

The total tortuosity is listed for protein release from EVAc matrices prepared with different molecular weight fractions of EVAc. γ-Globulin matrices were prepared with 119–180 μm particles and 40% protein loading [31]. BSA matrices were prepared with 106–150 μm particles and 25% protein loading [38].

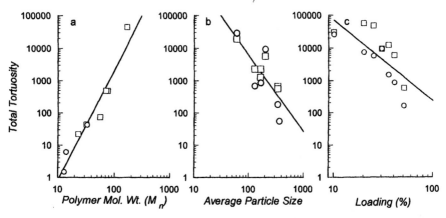

Figure 9.12 Tortuosity for protein release from EVAc matrices as a function of particle loading, particle size, and polymer molecular weight. The dependence of total tortuosity (*Fτ*) on various matrix fabrication parameters. (a) The tortuosity dependence on the molecular weight of the EVAc polymer used in the matrix system. The tortuosity values (*Fτ*) for γ-globulin at 40% loading and 119–180 μm particle size (squares) and BSA at 25% loading and 106–150 μm (circles) are plotted versus the number-averaged molecular weight of the EVAc. (b) The tortuosity dependence on the particle size of the macromolecule. Tortuosity values of BSA at 25% loading (squares) and BSA at 30% loading (circles) are plotted versus the average particle size. (c) The effect of loading on the tortuosity values. The *Fτ* values for γ-globulin at 150–250 μm (squares) and BSA at 150–250 μm (circles) are plotted versus the percentage loading. The lines are the power curve fits to the data. All symbols represent data selected from Tables 9.2 and 9.3; see table captions for original references.

Increasing the loading of protein in the matrix also increases the release rate of proteins (Table 9.2 and Figure 9.12c). When particle size is fixed, the tortuosity values generally decrease as the loading increases. Increasing the loading also provides protein molecules with less tortuous routes to the exterior of the matrix, thus increasing the diffusion coefficient and, therefore, the release rate [16, 36, 40, 41].

When particles of a fixed size are dispersed at a fixed loading, the molecular weight of EVAc can be changed to influence the rate of protein release (Table 9.2 and Figure 9.12). As the average molecular weight of the polymer in the matrix increases, the rate of protein release decreases [31, 38]. The rate of release correlates with mechanical properties of the polymer [31], suggesting that the mechanical properties of the polymer influence the structure of the pore network within the polymer. High molecular weight polymers have a low elastic modulus, forming a relatively non-deformable matrix, which limits the formation of small connecting channels. In low molecular weight polymers, the modulus is higher, the matrix is more deformable, and osmotic pressure within the pore space can cause the connecting pores to expand. Numerical simulations and mathematical models demonstrate that the ratio of small connecting pore size to pore size has a substantial influence on pore-to-pore transport rate [42, 43] .

To quantify the influence of polymer molecular weight, protein loading, and protein particle size on the rate of release, calculated overall tortuosities ($F\tau$) were compared for release from matrices where only one of these parameters was varied (Figure 9.12). The tortuosity was greatly influenced by the molecular weight of the polymer ($F\tau \propto M_n^{3,4}$; Figure 9.12a). Particle size and the loading also strongly influence the overall tortuosity ($F\tau \propto size^{-2.4}$ and $F\tau \propto loading^{-2.5}$; Figure 9.12b,c). Effective diffusion coefficients or overall tortuosities are useful for evaluating trends in the observed protein release rates. The empirical relationship between $F\tau$ and simple fabrication parameters (as shown in Figure 9.12) can be used to design the protein-releasing EVAc matrix, although there are protein-specific differences that are not yet completed determined.

To understand how microscopic properties of the material influence these phenomena, it is necessary to develop more complex models of protein release. When detailed information on the microgeometry of the porous network in the polymers is available (as it is for protein-loaded EVAc matrices [37]) or can be estimated accurately, detailed models of protein release can be developed. For example, percolation models of pore network topology (such as those described in Chapter 4, see Figure 4.20) were coupled with analytical models of pore-to-pore diffusion rates to predict the rate of diffusion of proteins from EVAc matrices [16]. Effective diffusion coefficients predicted using this approach agree with those estimated by measuring rates of protein release from the matrix (Figure 9.13).

Since the protein is loaded into the matrix at high concentration in solid particles, and protein molecules are confined to the water-filled pore space, it is possible for protein concentration within the pores to exceed protein solubility. The pseudo-steady-state model of Higuchi [10] (Equation 9-20) can be modified to account for this physical situation:

$$M_t = A\sqrt{D_{eff}c_s(2c_a - \varepsilon c_s)t} \qquad (9\text{-}24)$$

For water-soluble proteins, such as BSA, this expression does not compare well with experimental data, presumably because the solubility is comparable to the concentration of protein in the solid particles [44].

Incompleteness of Release.   For dispersed drug matrix systems in which the drug molecules do not dissolve in the polymer phase, release is often incomplete; less than 100% of the drug is released, even after very long incubations in water (see, e.g., Figure 9.11). Incomplete release occurs because some of the drug particles are isolated, or completely surrounded by polymer; the drug molecules within the volume have no pathway to the surface. The completeness of release can be modeled by percolation theory, which was introduced in the discussion on diffusion in heterogeneous media in Chapter 4. If the porosity is low, only pores that are near the material/water interface can be wetted. Since interior pore clusters are totally surrounded by material backbone, they will never be wetted. As the porosity of the material increases, large clusters of pores begin to form, and many

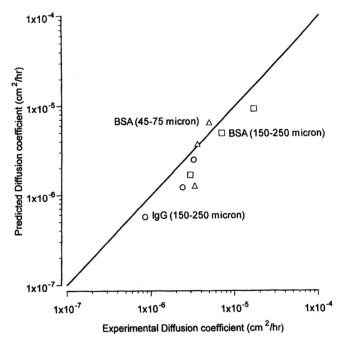

Figure 9.13 Effective diffusion coefficients for protein release from EVAc matrices. Effective diffusion coefficients predicted by percolation/diffusion models agree with diffusion coefficients estimated by measuring rates of protein release. From [16].

interior pores are now connected to the aqueous environment. Therefore, the fraction of pores that can be wetted, and the release of incorporated protein, increases.

To quantify this effect, two types of material porosity must be considered: total porosity and accessible porosity. Total porosity, ϵ, is the volume fraction of pores in the material; accessible porosity, ϵ_a, is the volume fraction of pores which are members of clusters that extend to the surface of the material. In desorption of solute from a porous material, only solute which is initially present in accessible pores can be released. The fractional volume of accessible pores, ϵ_a/ϵ, is denoted ϕ_a. As the intuitive model of porous materials—developed in the previous paragraph—suggests, both accessible porosity and fractional accessible porosity increase as the total porosity increases.

In materials of infinite extent, the above definitions remain valid. As noted previously (Chapter 4), for pore space topologies with a given coordination number, there exists a critical filling probability (porosity). In materials with filling probabilities above this critical value, the size of the largest cluster is comparable to the size of the lattice. The presence of this lattice spanning cluster does not require that the material be finite in extent; in fact, most analytical results in percolation theory assume that the lattice is infinite. For infinite materials, ϵ_a is the probability that any given site on the lattice belongs to the infinite cluster.

For Bethe lattices (see Chapter 4), the relationship of accessible porosity to total porosity can be analytically derived. For site percolation on a Bethe lattice with coordination number ζ, the accessible porosity is given by [45, 46]:

$$\varepsilon_a = \varepsilon - \varepsilon^* \left(\frac{1-\varepsilon}{1-\varepsilon^*}\right)^2 \tag{9-25}$$

where ϵ^* is the root of the equation:

$$\varepsilon^*(1-\varepsilon^*)^{\zeta-2} - \varepsilon(1-\varepsilon)^{\zeta-2} = 0 \tag{9-26}$$

As noted previously, the critical probability for the Bethe lattice is ϵ_c (equivalent to p_c defined in Chapter 4) $= 1/(\zeta-1)$. For lattices below this critical probability ($\epsilon < \epsilon_c$), the root of Equation 9-26 is $\epsilon^* = \epsilon$. The accessible porosity, from Equation 9-25, is zero, which indicates that a lattice spanning cluster is not present. For lattices above the critical probability ($\epsilon > \epsilon_c$) Equation 9-26 can be solved to find ϵ^*. Results for coordination numbers 3 and 4 are:

$$
\begin{aligned}
&\text{for } \zeta = 3 \qquad \epsilon^* = 1 - \epsilon \qquad\qquad\quad \text{when } \epsilon > \tfrac{1}{2}\\
&\text{for } \zeta = 4 \quad \epsilon^* = 1 - \tfrac{1}{2}\epsilon - \sqrt{\epsilon\left(1 - \tfrac{3}{4}\epsilon\right)} \quad \text{when } \epsilon > \tfrac{1}{3}
\end{aligned}
\tag{9-27}
$$

Solutions for other coordination numbers follow directly [45, 46]. Accessible porosity is found by substitution of the correct root of Equation 9-26 into Equation 9-25; the fraction of accessible porosity, ϕ_a, is found by dividing ϵ_a by the total porosity ϵ.

In Figure 9.14b, the fraction accessible porosity is plotted versus the total porosity for Bethe lattices of coordination numbers 3 and 7. For all coordination numbers, ϕ_a is zero for porosities less than the critical value. The critical porosity is indicated for each coordination number by the intercept of the curve with the x-axis. Above the critical porosity, ϕ_a rises sharply. In this transition region, the infinite, lattice-spanning cluster is growing and incorporating pores and smaller pore clusters that are isolated at lower porosities. At high porosities, ϕ_a becomes equal to unity, indicating that all the pores are members of the infinite cluster.

Even below the critical porosity, pore clusters exist. The clusters are finite in size and do not span the lattice. In the transition region—as the infinite cluster is incorporating more of these finite clusters—finite clusters still exist. For the Bethe lattice, the mean size of these finite clusters, S, depends on the lattice-filling probability (i.e., porosity) [45]:

$$S(\varepsilon) = \frac{1+\varepsilon^*}{1-(\zeta-1)\varepsilon^*} \tag{9-28}$$

where ϵ^* is again defined in Equation 9-26). Figure 9.14a also shows the mean cluster size for a Bethe lattice of coordination numbers 3 and 7. Below the critical porosity the mean cluster size increases with increasing porosity, exhibiting a singularity at the critical porosity. Above the critical porosity, the

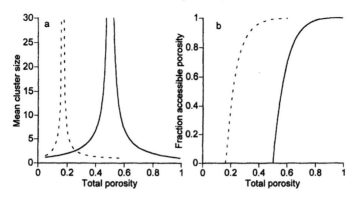

Figure 9.14 Fraction accessible porosity and mean cluster size for a Bethe lattice. Fraction accessible porosity and mean cluster size for Bethe lattices with $\zeta = 3$ (solid lines) and 7 (dashed lines). The fraction of accessible porosity (b), or the fraction of porosity that is part of an infinite cluster, is plotted versus the total porosity. The mean cluster size (a) exhibits a singularity at the critical porosity.

infinite cluster recruits smaller clusters and the mean size of finite clusters decreases.

The dependence of accessible porosity and cluster size distribution on lattice porosity has been evaluated for other lattices. Using Monte Carlo techniques on the cubic lattice, Kirkpatrick [47] identified a critical filling probability of 0.25. The quantitative dependence of accessible porosity above the critical probability is identical to a Bethe lattice with coordination number of 5 [48]. Winterfeld and co-workers [49, 50] examined three-dimensional Voronoi tesselations with a percolation threshold at 0.16; the accessible porosity for these chaotic composites is quantitatively well described as a Bethe lattice with coordination number 7 [48]. Similarly, the mean cluster size in a triangular lattice is well represented by a Bethe lattice of coordination number 4 [45]. Although the Bethe lattice is highly idealized—possessing no closed loops that are found in real porous materials—it can be used to approximate the properties of real materials. By selecting a Bethe lattice with an effective coordination number that fits the threshold behavior of a real material, many of the properties of the porous material can be analytically described [48, 51–54]. For example, the completeness of release from the dispersed protein/EVAc system is very similar to the accessible porosity defined on percolation lattices (Figure 9.15, see [16] for more details).

9.3 HYDROGEL DELIVERY SYSTEMS

Water-soluble polymers (see Appendix A.3) can be cross-linked to create materials, called hydrogels, that swell, but do not dissolve, in water. In these hydrogels, the rate of drug diffusion through the bulk material depends on the extent of cross-linking. Cross-links, or interconnections between polymer chains, con-

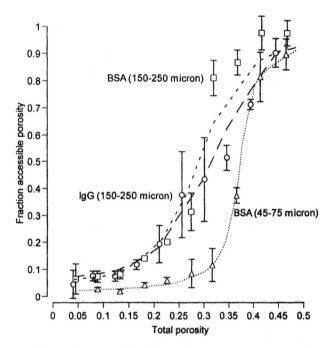

Figure 9.15   Total fraction released from EVAc/protein matrices. The total fraction released is plotted versus the total membrane porosity for three different systems: large particle (150–250 μm) BSA; large particle γ-globulin; and small particle (45–75 μm) BSA. Total porosity was determined by quantitative image analysis. Points represent the mean and standard deviation for five matrices. Solid lines indicate the model prediction assuming a Bethe lattice topology ($\zeta = 4$).

vert the ensemble of individual polymer chains into a macromolecular network. Individual polymer chains remain soluble in water, but cross-links (which can be covalent bonds or physical entanglements) prevent individual molecules from dissolving in the aqueous phase. Instead, the material swells as water diffuses into the interstices of the network; this swelling is limited by osmotic forces and the physical integrity of the network itself. In this manner, the extent of cross-linking determines the extent of swelling and, therefore, the distance between chains within the cross-linked network [55, 56] (Figure 9.16). When other molecules (e.g., entrapped drugs) are also diffusing within the network, the rate of diffusion can depend on the interchain separation and the size of the diffusing drug (Figure 9.17). The drug release characteristics of hydrogels can often be controlled by copolymerization of compatible monomers [58]. Since the rate of drug diffusion in hydrogels can be controlled, these materials are potentially useful as drug delivery systems. In addition, many of the water-soluble polymers that are used as the basis of hydrogels are biocompatible.

Hydrogel drug delivery systems can be used in the hydrated state (as in Figure 9.17) or dried before use. Dried hydrogels absorb water from the exter-

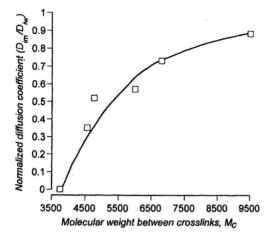

Figure 9.16
Diffusion of albumin in PVA hydrogels with different cross-linking density. The diffusion coefficient for bovine serum albumin at 37 °C was measured in poly(vinyl alcohol) hydrogels prepared with different cross-linking densities. Adapted from [56].

nal environment. Most drug molecules diffuse very slowly in the dried polymer network, so release of agents from these systems follows the penetration of water into the polymer. In some situations, hydrogel materials can be designed so that the rate of water entry controls the release process. Swellable hydrogels are frequently composed of ionic polymers; therefore, the dynamics of swelling are influenced by the ionic milieu (Figure 9.18). This property may permit the design of drug delivery systems that respond in predetermined ways to changes in the local environment.

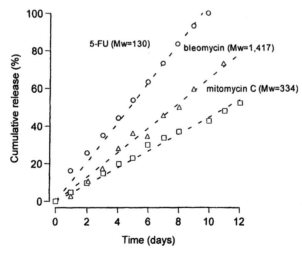

Figure 9.17 Release of anticancer drugs from poly(HEMA)–collagen hydrogels. Controlled release of 5-fluorouracil (M_w 130), mitomycin C (M_w 334), and bleomycin (M_2 1,417) into phosphate-buffered water at 37 °C from p(HEMA) polymerized in the presence of drug and 5% collagen. The hydrogels were hydrated (∼ 40% water) throughout the experiment. Adapted from [57].

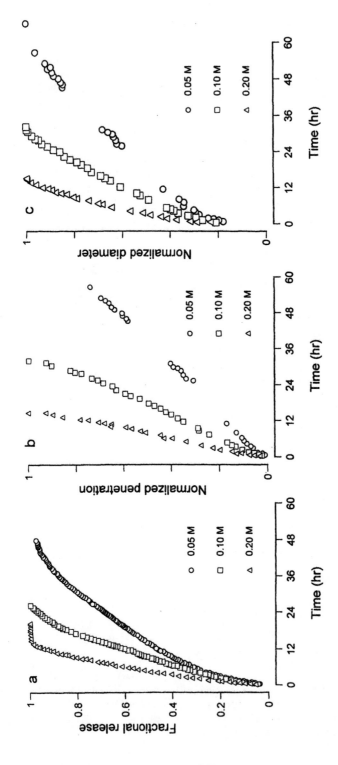

Figure 9.18 Release of oxprenolol HCl from hydrophobic polyelectrolyte beads. Controlled release of oxprenolol HCl from ~ 1 mm diameter beads of poly[(methyl methacrylate)-*co*-(methacrylic acid)] into phosphate-buffered water of different buffer concentration at pH 7.4 and 37 °C. Adapted from Kim and Lee, *Pharmaceutical Research*, 1992, **9**, 1268–1274.

9.4 DEGRADABLE DELIVERY SYSTEMS

9.4.1 Introduction

Drug delivery systems based on biodegradable polymers have the potential advantage that the supporting matrix will dissolve after drug release, so that no residual material remains in the tissue. This advantage must be balanced by the potential disadvantage that degrading polymers introduce an additional level of complexity into the design of useful materials. In addition, polymer degradation and erosion releases large quantities of potentially harmful polymer degradation products into the body. Therefore, the discovery of degradable polymers that are useful as drug delivery vehicles has focused on materials that (i) degrade in a controllable fashion and (ii) degrade into naturally occurring or inert chemicals.

The physical process of dissolution of a polymer matrix or microsphere, in which a solid material slowly losses mass and eventually disappears, is called *bioerosion*. The mechanism of bioerosion may be simple: e.g., the solid polymer may erode by dissolution of the individual polymer chains within the matrix. On the other hand, in many cases the polymer chains within the matrix must change to permit bioerosion. For example, the molecular weight of the polymer may decrease within the matrix following placement within the biological environment. This process, called *biodegradation*, may occur enzymatically, relying on catalysts present within the environment or embedded within the polymer itself, or hydrolytically, if polymers that are susceptible to hydrolytic breakdown are used. As biodegradation proceeds, the molecular weight of the polymer decreases. When the constituent polymer molecules become sufficiently small, they dissolve and the polymer matrix erodes.

Synthetic biodegradable polymers have been used as controlled-release drug delivery systems since 1970, when cyclazocine was released from a poly(-lactic acid) implant [59]. While copolymers of lactic acid and glycolic acid—poly(lactide-*co*-glycolide) or pLGA—continue to be the most widely used biodegradable polymers [60, 61], drug-releasing devices have now been formed from many different classes of biodegradable polymers including polyamides [62], poly(*ortho* esters) [63], polyesters [64], poly(iminocarbonates) [65], polyurethanes [66], poly(organophosphazenes) [67], polyphosphates and polyphosphonates [68], pseudo-poly(amino acids) [69], and polyanhydrides [70–72]. Properties of the material can be manipulated through chemical features of the polymer. For example, molecular weight and copolymerization are used to control properties of the pLGA system. Polyanhydride polymers contain the most hydrolytically unstable polymer linkage; therefore, materials based on polyanhydride chemistry should degrade rapidly. However, chemical stability of the polymer is only one of the factors contributing to overall erosion of the material; polyanhydride matrices with controlled hydrophobicity can be produced by copolymerization of hydrophobic and hydrophilic monomers. Proper monomer selection leads to slow water penetration and controlled polymer erosion from the matrix surface [70, 72].

9.4.2 Mechanisms of Degradation and Erosion

Degradation and disappearance of a biodegradable polymer matrix occurs in a sequence of steps. Some of these steps are understood reasonably well; as a consequence mathematical expressions can be used to describe some aspects of polymer degradation and drug release. These mechanisms will be illustrated using the pLGA system.

Most of the degradable polymers considered for use in pharmaceutical applications, including pLGA, degrade by hydrolysis (see Appendix A.2), so water is required for the degradation reaction. Since the material is initially dehydrated, water must enter before degradation. The rate of water penetration depends on the degree of hydrophobicity and morphology of the polymer matrix. Figure 9.19a shows water uptake into a thin film of pLGA. Water penetration occurs initially by diffusion of water molecules through either the bulk of the polymer or through voids or pore space in the material. If water enters the bulk polymer phase, the diffusion coefficient for solutes in that hydrated phase increases as predicted by free volume theory:

$$\ln\left(\frac{D_{i:p}}{D_{i,\text{water}}}\right) = \frac{kV_s}{V_f}\frac{1-H}{H} \tag{9-29}$$

where k is a constant characterizing the polymeric material, V_s is the volume of the solute i, V_f is the free volume of water in the polymer phase, and H is the fractional degree of hydration [73].

Polymer molecules that are exposed to water will hydrolyze, causing the average molecular weight of the polymer to decrease with time. For polyester materials such as pLGA, the kinetics of degradation are first order [74]:

$$\frac{M_n}{M_n^0} = \exp(-kt[\text{H}_2\text{O}]\,[\text{pLGA}]) \tag{9-30}$$

where M_n is the number average molecular weight, which varies with time, and M_n^0 is the initial value of M_n. This relationship is consistent with experimental data, as shown in Figure 9.19b.

Eventually, small polymer units, either monomers or short oligomers, are formed; these smaller units can dissolve in water and diffuse out of the matrix. For pLGA, only oligomers with molecular weight less than several hundred are sufficiently soluble in water for diffusion. For the system illustrated in Figure 9.19, this reduction in molecular weight requires 10 days. The disappearance of monomer/oligomer from the matrix should occur by diffusion-controlled desorption, and therefore can be described by Equation 9-18 or 9-19. As the small polymer fragments leave the matrix, additional channels for water entry and diffusion are created. Therefore, the diffusion coefficient increases with time, making prediction of weight loss based on Equation 9-18 difficult. Slow water penetration (Figure 9.19a) delays the physical erosion of the material, creating a substantial lag period between immersion of the matrix in buffered saline and release of the entrapped drug (Figure 9.19). Because of the complexity of this degradation/erosion process,

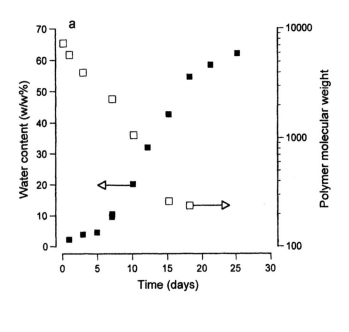

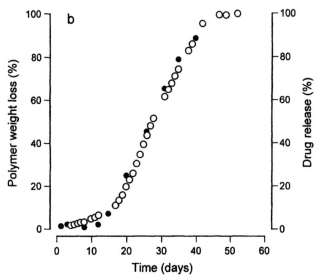

Figure 9.19 Changes in a pLGA system during degradation and drug release. The figure shows (a) the uptake of water (filled squares) and the decrease in molecular weight (open squares); and (b) the loss of polymer mass (filled circles) and the release of drugs (open circles) for a 18,000 weight-average molecular weight pLGA 50:50 copolymer [74].

the overall kinetics of degradation and release from a macroscopic piece of material are difficult to describe and predict.

In principle, matrices formed from degradable polymers disappear in one of two idealized patterns: surface erosion or bulk erosion (Figure 9.20). In bulk erosion (such as occurs with pLGA materials), the polymer disappears uniformly throughout the material: a microporous matrix eventually becomes spongy, with the water-filled holes becoming larger until the matrix is no longer mechanically stable. During surface erosion, the polymer disappears from the surface, so that the matrix becomes progressively smaller with time. In designing degradable polymers for drug delivery, materials that display surface erosion are usually preferred, since drug release from the slowly shrinking matrix should be more predictable. In addition, it is sometimes possible to design drug delivery systems in which the kinetics of erosion control the rate of drug release from the matrix. Surface erosion, the more orderly of the two idealized mechanisms, potentially provides a constant rate of polymer erosion (for materials where the surface area remains relatively constant during the erosion process). In reality, these idealized patterns of erosion are observed in only a few special cases, with most materials degrading by a combination of processes that appears intermediate to these two extremes. When drug molecules are dispersed or dissolved within the polymer matrix, this situation is further complicated, since the presence of drug molecules often influence the rate of water penetration. In addition, release of water-soluble drugs from the matrix can create additional routes for water entry, while degradation/erosion of the polymer matrix provides additional pathways for drug diffusion.

A full model for drug release from degradable materials must account for the complex physical changes that occur in a matrix during degradation and erosion. Some new models can account for the influence of polymer crystallinity and morphology on local rates of water penetration and polymer degradation [75, 76]. Consider the cross-section of a polymer matrix shown in Figure 9.20c; the polymer matrix is assumed to consist of a regular, three-dimensional array of lattice points (the individual lattice points would be much smaller than the matrix; for simplicity, only four layers are shown in the figure). Initially, each lattice site is randomly assigned a characteristic property; for example, the site is assumed to be either crystalline polymer or amorphous polymer with drug. The characteristics for each site are assigned randomly by a statistical process that insures that the overall properties of the matrix matches the system being modeled. The process of degradation/erosion is simulated by assuming that the matrix is immersed in an aqueous environment. Immersion exposes the outer layer of lattice sites to water; these sites are now susceptible to degradation. Since hydrolysis of polymer bonds usually occurs by first-order kinetics (see Equation 9-30), the lifetime of a polymer lattice site can be predicted from Poisson statistics:

$$t_{lattice} = \frac{1}{\lambda n} \ln(1 - \varepsilon_r) \tag{9-31}$$

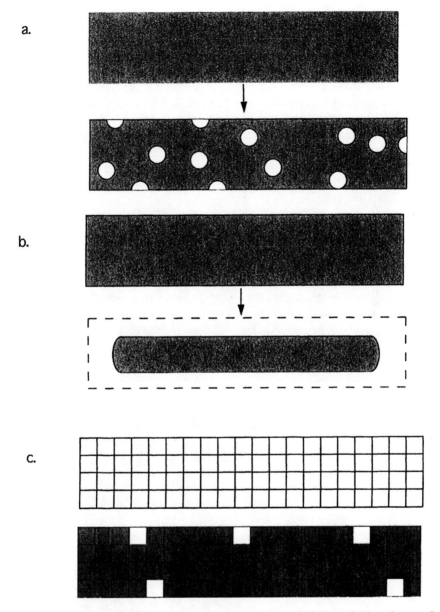

Figure 9.20 Idealized patterns of erosion for matrices of biodegradable polymers. In bulk erosion (a), the degradation or erosion events occur more uniformly throughout the matrix. It is generally assumed that surface erosion can be better controlled, offers better control over the rate of release of incorporated compounds, and is preferred for many applications, particularly for drug delivery. In surface erosion (b), the polymer matrix degrades heterogeneously. (c) Model of erosion process in a semi-crystalline polymer. White squares indicate water-filled pore space with dissolved drug, black squares indicate drug-free regions of crystalline polymer, and gray squares indicate regions of drug-loaded amorphous polymer.

were t_{lattice} is the lifetime of the lattice site after exposure to water, λ is a parameter which characterizes the susceptibility of the lattice site to degradation (i.e., $1/\lambda$ is the average lifetime for a lattice site of a particular type), n is the number of elements in the grid, and ϵ_r is a random variable uniformly distributed between 0 and 1. Figure 9.21 shows an example of polymer erosion simulated for a material with a crystallinity of 35% (i.e., 35% of the lattice sites are initially crystalline), $n = 100$, and λ equal to 10^{-8}s^{-1} for crystalline sites and 10^{-6}s^{-1} for amorphous sites. This simulation agrees reasonably well with microscopic observations of erosion of polyanhydride copolymers of carboxyphenoxypropane (CPP) and sebacic acid (SA) [75]. The rate of release of degraded polymer monomers or ecapsulated drug can be predicted by applying the diffusion equation within the complex, time-variant geometry determined by this simulation [75, 76].

9.4.3 Examples of Drug Release from Degradable Materials

In spite of the complexity of the release process, the release of drugs from matrices of biodegradable polymers often has kinetics similar to diffusion from non-degradable matrices. Figure 9.22 shows the controlled release of three different anticancer drugs from matrices composed of a polyanhydride polymer (20:80 pCPP/SA, see Appendix A.2 for the chemical structure); except for the encapsulated drug, the matrices were identical. Both BCNU and 4-hydroxycyclophosphamide (4-HC) were released by diffusion; most of the drug was released before substantial degradation of the matrix had occurred.

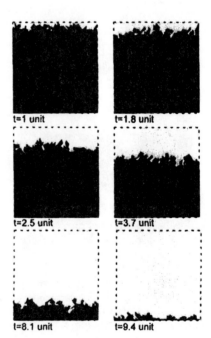

Figure 9.21
Simulation of polymer matrix erosion. From [75].

The data are well represented by models of diffusional release (Equation 9.19 was used for the solid lines in Figure 9.22a and 9.22b). Paclitaxel, however, is much less soluble in water than the other two drugs and it is released more slowly from the matrix. In certain situations, such as release of sparingly soluble drugs or agents incorporated at low loadings, the release of incorporated drug molecules appears to follow the kinetics of polymer degradation. This observation suggests that the degradation/erosion process controls the rate of drug release from the matrix. This phenomenon has been demonstrated most clearly with the polyanhydride copolymers pCPP/SA (Figure 9.23).

A number of groups have produced biodegradable materials based on water-soluble polymers [78, 79]. These degradable materials combine the advantages of a hydrogel (i.e., control of molecular diffusion) with the advantages of degradability (i.e., disappearance after implantation). Degradation is usually provided by introduction of ester linkages into the polymer chain; these ester linkages can be hydrolyzed, resulting in chain scission and dissolution of the resulting low molecular weight monomers, macromers, or oligomers. When the time period for degradation is much longer than that for release of the encapsulated drug, the rate of drug release does not depend on degradation and can be analyzed by the same approach outlined above for hydrogels. When the rate of degradation is comparable, the same difficulties encountered with biodegradable matrix systems are present and analysis requires a description of the degradation process.

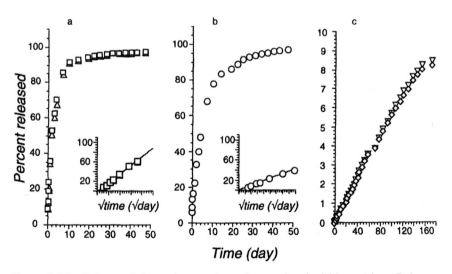

Figure 9.22 Release of chemotherapy drugs from polyanhydride matrices. Release of anticancer drugs from a pCPP/SA polyanhydride matrix. (a) BCNU, (b) 4-hydroxy-cyclophosphamide, and (c) paclitaxel. The insets in panels (a) and (b) show diffusion-controlled release (i.e., the percentage released is proportional with respect to the square root of time); D_{eff} is 7×10^{-10} and 3×10^{-10} cm^2/s, respectively. Data from [77].

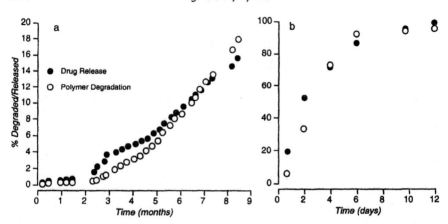

Figure 9.23   Release of *p*-nitrolaniline from matrices of polyanhydride copolymers. PCPP/SA matrices were loaded with 10% *p*-nitroaniline. Release and polymer erosion are shown for matrices with 100% CPP/0% SA (a) and 21% CPP/79% SA (b). From [70].

An interesting new approach combines the principles of drug modification, described in Chapter 8, with release from biodegradable hydrogels. In this situation, the agent is covalently coupled, through a biodegradable linkage, to the hydrogel network. The rate of release from this system can be determined by the solution of a modified form of the diffusion equation [80], Equation 9-1:

$$\frac{\partial c_{\mathrm{p}}}{\partial t} = \frac{\partial}{\partial x}\left(D_{i:\mathrm{p}}\,\frac{\partial c_{\mathrm{p}}}{\partial x}\right) + R_{\mathrm{form}}(x, t) \qquad (9\text{-}32)$$

where R_{form} is the local rate of formation of diffusible drug (or, in other words, the rate of generation of drug by cleavage of the covalently bound drug from the hydrogel network) within the material. This equation applies to any system in which a diffusible drug, initially present within a polymer film, is released by breakdown of drug–polymer bonds. If the cleavage of drug from the polymer occurs by a first-order process, then the rate of drug formation within the matrix is given by:

$$R_{\mathrm{form}} = kc_{\mathrm{p}}^{\mathrm{bound}} = k\left(c_0^{\mathrm{bound}}e^{-kt}\right) \qquad (9\text{-}33)$$

where k is the first-order rate constant for drug cleavage from the polymer and c_0^{bound} is the initial concentration of immobilized drug within the material. This approach has recently been used to model the release of covalently coupled proteins from a biodegradable hydrogel [79].

9.5 PARTICULATE DELIVERY SYSTEMS

While implantable drug delivery systems may be useful for certain applications, in many cases an injectable or ingestable delivery system is desired. For that reason, polymer particulates represent a potentially important class of drug delivery systems. Because of their small size, polymer particulates can be injected into the desired tissue site or into the blood stream. Several different classes of particulates have been examined for use as drug delivery vehicles including microcapsules, microspheres, and nanospheres (Figure 9.24). Microcapsules consist of a polymer shell enclosing a drug-loaded core. Microspheres are homogeneous particles in which the drug of interest is dispersed or dissolved within the solid polymer phase. Nanospheres are much smaller then microcapsules or microspheres. Nanospheres are usually homogeneous particles, similar to microspheres, but they are frequently modified at the surface to increase their stability in the body or to provide targeting capability.

A variety of methods have been reported for producing microspheres including phase separation by polymer/polymer incompatibility and coacervation [81]; solvent evaporation or solvent removal [82]; hot-melt microencapsulation; spray drying; interfacial polymerization; and supercritical fluid-processing techniques (such as the gas antisolvent spray precipitation process [83] or rapid expansion of supercritical fluids [84]). The characteristics of the most important of these methods have been reviewed [85, 86].

The rate of agent release from a preparation of microspheres depends on an intimidating number of variables including characteristics of the drug (size and solubility), properties of the polymer (composition and molecular weight), and the method of particle production. The rate of release can usually be adjusted by changing these properties. Release from pLGA microparticles, for example, depends on both composition and molecular weight, because these properties influence the rate of water entry into the particles and, therefore, the overall rate of degradation (see Appendix A.2).

Often, the role of the particulate delivery system is sustained release of the agent, in order to reduce the frequency of drug administration. In this situation, a constant rate of drug release is usually desired. For example, microparticle delivery systems (average diameter $\sim 20\,\mu$m) for leuprolide acetate were produced from pLGA (75:25) with an average molecular weight of 14 kDa [87]; these particles release the peptide hormone for $\sim$1 month, providing reasonably constant release throughout the period (Figure 9.25). This technology is the basis of the Lupron Depot®, a successful clinical product for the treatment of prostatic cancer, precocious puberty, and endometriosis. These techniques are particularly valuable for the release of proteins and peptides, since they allow long-term delivery of agents that are otherwise cleared from the circulation rapidly [88].

In other applications, different patterns of release are desirable. Microparticle vaccines are often designed to mimic a pattern of vaccine "boosts" that elicits optimal immunity. This approach has been tested for HIV vaccines based on recombinant gp120 [89]. A malaria vaccine has been

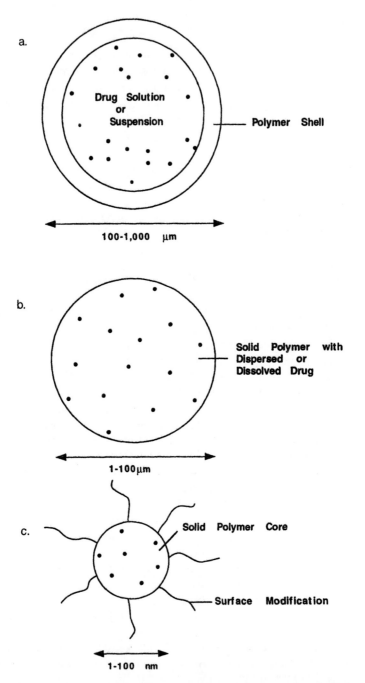

Figure 9.24 Particles for drug delivery. Particulate delivery systems include (a) microcapsules, (b) microparticles, and (c) surface-modified nanoparticles. In the nanoparticles, the drug is entrapped in the solid polymer core.

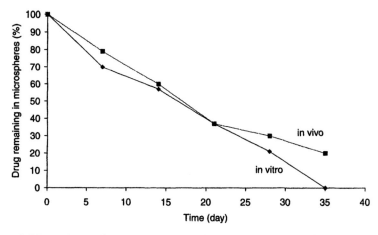

Figure 9.25   Release of leuprolide acetate from pLGA. From [87]. Release in *in vitro* test systems is similar to release in the body.

formulated by mixing microparticles that were prepared from different poly-(lactic acid) PLA and pLGA copolymers; each of the materials provided a burst of antigen release at different times (Figure 9.26).

The useful attributes of microparticles and drug modification have been combined in a recent approach. Long-circulating nanospheres were produced with a core of pLGA or polycaprolactone and PEG at the surface [90]. The nanospheres were produced by synthesizing diblock copolymers of PEG-R, where R was, for example, pLGA. These copolymers were dissolved in organic solvent, added to an aqueous phase to form an emulsion, and the solvent was evaporated. No surfactants were necessary because of the presence of PEG, which migrated to the surface of the emulsion. The nanospheres produced were 140 ± 10 nm in diameter, and released lidocaine over a period of ~ 4–6 h. Nanospheres without PEG were quickly cleared from the circulation, with > 90% eliminated with 5 min. PEG-containing nanospheres were cleared more slowly, with 20–50% remaining after 5 min.

Liposomes can also be considered particulate delivery systems, but in this text they are considered together with drug-modification methods in Chapter 8.

9.6 RESPONSIVE DELIVERY SYSTEMS

A central problem in drug delivery is matching the administration of the drug with the biological process that is under treatment. In many situations, particularly when the concentration of drug necessary for action is not much less than the concentration that causes toxicity, regulation of drug delivery is critical for success. The search for "smart" methods, in which drug delivery is matched with an underlying biological process, requires delivery systems that can target and respond to the biological process.

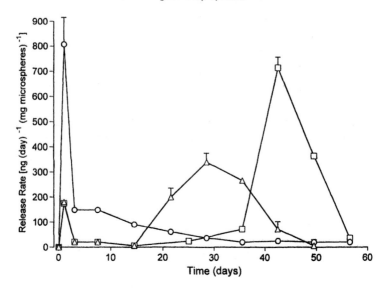

Figure 9.26   Release of a malarial antigen from pLA and pLGA microspheres. Altering the composition and molecular weight of the polymer can influence release from microspheres. Here, the malarial antigen was released from three different formulations of biodegradable polyesters.

The simplest responsive delivery systems use human intelligence. It is possible to argue that all drug delivery is responsive. For example, a migraine headache sufferer takes a second dose of ibuprofen because the biological process signals her that more drug is needed. In this example, the "smart" element (the suffering patient) is activated by pain. Diabetics adjust their insulin doses throughout the day using a variety of biological signals; they might take additional insulin with a heavy meal or adjust doses for exercise; diabetics can also measure glucose levels in the blood or urine in order to evaluate their need for insulin more accurately. The need for regulation of drug delivery is often so important that hospitalization is required (i.e., the intelligence of multiple humans is required to adjust the doses). Heparin is administered by computer-controlled infusion; the rate of heparin delivery is regulated by modifying the rate of infusion in order to maintain a desired anticoagulant activity in the blood, which is measured by testing blood samples. Modern responsive delivery systems propose to pre-insert human intelligence into the drug delivery device, so that less manipulation or human involvement is required. Responsive delivery systems can potentially control the disease process more efficiently, because drug doses are maintained at optimal levels for a greater fraction of the treatment period.

Biological recognition is the basis of some of the most promising approaches for responsive delivery systems. By introducing biological signals into the material, drug release can be initiated by biological events that occur after placement of the drug delivery device into the tissue. This approach has

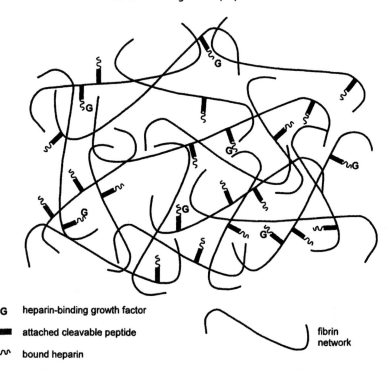

G   heparin-binding growth factor

■   attached cleavable peptide

∿   bound heparin

fibrin
network

Figure 9.27 Controlled release activated by cellular infiltration and enzyme activity. Adapted from [93]. The gel consists of a fibrin base with covalently bound bidomain peptides. The peptide has a domain that confers heparin binding and a domain that is susceptible to enzymatic degradation. The agent for release is a growth factor that binds with high affinity to heparin. Release is initiated by cellular invasion of the gel and local secretion of an enzyme that cleaves the peptide.

been explored most completely for the delivery of insulin, which is ideally released only in response to increased glucose concentration in the blood. Polymer-based materials have been designed that release insulin after glucose binding to lectins [91] or glucose conversion catalyzed by specific enzymes [92]. In a new approach [93], drug delivery materials are engineered from a polymeric base material, which is chemically modified to include peptides that are susceptible to cleavage by enzymes released from certain cells (Figure 9.27). As cells invade the material, they release the enzyme, which causes degradation of the peptide and release of the drug; drug delivery is therefore activated by a biological process that occurs locally in the tissue.

SUMMARY

- Polymeric membranes can be used to control the rate of release. Reservoir and transdermal devices are conceptually simple; the rate of drug release can be predicted by simple mathematical expressions. However, many drugs are not suitable for use in membrane-type devices.

- Matrix-type delivery systems are simple to make; release is usually controlled by diffusion of drug through the polymer matrix. Mathematical descriptions of release are more complicated than are obtained for membrane-type devices, and it is difficult to produce devices that provide a constant rate of release. However, these materials are versatile and almost any compound can be formulated into a controlled-release matrix.

- Degradable polymers, particularly polymers that degrade by hydrolysis, are appealing materials for use in clinical medicine because they disappear after implantation. Degradation is often difficult to control; families of related polymers provide the greatest versatility in design.

REFERENCES

1. Folkman, J. and D. Long, The use of silicone rubber as a carrier for prolonged drug therapy. *Journal of Surgical Research*, 1964, **4**, 139–142.
2. Frank, M.L., *et al.*, Characteristics and attitudes of early contraceptive implant acceptors. *Family Planning Prespectives*, 1992, **24**, 208.
3. Kuu, W.Y., R.W. Wood, and T.J. Roseman, Factors influencing the kinetics of solute release, in A. Kydonieus, ed., *Treatise on Controlled Drug Delivery*. New York: Marcel Dekker, 1992, pp 37–154.
4. Various, *Physicians' Desk Reference*. Montvale, NJ: Medical Economics Co., 1998.
5. Crank, J., *The Mathematics of Diffusion*. 2nd ed. Oxford: Oxford University Press, 1975, 414 pp.
6. Abramowitz, M. and I.A. Stegun, *Handbook of Mathematical Functions with Formulas, Graphs, and Mathematical Tables*. Washington, DC: National Bureau of Standards, 1964, 1046 pp.
7. Saltzman, W.M. and L.B. Tena, Spermicide permeation through biocompatible polymers. *Contraception* 1991, **43**, 497–505.
8. Reinhard, C., *et al.*, Polymeric controlled release of dexamethasone in normal rat brain. *Journal of Controlled Release*, 1991, **16**, 331–340.
9. Rhine, W., *et al.*, A new approach to achieve zero-order release kinetics from diffusion-controlled polymer matrix systems, in R.W. Baker, *Controlled Release of Bioactive Materials*. New York: Academic Press, 1980, pp. 177-187.
10. Higuchi, T., Rate of release of medicaments from ointment bases containing drugs in suspension. *Journal of Pharmaceutical Sciences*, 1961, **50**(10), 874–875.
11. Marson, F., Antifouling paints. I. Theoretical approach to leaching of soluble pigments from insoluble paint reticles. *Journal of Applied Chemistry*, 1969, **19**, 93.
12. Gregorian, R.S. and C.C. Kirk, Process for forming crosslinked oriented, microporous polyolefin film. U.S. Patent 3376238, 1968.
13. Freese, A., *et al.*, Controlled release of dopamine from a polymeric brain implant: *in vitro* characterization. *Experimental Neurology*, 1989, **103**, 234–238.
14. During, M.J., *et al.*, Controlled release of dopamine from a polymeric brain implant: *in vivo* characterization. *Annals of Neurology*, 1989, **25**, 351–356.
15. Langer, R. and J. Folkman, Polymers for the sustained release of proteins and other macromolecules. *Nature*, 1976, **263**, 797–800.
16. Saltzman, W.M. and R. Langer, Transport rates of proteins in porous polymers with known microgeometry. *Biophysical Journal*, 1989, **55**, 163–171.
17. Luo, D., *et al.*, Controlled DNA delivery systems. *Pharmaceutical Research*, 1999, **16**(8), 1300–1308.

18. Stannett, V.T., *et al.*, Recent advances in membrane science and technology. *Advances in Polymer Science*, 1979, **32**, 69–121.
19. Brown, L., *et al.*, Controlled release of insulin from polymer matrices: control of diabetes in rats. *Diabetes*, 1986, **35**, 692–697.
20. Fischel-Ghodsian, F., *et al.*, Enzymatically controlled drug delivery. *Proceedings of the National Academy of Sciences USA*, 1988, **85**, 2403–2406.
21. Edelman, E., D. Adams, and M. Karnovsky, Effect of controlled adventitial heparin delivery on smooth muscle cell proliferation following endothelial injury. *Proceedings of the National Academy of Sciences USA*, 1990, **87**, 3773–3777.
22. Powell, E.M., M.R. Sobarzo, and W.M. Saltzman, Controlled release of nerve growth factor from a polymeric implant. *Brain Research*, 1990, **515**, 309–311.
23. Murray, J., *et al.*, A micro sustained release system for epidermal growth factor. *In Vitro*, 1983, **19**, 743–748.
24. Beaty, C.E. and W.M. Saltzman, Controlled growth factor delivery induces differential neurite outgrowth in three-dimensional cell cultures. *Journal of Controlled Release*, 1993, **24**, 15–23.
25. Hoffman, D., L. Wahlberg, and P. Aebischer, NGF released from a polymer matrix prevents loss of ChAT expression in basal forebrain neurons following a fimbria-fornix lesion. *Experimental Neurology*, 1990, **110**, 39–44.
26. Lee, A. and R. Langer, Shark cartilage contains inhibitors of tumor angiogenesis. *Science*, 1983, **221**, 1185–1187.
27. Radomsky, M.L., *et al.*, Macromolecules released from polymers: diffusion into unstirred fluids. *Biomaterials*, 1990, **11**, 619–624.
28. Radomsky, M.L., *et al.*, Controlled vaginal delivery of antibodies in the mouse. *Biology of Reproduction*, 1992, **47**, 133–140.
29. Sherwood, J.K., R.B. Dause, and W.M. Saltzman, Controlled antibody delivery systems. *Bio/Technology*, 1992, **10**, 1446–1449.
30. Saltzman, W.M., Antibodies for treating and preventing disease: the potential role of polymeric controlled release. *Critical Reviews in Therapeutic Drug Carrier Systems*, 1993, **10**(2), 111–142.
31. Saltzman, W.M., *et al.*, Controlled antibody release from a matrix of poly(ethylene-co-vinyl acetate) fractionated with a supercritical fluid. *Journal of Applied Polymer Science*, 1993, **48**, 1493–1500.
32. Preis, I. and R.S. Langer, A single-step immunization by sustained antigen release. *Journal of Immunological Methods*, 1979, **28**, 193–197.
33. Wyatt, T.L., *et al.*, Antigen-releasing polymer rings and microspheres stimulate mucosal immunity in the vagina. *Journal of Controlled Release*, 1998, **50**, 93–102.
34. Sherwood, J.K., *et al.*, Controlled release of antibodies for sustained topical passive immunoprotection of female mice against genital herpes. *Nature Biotechnology*, 1996, **14**, 468–471.
35. Mahoney, M.J. and W.M. Saltzman, Millimeter-scale positioning of a nerve-growth-factor source and biological activity in the brain. *Proceedings of the National Academy of Sciences USA*, 1999, **96**, 4536–4539.
36. Siegel, R. and R. Langer, Controlled release of polypeptides and other macromolecules. *Pharmaceutical Research*, 1984, 2–10.
37. Saltzman, W.M., S.H. Pasternak, and R. Langer, Quantitative image analysis for developing microstructural descriptions of heterogeneous materials. *Chemical Engineering Science*, 1987, **42**, 1989–2004.

38. Hsu, T. and R. Langer, Polymers for the controlled release of macromolecules: Effect of molecular weight of ethylene-vinyl acetate copolymer. *Journal of Biomedical Materials Research*, 1985, **19**, 445–460.
39. Bawa, R., *et al.*, An explanation for the controlled release of macromolecules from polymers. *Journal of Controlled Release*, 1985, **1**, 259–267.
40. Rhine, W.D., D.S.T. Hsieh, and R. Langer, Polymers for sustained macromolecular release: procedures to fabricate reproducible delivery systems and control release kinetics. *Journal of Pharmaceutical Science*, 1980, **69**, 265–270.
41. Miller, E. and N. Peppas, Diffusional release of water-soluble bioactive agents from ethylene-vinyl acetate copolymers. *Chemical Engineering Communication*, 1983, **22**, 303–315.
42. Siegel, R. and R. Langer, A new Monte Carlo approach to diffusion in constricted porous geometries. *Journal of Colloid and Interface Science*, 1986, **109**, 426–440.
43. Ballal, G. and K. Zygourakis, Diffusion in particles with varying cross-section. *Chemical Engineering Science*, 1985, **40**, 1477–1483.
44. Zhang, X., *et al.*, Controlled release of albumin from biodegradable poly (DL-lactide) cylinders. *Journal of Controlled Release*, 1993, **25**, 61–69.
45. Fisher, M. and J. Essam, Some cluster size and percolation problems. *Journal of Mathematical Physics*, 1961, **2**, 609–619.
46. Mohanty, K., J. Ottino, and H. Davis, Reaction and transport in disordered composite media: Introduction of percolation concepts. *Chemical Engineering Science*, 1982, **37**, 905–924.
47. Kirkpatrick, S., Percolation and conduction. *Reviews of Modern Physics*, 1973, **45**, 574–588.
48. Reyes, S. and K. Jensen, Estimation of effective transport coefficient in porous solids based on percolation concepts. *Chemical Engineering Science*, 1985, **40**, 1723–1734.
49. Winterfeld, P., L. Scriven, and H. Davis, Percolation and conductivity of random two dimensional composites. *Journal of Physics C: Solid State Physics*, 1981, **14**, 2361–2376.
50. Winterfeld, P.H., Percolation and conduction phenomena in disordered composite media, PhD thesis, University of Minnesota, 1981.
51. Larson, R., H. Davis, and L. Scriven, Displacement of residual nonwetting fluid from porous media. *Chemical Engineering Science*, 1981, **36**, 75–85.
52. Larson, R., L. Scriven, and H. Davis, Percolation theory of two phase flow in porous media. *Chemical Engineering Science*, 1981, **36**, 57–73.
53. Reyes, S. and K. Jensen, Percolation concepts in modelling of gas–solid reactions— I. Application to char gasification in the kinetic regime. *Chemical Engineering Science*, 1986, **41**, 333–343.
54. Reyes, S. and K. Jensen, Percolation concepts in modelling of gas–solid reactions— II. Applications to char gasification in the diffusion regime. *Chemical Engineering Science*, 1986, **41**, 345–354.
55. Peppas, N. and D. Meadows, Macromolecular structure and solute diffusion in membranes: An overview of recent theories. *Journal of Membrane Science*, 1983, **16**, 361–377.
56. Reinhart, C. and N. Peppas, Solute diffusion in swollen membranes. Part II. Influence of crosslinking on diffusive properties. *Journal of Membrane Science*, 1984, **18**, 227–239.
57. Jeyanthi, R. and K. Rao, Controlled release of anticancer drugs from collagen- poly(hema) hydrogel matrices. *Journal of Controlled Release*, 1990, **13**, 91–98.

58. Antonsen, K.P., *et al.*, Controlled release of proteins from 2-hydroxyethyl methacrylate copolymer gels. *Biomat., Art. Cells & Immob. Biotech.*, 1993, **21**(1), 1–22.

59. Yolles, S., J.E. Eldridge, and J.H.R. Woodland, Sustained delivery of drugs from polymer/drug mixtures. *Polymer News*, 1970, **1**, 9.

60. Lewis, D.H., Controlled release of bioactive agents from lactide/glycolide polymers, in M. Chasin and R. Langer, *Biodegradable Polymers as Drug Delivery Systems*, New York: Marcel Dekker, 1990, pp. 1–42.

61. Sanders, L., *et al.*, Prolonged controlled release of Naferlin, a luteinizing hormone-releasing hormone analogue, from biodegradable polymeric implants. *Journal of Pharmaceutical Science*, 1986, **75**, 356–360.

62. Sidman, K.R., *et al.*, Biodegradable, implantable sustained release systems based on glutamic acid copolymers, *Journal of Membrane Science*, 1980, **7**, 277–291.

63. Heller, J., Development of poly(ortho esters): a historical overview. *Biomaterials*, 1990, **11**, 659–665.

64. Pitt, C.G., T.A. Marks, and A. Schlinder, Biodegradable drug delivery systems based on aliphatic polymers: application to contraceptives and narcotic antagonists, in R.W. Baker, *Controlled Release of Bioactive Compounds*, New York: Academic Press, 1980, pp. 19–43.

65. Pulapura, S., C. Li, and J. Kohn, Structure–property relationships for the design of polyiminocarbonates. *Biomaterials*, 1990, **11**, 666–678.

66. Nathan, A., *et al.*, Copolymers of lysae and polyethylene glycol. *Bioconjugate Chemistry*, 1993, **4**, 54–62.

67. Allcock, H.R., Polyphosphazenes as new biomedical and bioactive materials, in M. Chasin and R. Langer, *Biodegradable Polymers as Drug Delivery Systems*, New York: Marcel Dekker, 1990 pp. 163–193.

68. Richards, M., *et al.*, Evaluation of polyphosphates and polyphosphonates as degradable materials. *Journal of Biomedical Materials Research*, 1991, **25**, 1151–1167.

69. Kohn, J., Pseudopoly(amino acids), in M. Chasin and R. Langer, *Biodegradable Polymers as Drug Delivery Systems*, New York: Marcel Dekker, 1990, pp. 195–229.

70. Leong, K., B. Brott, and R. Langer, Bioerodible polyanhydrides as drug-carrier matrices. I: Characterization, degradation and release characteristics. *Journal of Biomedical Materials Research*, 1985, **19**, 941–955.

71. Rosen, H., *et al.*, Bioerodible polyanhydrides for controlled drug delivery. *Biomaterials*, 1983, **4**, 131–133.

72. Tamada, J. and R. Langer, Review: The development of polyanhydrides for drug delivery. *Journal of Biomaterials Science Polymer Edition*, 1992, **3**(4), 315–353.

73. Yasuda, H., *et al.*, Permeability of solutes through hydrated polymer membranes. III. Theoretical background for the selectivity of dialysis. *Makromol. Chem.*, 1969, **126**, 177–186.

74. Shah, S.S., Y. Cha, and C.G. Pitt, Poly(glycolic acid-co-DL-lactic acid): diffusion or degradation controlled drug delivery. *Journal of Controlled Release*, 1992, **18**, 261–270.

75. Gopferich, A. and R. Langer, Modeling of polymer erosion. *Macromolecules*, 1993, **26**, 4105–4112.

76. Gopferich, A. and R. Langer, The influence of microstructure and monomer properties on the erosion mechanism of a class of polyanhydrides. *Journal of Polymer Science: Part A: Polymer Chemistry*, 1993, **31**, 2445–2458.

77. Fung, L.K., *et al.*, Pharmacokinetics of interstitial delivery of carmustine, 4-hydro-peroxycyclophosphamide, and paclitaxel from a biodegradable polymer implant in the monkey brain. *Cancer Research*, 1998, **58**, 672–684.
78. West, J.L. and J.A. Hubbell, Photopolymerized hydrogel materials for drug delivery applications. *Reactive Polymers*, 1995, **25**, 139–147.
79. Zhao, X. and J. Harris, Novel degradable poly(ethylene glycol) hydrogels for controlled release of protein. *Journal of Pharmaceutical Sciences*, 1998, **87**(11), 1450–1458.
80. Pitt, C. and S. Shah, The kinetics of drug cleavage and release from matrixes containing covalent polymer–drug conjugates. *Journal of Controlled Release*, 1995, **33**, 391–395.
81. Thomasin, C., H.P. Merkle, and B.A. Gander, Physico-chemical parameters governing protein microencapsulation into biodegradable polyesters by coacervation. *International Journal of Pharmaceutics*, 1997, **147**, 173–186.
82. Mathiowitz, E., *et al.*, Polyanhydride microspheres as drug carriers. II. Microencapsulation by solvent removal. *Journal of Applied Polymer Science*, 1988, **35**, 755–774.
83. Randolph, T.W., *et al.*, Sub-micrometer-sized biodegradable particles of poly(L-lactic acid) via the gas antisolvent spray precipitation process. *Biotechnology Progress*, 1993, **9**, 429–435.
84. Tom, J.W. and P.G. Debenedetti, Formation of bioerodible polymeric microspheres and microparticles by rapid expansion of supercritical solutions. *Biotechnological Progress*, 1991, **7**, 403–411.
85. Cleland, J. Protein delivery from biodegradable microspheres, in L. Sanders and W. Hendren, *Protein Delivery–Physical Systems*, New York: Plenum Press, 1997, pp. 1–43.
86. Okada, H. and H. Toguchi, Biodegradable microspheres in drug delivery. *Critical Reviews in Therapeutic Drug Carrier Systems*, 1995, **12**(1), 1–99.
87. Ogawa, Y., *et al.*, *In vivo* release profiles of leuprolide acetate from microcapsules prepared with polylactic acids or copoly(lactic/glycolic) acids and *in vivo* degradation of these polymers. *Chem. Pharm. Bull. (Japan)*, 1988, **36**, 2576–2581.
88. Putney, S.D. and P.A. Burke, Improving protein therapeutics with sustained-release formulations. *Nature Biotechnology*, 1998., **16**, 153–157.
89. Cleland, J., *et al.*, Development of a single-shot subunit vaccine for HIV-1 2. Defining optimal autoboost characteristics to maximize the humoral immune response. *Journal of Pharmaceutical Sciences*, 1996, **85**(12), 1346–1348.
90. Gref, R., *et al.*, Biodegradable long-circulating polymeric nanospheres. *Science*, 1994, **263**, 1600–1603.
91. Liu, F., *et al.*, Glucose-induced release of glycosylpoly(ethylene glycol) insulin bound to a soluble conjugate of concanavalin A. *Bioconjugate Chemistry*, 1997, **8**(5), 664–672.
92. Kost, J., *et al.*, Glucose sensitive membranes containing glucose oxidase: activity, swelling, and permeability studies. *Journal of Biomedical Materials Research*, 1984, **19**, 1117–1133.
93. Sakiyama-Elbert, S.E. and J.A. Hubbell, Development of fibrin derivatives for controlled release of heparin-binding growth factors. *Journal of Controlled Release*, 2000, **65**, 389–402.

10

Case Studies in Drug Delivery

Most of the change we think we see in life
Is due to truths being in and out of favour.
Robert Frost, The Black Cottage (1914)

This chapter illustrates the concepts presented throughout the book through the examination of three different clinical scenarios in which new methods for drug delivery are needed. Many agents cannot be administered orally because of poor absorption from the intestinal tract into the blood, yet they are rapidly eliminated once they enter the blood stream. Section 10.1[1] describes a controlled delivery system that produces prolonged levels of such an antiviral agent in the blood. The brain is protected from changes in blood chemistry by the blood–brain barrier; this natural defense mechanism makes drug delivery to the brain difficult. Section 10.2 describes one method for achieving prolonged concentrations of an active agent within a region of the brain. Finally, some agents are active on the skin or mucosal surfaces but must be present for long periods. Section 10.3 presents a method for prolonging the residence time of macromolecules on a mucosal surface.

The three problems incorporate both aspects of the drug delivery challenge: design of methods or materials for introducing drugs into the body and optimization of the design to integrate the delivery system with the body's natural mechanisms for distributing and eliminating foreign agents.

10.1 CONTROLLED DELIVERY OF SYSTEMIC THERAPY

Viral diseases are a significant cause of disability and death. Many deadly viral diseases—such as smallpox and polio—are now under control, due largely to the development of protective vaccines, but vaccines for certain viral illnesses

---

1. Sections 10.1 and 10.3 were co-authored by Rebecca K. Willits and W. Mark Saltzman. Section 10.2 has appeared elsewhere in a different form [1].

have been difficult to develop. The development of an AIDS vaccine has been a priority among biomedical research efforts since the mid-1980s; tremendous energy and resources have been invested in this pursuit, but clinical progress towards a vaccine has been slow.

On the other hand, antiviral therapies have had a significant impact on the clinical care of AIDS patients, particularly since multi-drug regimens targeting the retroviral reverse transcriptase and protease enzymes were developed in the late 1990s. However, many patients cannot tolerate aggressive antiviral therapy and development of drug-resistant viral strains is a persistent problem. Therefore, the search for alternative methods for blocking retroviral infections continues.

10.1.1 Use of T-20 to Block HIV Replication in Humans

Glycoproteins in the viral envelope permit human immunodeficiency virus type 1 (HIV-1) to recognize, fuse with, and enter human cells. The surface protein gp120 binds to a receptor, called CD4, and chemokine co-receptors on cells of the immune system; agents that block gp120–CD4 binding prevent viral infection of cultured cells. Some agents that block gp120 binding to the CD4 receptor have been tested in AIDS patients including dextran sulfate, soluble CD4, and CD4–IgG. Patients receiving dextran sulfate experienced an increase in viral titers and significant toxicity [2]. Therapies involving CD4 required high doses and produced inconsistent results [3]. The reasons for failure of gp120 blockade are not known; these set-backs have encouraged investigators to explore other targets for inhibiting viral entry.

The gp120-mediated binding event brings the virus in close proximity with the cell surface; fusion with the cell is facilitated by a conformational change in a second subunit of the receptor, gp41 (Figure 10.1). Therefore, agents that block this gp41-mediated fusion event should permit viral binding, but prohibit viral entry into the cell. T-20 is a 36 amino acid synthetic peptide, which was designed by analysis of the gp41 protein structure, that competitively inhibits the conformational change in gp41 and, therefore, blocks viral fusion. In cell-culture experiments, T-20 inhibits virus infection at concentrations as low as 2 ng/mL. Based on this activity, T-20 has been tested for activity as an anti-retroviral agent in patients with AIDS [4].

10.1.2 Pharmacokinetics of T-20 after Intravenous Administration

The T-20 peptide was administered to patients via intravenous injection [4]. Because the half-life of the peptide is short ($t_{1/2} = 1.8$ h), and because it was suspected that several weeks of antiviral therapy would be needed, patients were given intravenous infusions of T-20 every 12 h for 14 days at doses ranging from 3 to 100 mg per dose. Expected plasma concentrations for T-20, based on the average $t_{1/2}$ and V_d measured in AIDS patients, over the first week of treatment are shown in Figure 10.2.

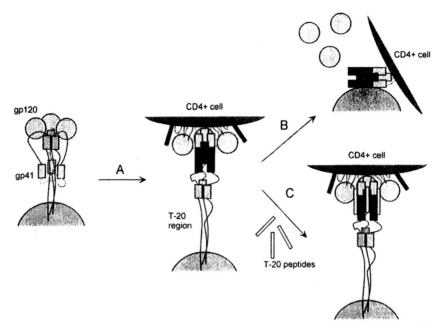

Figure 10.1   Proposed mechanism of anti-retroviral activity of peptide T-20. Based on a drawing in [4]. During step A, the virus gp120 binds to CD4 on the target cell and a conformational change in gp41 exposes the fusion peptide, which inserts itself into the cell membrane. In the absence of T-20 peptide (step B), further conformational changes in gp41 bring the cell and virus surfaces together. Fusion and viral entry can now occur. T-20 peptide binds to a region of the gp41 structure (step C), preventing the conformational changes that allow viral fusion and entry.

The highest dose T-20 regimen (100 mg, twice daily) was effective at reducing viral load in patients. Viral concentration dropped by 99% over the 14-day treatment period: this corresponds to a half-life of viral clearance of ~ 2 days.

10.1.3 Design of a Controlled-Delivery System for T-20

Required Rate of Release. Controlled-release delivery systems are an alternative to frequent intravenous administration. Here, we consider the design of an implantable controlled-release matrix for T-20 delivery to human patients. The objective is to provide sustained plasma levels after a single administration. A one-compartment pharmacokinetic model provides a useful means for estimating the required rate of release from the matrix. Equation 7-1 can be modified to account for a continuous release of drug into the central compartment:

$$\dot{M} - kV_{\mathrm{d}}c = \frac{\mathrm{d}c}{\mathrm{d}t}V_{\mathrm{d}} \tag{10-1}$$

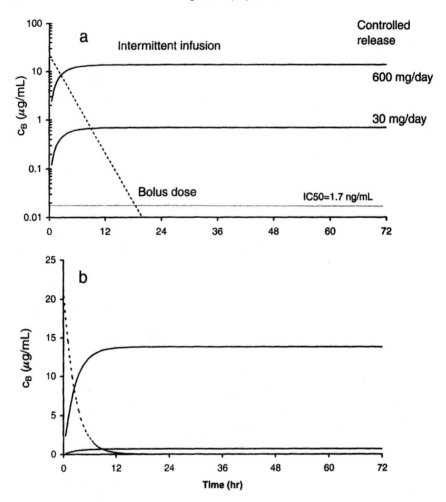

Figure 10.2   Pharmacokinetics of intravenous T-20 administration. The graphs show the expected changes in plasma concentration vs time after intravenous administration of a single injection (dashed line) or multiple injections (solid line) of intravenous T-20 (100mg/dose). The curves are based on the half-life (1.8 h) and volume of distribution (4.7 L) measured in 17 human volunteers [4] using a one-compartment model (see Equation 7-3). Because of its rapid elimination, multiple doses are needed to maintain the peptide level in the effective range.

where $\dot{M}$ is the rate of release into the plasma. Assuming that the initial concentration is zero, this first-order differential equation yields:

$$c(t) = \frac{\dot{M}}{V_d k}\left(1 - e^{-kt}\right) \tag{10-2}$$

Plasma concentration at steady state is obtained from Equation 10-2 in the limit as $t \to \infty$:

$$c_{SS} = \frac{\dot{M}}{kV_d} \tag{10-3}$$

Plasma concentrations based on Equation 10-2 are shown in Figure 10.2 for two different release rates, 30 and 600 mg/day. These release rates provide steady-state concentrations that encompass most of the range expected from the twice daily infusion. The steady-state concentration is reached in $\sim 6\,\mathrm{h}$.

Design of a Matrix-type Delivery Device. Assume that the agent is delivered from an implanted polymer matrix, similar to the matrices described in Chapter 9 (see Figure 9.6). Implanted matrices of poly(ethylene-*co*-vinyl acetate) (EVAc) can be loaded with up to 50% drug by weight. A high loading will be necessary in this case, because our anticipated daily dose of T-20 is 30 to 600 mg/day. A device that lasts 14 days must weigh at least 840 mg to accommodate the T-20; this could be achieved by a 2-mm thick disk with a radius of 1.2 cm. For comparison, the Norplant® system for hormonal contraception, which is implanted subcutaneously in the arm, consists of six long cylinders, each being 2.4 mm in diameter and 34 mm in length. Norplant®, therefore, has a total volume that is slightly greater than the minimum-sized, disk-shaped device proposed here.

In most situations, the rate of release of agent from the matrix depends on the rate of agent diffusion through the polymer matrix; diffusion is characterized by an effective diffusion coefficient D_{eff}, which was defined in Equation 9-21. As the agent is released from the matrix, it accumulates in the tissue surrounding the implant. Therefore, the concentration of agent in the tissue surrounding the implant changes with time. The overall process of release from the matrix can be described by the equations for transient diffusion in the matrix:

$$\frac{\partial c_p}{\partial t} = D_{eff} \frac{\partial^2 c_p}{\partial x^2} \tag{10-4}$$

subject to the following initial and boundary conditions:

$$\begin{aligned}
c_p &= c_{p0} & t &= 0 & 0 &\le x \le L \\
c_p &= c_Q(t) & t &> 0 & x &= 0, L
\end{aligned} \tag{10-5}$$

where L is the thickness of the matrix, c_p is the agent concentration in the matrix, c_{p0} is the initial concentration of agent within the matrix, and c_Q is the concentration of agent in the tissue surrounding the implant (Figure 10.3). To simplify the analysis, the tissue surrounding the implant is assumed to be well mixed (which will be reasonable as long as agent diffusion in the tissue space is rapid compared to diffusion within the matrix).

The solution to differential Equation 10-4, subject to conditions in Equation 10-5, is (see [5], p. 102):

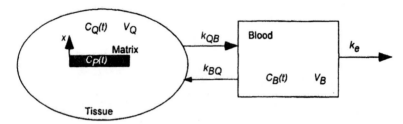

Figure 10.3   Compartment model for controlled delivery of T-20 by implanted matrices. The T-20 peptide is released from an implanted matrix. The agent accumulates in the local tissue and permeates through capillary walls to enter the blood.

$$c_{\mathrm{p}}(t) = \frac{2}{\pi} \sum_{n=0}^{\infty}$$

$$\left\{ \begin{array}{l} \left[\dfrac{(1-(-1^n)}{n} \exp\!\left(-\dfrac{D_{\mathrm{eff}} n^2 \pi^2 t}{L^2}\right) \sin\!\left(\dfrac{n\pi x}{L}\right) \right] c_{\mathrm{p}0} \\[2ex] + \dfrac{2D_{\mathrm{eff}}}{L^2} n\pi(1-(-1)^n) \sin\!\left(\dfrac{n\pi x}{L}\right) \int_0^1 \exp\!\left(-\dfrac{D_{\mathrm{eff}} n^2 \pi^2 (t-\lambda)}{L^2}\right) c_Q(\lambda)\mathrm{d}\lambda \end{array} \right\}$$

$$(10\text{-}6)$$

For a matrix with cross-sectional area A, the total quantity of agent remaining in the matrix after time t, M_t, can be found by integrating the concentration profile, Equation 10-6:

$$M_t = A \int_0^L c_{\mathrm{p}}(t)\mathrm{d}x \qquad (10\text{-}7)$$

which will decrease from the initial quantity of agent in the matrix M_{t0}, or $ALc_{\mathrm{p}0}$, to zero as the agent is released.

The functional dependence of M_t on time can be found by substituting Equation 10-6 into the integral Equation 10-7, but this integral cannot be evaluated until the function $c_Q(t)$ is known. The release from matrices into an infinite reservoir, in which $c_Q(t) = 0$, was discussed in Chapter 9. To achieve release over a one-month period, the matrix must be designed so that D_{eff} is $\sim 10^{-9}$ cm^2/s (Figure 9.7).[2]

Prediction of Blood Concentrations after Implantation of Controlled-release System.   To relate agent release from the matrix to concentrations in the tissue, a simple compartmental model is employed (Figure 10.3). An additional relationship between agent release from the matrix and

2. Note that the calculation used to construct Figure 9.7 assumed a matrix thickness of 1 mm, whereas a thicker matrix may be required in order to achieve the necessary amount of drug.

concentration in the tissue surrounding the matrix can be found from a mass balance on the tissue:

$$V_Q c_Q(t) + \int_0^t [k_{QB} c_Q(t) - k_{BQ} c_B(t)]dt = M_{t0} - A \int_0^L c_p(t)dx \qquad (10\text{-}8)$$

where V_Q is the volume of the tissue surrounding the implant (i.e., the volume through which released agent in the tissue is uniformly distributed), c_B is the concentration in the blood, and k_{BQ} and k_{QB} are rate constants for transfer of agent between the tissue and the blood circulating through the tissue. A mass balance on the blood provides an additional relationship:

$$V_B c_B(t) - \int_0^t [k_{QB} c_Q(t) - k_{BQ} c_B(t)]dt = - \int_0^t k_e c_B(t)dt \qquad (10\text{-}9)$$

where k_e is a first-order rate constant for elimination of agent from the blood. Substitution of Equation 10-6 into Equation 10-8 gives an integral equation in terms of the two unknown functions c_B and c_Q; this equation must be solved simultaneously with Equation 10-9, which is also an integral equation in terms of the same two unknown functions. These equations can be solved by using numerical methods to estimate the values of the integrals.

The objective in this application is prolonged maintenance of T-20 in the blood after implantation. Figure 10.4 shows T-20 concentrations after implantation of a controlled-release matrix as a function of the parameters D_{eff} (which is designed into the polymer matrix) and k_{QB}. The elimination rate constant, k_e, was determined from the half-life for disappearance of the agent from the plasma, $\ln(2)/t_{1/2} = 1.8\,h$ (see Figure 10.2).

The effect of transfer coefficient—the rate of movement from the subcutaneous compartment to the blood—is illustrated in Figure 10.4. For small values of k_{BQ} ($10^{-6}\,s^{-1}$, which corresponds to a very long subcutaneous tissue-to-blood transfer half-life, 200 h) drug concentrations remain relatively constant throughout the 14-day period. For larger values of k_{BQ}—illustrated here as $10^{-4}\,s^{-1}$, which corresponds to a transfer half-life of 2 h—the concentration in the subcutaneous compartment drops more rapidly and blood concentrations are higher. Assuming that the actual transfer coefficient is $\sim 10^{-4}\,s^{-1}$, the controlled-release system provides concentrations in the blood (in the range 0.1–0.2 μg/mL) throughout the two-week treatment period; this concentration is slightly lower than the most effective concentration after repeated bolus administration, which was produced by high-dose treatment in patients (Figure 10.2).

10.1.4 Conclusion

Controlled-release matrices are capable of releasing agents for prolonged periods, as described in Chapter 9; this example illustrates one possible application of the technology. The calculations suggest that reasonably constant concen-

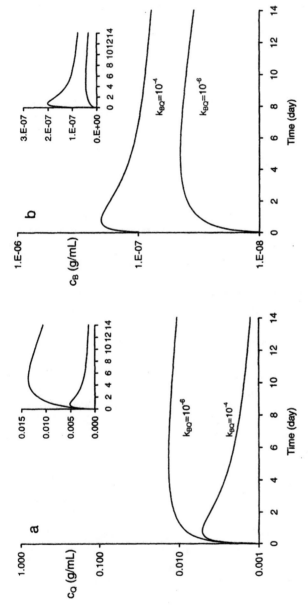

Figure 10.4 Predicted plasma concentrations of T-20 after implantation of controlled release matrix. T-20 peptide is released from an implanted matrix; resulting concentrations are shown for the subcutaneous compartment (a) and the blood (b). The matrix thickness is 2 mm, c_{p0} is 1 g/mL, and D_{eff} is 10^{-9} cm^2/s. The transfer coefficients—k_{BQ} and k_{QB}—are assumed to be equal; results are shown for two different values (10^{-6} and 10^{-4} s^{-1}). The larger graphs show a semi-log plot of concentration versus time; insets show the same curves as a linear plot.

trations can be achieved for prolonged periods; this prediction is based on the assumption of a simple expression for the rate of exchange between the subcutaneous interstitial space and the blood. Perhaps the most significant obstacle to this particular system is the large quantity of agent that must be administered by twice daily doses in order to be effective. Unless the agent is effective at a lower concentration when delivered continuously, the implant will be large. This approach may be more practical for more potent agents.

10.2 IMPLANTS FOR LOCAL DRUG DELIVERY

Traditional methods for delivering drugs to the brain have several disadvantages. Many drugs, particularly water-soluble or high molecular weight compounds, do not enter the brain following systemic administration because they penetrate the blood–brain barrier (BBB) very slowly (see Figure 5.32). This barrier to transport limits the number of drug molecules that are candidates for treating brain disease. Therefore, a number of strategies have been developed for the delivery of protein therapeutic agents to the central nervous system (CNS) [6].

10.2.1 Drug Delivery to the Brain

Since the BBB is generally permeable to lipid-soluble compounds, which can dissolve and diffuse through endothelial cell membranes [7, 8], a common approach for enhancing brain delivery of compounds is to modify them chemically so that their lipid solubility is enhanced. An extension of this approach involves the chemical modification of drugs by conjuction to methyldihydropyridine [9]. The chemically modified drug is permeable in brain capillaries, but immediately converted into an impermeable salt by enzymes within the brain tissue. Unfortunately, these lipidization approaches do not appear to be useful for peptides larger than 1,000. Another approach for increasing permeability is the entrapment of peptides or proteins in liposomes [10], which have not yet been used successfully to deliver drugs to the brain, probably due to their limited stability in the plasma and rapid uptake at other tissue sites.

Several strategies for increasing the permeability of the brain capillaries to proteins have been developed. The permeability of the BBB can be transiently increased by intra-arterial injection of the solutions with high osmolarity, which disrupts inter-endothelial tight junctions [11]. Certain protein modifications, such as cationization by hexamethyldiamine [12] and anionization by succinylation [13], produce enhanced uptake in the brain. Modification of drugs [14] and proteins [15] by linkage to an anti-transferrin receptor antibody also appears to enhance transport into the brain. This approach depends on receptor-mediated transcytosis of transferrin–receptor complexes by brain endothelial cells; substantial uptake also occurs in the liver.

Some protein drugs have been administered intranasally [16], an approach that frequently leads to significant systemic bioavailability. Although the intranasal approach has been attempted for drug delivery to the brain, it has only

been successfully demonstrated for small, lipid soluble compounds like progesterone. Intraventricular therapy, in which agents are administered directly into
the cerebrospinal fluid (CSF) of the ventricles, results in high concentrations
within the brain tissue, but only in regions immediately surrounding the ventricles [17]. Because the agent must diffuse into the brain parenchyma from the
ventricles, and because of the high rate of clearance of agents in the CNS into
the peripheral circulation, this strategy has not been used to deliver agents deep
into the brain.

An alternative approach is direct delivery of agents into the tissue by controlled release or infusion; both techniques provide sustained drug delivery. In
addition, polymeric controlled-release devices or catheter systems can be
implanted at specific sites, localizing therapy to a brain region. Because these
methods provide a localized and continuous source of active drug molecules, the
total drug dose can be less than needed with systemic administration. With
polymeric controlled release, the implants can also be designed to protect unre-
leased drug from degradation in the body and to permit localization of extremely high doses (up to the solubility of the drug) at precisely defined locations in
the brain. Infusion systems require periodic refilling; drug is usually stored in a
liquid reservoir at body temperature and many drugs are not stable under these
conditions. In this example, we will consider the use of controlled-release
implants for delivery of chemotherapy directly at the site of a brain tumor.

10.2.2 Design of Implanted Controlled-Delivery Systems
for Chemotherapy

The kinetics of drug release from a controlled-release system are frequently
characterized by measuring the amount of drug released from the matrix into a
well-stirred reservoir of phosphate buffered water or saline at 37 °C.
Controlled-release profiles for some representative anticancer agents are
shown in Figure 10.5; all of the agents selected for these studies—1,3-bis(2-
chloroethyl)-1-nitrosourea (BCNU), 4-HC, cisplatin, and taxol—are used clinically for chemotherapy of brain tumors. The controlled-release period can vary
from several days to many months, depending on the drug and polymer cho-
sen. Therefore, the delivery system can be tailored to the therapeutic situation
by selection of implant properties.

The release of drug molecules from polymer matrices can be regulated by
diffusion of drug through the polymer matrix or degradation of the polymer
matrix. In many cases—including the release of BCNU, cisplatin, and 4-HC
from the degradable matrices shown in Figure 10.5—drug release from biodegradable polymers appears to be diffusion regulated, probably because the
degradation time is much longer than the time required for drug molecules
to diffuse through the polymer. Only diffusion-regulated release is discussed
here, since most degradable polymers provide release kinetics that are consistent with diffusion. In a few special cases linear release, which appears to
correlate with the polymer degradation rate, can be achieved, however; this
might be the case for paclitaxel release from the biodegradable matrix (Figure

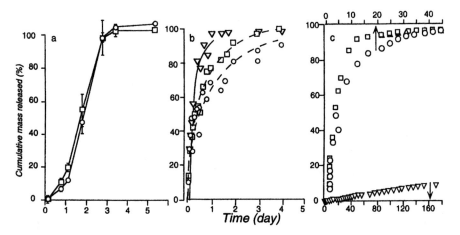

Figure 10.5 Controlled release of anticancer compounds from polymeric matrices.
(a) Release of cisplatin (circles) from p(FAD/SA) initially containing 10% drug.
Similar results have been obtained for BCNU, MTX, and a variety of other compounds (see [18, 19] for details). (b) Release of BCNU from EVAc (circles), p(CPP/SA) (squares), and p(FAD/SA) (triangles) matrices initially containing 20% drug.
(c) Release of BCNU (squares), 4-HC (circles), and paclitaxel (triangles) from p(CPP/SA) matrices initially containing 20% drug. Note that panel c has two time axes: the lower axis applies to the release of paclitaxel and the upper axis applies to the release of BCNU and 4-HC

10.5c), although the exceedingly low solubility of paclitaxel in water may also contribute substantially to the slowness of release.

For diffusion-mediated release, the amount of drug released from the polymer is proportional to the concentration gradient of the drug in the polymer. By performing a mass balance for drug within a differential volume element in the polymer, the concentration of drug within the polymer as a function of position and time can be described as above (see Equation 10-4):

$$\frac{\partial c_p}{\partial t} = D_{i:p} \nabla^2 c_p \qquad (10\text{-}10)$$

where c_p is the local concentration of drug in the polymer, $D_{i:p}$ is the diffusion coefficient of the drug in the polymer matrix, and t is the time following immersion in the reservoir. This equation can be solved, with appropriate boundary and initial conditions, to obtain the cumulative mass of drug released as a function of time, as in the first case study (Section 10.1). A useful approximate solution, which is valid for the initial 60% of release, is (recall section 9.2):

$$M_t = 4M_o \sqrt{\frac{D_{i:p} t}{\pi L^2}} \qquad (10\text{-}11)$$

where M_t is the cumulative mass of drug released from the matrix, M_o is the initial mass of drug in the matrix, and L is the thickness of the implant. By comparing Equation 10-11 to the experimentally determined profiles, the rate of diffusion of the agent in the polymer matrix can be estimated (Table 10.1).

The design of these chemotherapy delivery systems could be modified to achieve other release rates by selection of materials and methods of production, as discussed in Chapter 9.

10.2.3 Drug Transport after Release from the Implant

Bypassing the BBB is necessary, but not sufficient, for effective drug delivery. Consider the consequences of implanting a delivery system, such as the ones characterized above, within the brain. Molecules released into the interstitial fluid in the brain extracellular space must penetrate into the brain tissue to reach tumor cells distant from the implanted device. Before these drug molecules can reach the target site, however, they might be eliminated from the interstitium by partitioning into brain capillaries or cells, entering the CSF, or being inactivated by extracellular enzymes. Elimination always accompanies dispersion; therefore, regardless of the design of the delivery system, one must understand the dynamics of both processes in order to predict the spatial pattern of drug distribution after delivery. Although this case study focuses on drug transport in the context of polymeric controlled release to the brain, many of these issues apply to other novel drug delivery strategies (see [20] for further examples in the CNS).

Assume that the drug diffuses through, and is simultaneously eliminated from, tissue in the vicinity of an implant releasing active molecules. The polymer implant is surrounded by biological tissue, composed of cells and an extracellular space (ECS) filled with extracellular fluid (ECF). Immediately

Table 10.1   Diffusion coefficients for chemotherapy drug release

Drug	Polymer	Initial loading (%)	$D_{I,p}$ (cm^2/s)
Cisplatin	P(FAD/SA)	10	6.8×10^{-9}
BCNU	EVAc	20	1.6×10^{-8}
BCNU	P(FAD/SA)	20	6.9×10^{-8}
BCNU	P(CPP/SA)	20	2.3×10^{-8} (panel b)
			2.0×10^{-8} (panel c)
4-HC	P(CPP/SA)	20	3.1×10^{-10}
Taxol	P(CPP/SA)	20	n.a.

The diffusion coefficients were obtained by comparing the experimental data shown in Figure 10.5 to Equation 10-11 and determining the best value of the diffusion coefficient to represent the data. This technique is discussed in more detail elsewhere. FAD = fatty acid dimer, SA = sebacic acid, CPP = carboxyphenoxypropane, n.a. = not applicable.

following implantation, drug molecules escape from the polymer and penetrate the tissue. Once in the brain tissue, drug molecules (i) diffuse through the tortuous ECS in the tissue, (ii) diffuse across semipermeable tissue capillaries to enter the systemic circulation, and therefore are removed from the brain tissue, (iii) diffuse across cell membranes, by passive, active, or facilitated transport paths, to enter the intracellular space, (iv) transform, spontaneously or by an enzyme-mediated pathway, into other compounds, and (v) bind to fixed elements in the tissue. All of these events influence drug therapy: diffusion through the ECS is the primary mechanism of drug distribution in brain tissue; elimination of the drug occurs when it is removed from the ECF or transformed; and binding or internalization may slow the progress of the drug through the tissue.

A mass balance on a differential volume element in the tissue [21] gives a general equation describing drug transport in the region near the polymer [17]:

$$\frac{\partial c_t}{\partial t} + \bar{v} \cdot \nabla c_t = D_b \nabla^2 c_t + R_e(c_t) - \frac{\partial B}{\partial t} \tag{10-12}$$

where c_t is the concentration of the diffusible drug in the tissue surrounding the implant (g/cm^3 tissue), $\bar{v}$ is the fluid velocity (cm/s), D_b is the diffusion coefficient of the drug in the tissue (cm^2/s), $R_e(c)$ is the rate of drug elimination from the tissue (g/s $\cdot$ cm^3 tissue), B is the concentration of drug bound or internalized in cells (g/cm^3 tissue), and t is the time following implantation. (A different form of this equation, which accounts more rigorously for transfer of drug between different phases in the tissue, is also available [22].) In deriving this equation, the conventions developed by Nicholson [23], based on volume averaging in a complex medium, and Blasberg *et al.* [17] were combined. In this version of the equations, the concentrations c_t and B and the elimination rate $R_e(c_t)$ are defined per unit volume of tissue. D_b is an effective diffusion coefficient, which must be corrected for the tortuosity of the ECS as in Chapter 4.

When the binding reactions are rapid, the amount of intracellular or bound drug can be assumed to be directly proportional, with an equilibrium coefficient K_{bind}, to the amount of drug available for internalization or binding:

$$B = K_{bind} c_t \tag{10-13}$$

Substitution of Equation 10-13 into Equation 10-12 yields, with some simplification:

$$\frac{\partial c_t}{\partial t} = \frac{1}{1 + K_{bind}} \left(D_b \nabla^2 c_t + R_e(c_t) - \bar{v} \cdot \nabla c_t \right) \tag{10-14}$$

The drug elimination rate, $R_e(c_t)$, can be expanded into the following terms:

$$R_e(c_t) = k_{bbb} \left(\frac{c_t}{\varepsilon_{ecs}} - c_{plasma} \right) + \frac{V_{max} c_t}{K_m + c_t} + k_{ne} c_t \tag{10-15}$$

where k_{bbb} is the permeability of the BBB (defined on the basis of concentration in the ECS), c_{plasma} is the concentration of drug in the blood plasma, V_{max}

and K_m are Michaelis–Menten constants, and k_{ne} is a first-order rate constant for drug elimination due to non-enzymatic reactions. For any particular drug, some of these rate constants may be very small, reflecting the relative importance of each mechanism of drug elimination. If it is assumed that the permeability of the BBB is low ($C_{pl} \ll C$) and the concentration of drug in the brain is sufficiently low so that any enzymatic reactions are in the first-order regime ($C \ll K_m$), Equation 10-15 can be reduced to:

$$-R_e(c_t) = \frac{k_{bbb}}{\varepsilon_{ecs}} c_t + \frac{V_{max}}{K_m} c_t + k_{ne} c_t = k_{app} c_t \qquad (10\text{-}16)$$

where k_{app} is a lumped first-order rate constant. With these assumptions, Equation 10-14 can be simplified by definition of an apparent diffusion coefficient, D^*, and an apparent first-order elimination constant, k^*:

$$\frac{\partial c_t}{\partial t} = D^* \nabla^2 c_t + k^* c_t - \frac{\bar{v} \cdot \nabla c_t}{1 + K_{bind}} \quad \text{where } k^* = \frac{k_{app}}{1 + K_{bind}} \quad \text{and } D^* = \frac{D_b}{1 + K_{bind}}$$
$$(10\text{-}17)$$

Boundary and initial conditions are required for solution of Equation 10-17. If a spherical implant of radius R is implanted into a homogeneous region of the brain, at a site sufficiently far from anatomical boundaries, the following assumptions are reasonable:

$$c_t = 0 \quad \text{for} \quad t = 0; \; r > R$$
$$c_t = c_i \quad \text{for} \quad t > 0; \; r = R \qquad (10\text{-}18)$$
$$c_t = 0 \quad \text{for} \quad t > 0; \; r \to \infty$$

In many situations, drug transport due to bulk flow can be neglected. This assumption ($\bar{v}$ is zero) is common in previous studies of drug distribution in brain tissue [17]. For example, in a study of cisplatin distribution after continuous infusion into the brain, the effects of bulk flow were found to be small, except within 0.5 mm of the site of infusion [24]. In the cases considered here, since drug molecules enter the tissue by diffusion from the polymer implant, not by pressure-driven flow of a fluid, no flow should be introduced by the presence of the polymer. With fluid convection assumed to be negligible, the general governing equation in the tissue, Equation 10-17, reduces to:

$$\frac{\partial c_t}{\partial t} = D^* \nabla^2 c_t + k^* c_t \qquad (10\text{-}19)$$

The no-flow assumption may be inappropriate in certain situations. In brain tumors, edema and fluid movement are components of the disease and some drugs can elicit cytotoxic edema. Certain drug/polymer combinations can release drugs in sufficient quantity to create density-induced fluid convection.

Equation 10-19 with the conditions provided in Equation 10-18 can be solved by Laplace transform techniques [20] to yield:

$$\frac{c_t}{c_i} = \frac{1}{2\zeta}\left\{\exp[-\phi(\zeta-1)]\mathrm{erfc}\left[\frac{\zeta-1}{2\sqrt{\tau}} - \phi\sqrt{\tau}\right] + \exp[\phi(\zeta-1)]\mathrm{erfc}\left[\frac{\zeta-1}{2\sqrt{\tau}} + \phi\sqrt{\tau}\right]\right\}$$

(10-20)

where the dimensionless variables are defined as follows:

$$\zeta = \frac{r}{R} \qquad \tau = \frac{D^*t}{R^2} \qquad \phi = R\sqrt{\frac{k^*}{D^*}}$$

(10-21)

The differential equation also has a steady-state solution:

$$\frac{c_t}{c_i} = \frac{1}{\zeta}\exp[-\phi(\zeta-1)]$$

(10-22)

Figure 10.6 shows concentration profiles calculated using Equations 10-20 and 10-22. Using reasonable values for all of the parameters, steady state is reached approximately 1 h after implantation of the delivery device. The time required to achieve steady state depends on the rate of diffusion and elimination, as previously described [25], but will be significantly less than 24 h for most drug molecules.

10.2.4 Application of Diffusion–Elimination Models to Intracranial BCNU Delivery Systems

This analysis, which assumes that local drug concentrations are governed by the rates of diffusion and first-order elimination in the tissue, agrees well with experimental concentration profiles obtained after implantation of controlled-release polymers (Figure 10.7). At 3, 7, and 14 days after implantation of a BCNU-releasing implant, the concentration profile at the site of the implant was very similar. The parameter values (obtained by fitting Equation 10.22 to the experimental data) were consistent with parameters obtained using other methods [22], suggesting that diffusion and first-order elimination were sufficient to account for the pattern of drug concentration observed during this period. Parameter values were similar at 3, 7, and 14 days, indicating that the rates of drug release, dispersion, and elimination did not change during this period. This equation has been compared to concentration profiles measured for a variety of molecules delivered by polymer implants to the brain—dexamethasone [25, 26], molecular weight fractions of dextran [27], nerve growth factor in rats [28], and BCNU in rats [22], rabbits [29], and monkeys [30]. In all of these cases, the steady-state diffusion–elimination model appears to capture most of the important features of drug transport.

This model suggests guidelines for the design of intracranial delivery systems. Table 10.2 lists the important physical and biological characteristics of some of a few compounds that have been considered for interstitial delivery to treat brain tumors. When the implant is surrounded by tissue, the maximum rate of drug release is determined by the solubility of the drug, c_s, and the rate of diffusive transport through the tissue:

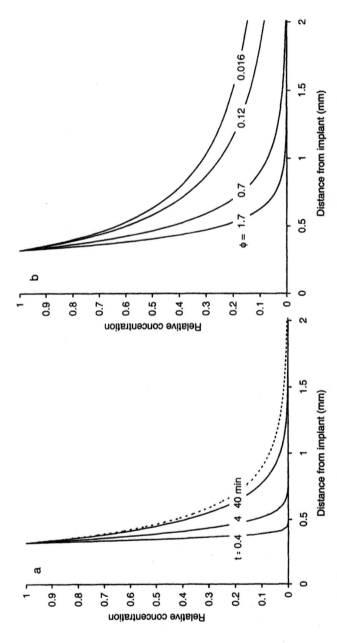

Figure 10.6 Concentration profiles after implantation of a spherical drug-releasing implant. (a) Solid lines represented the transient solution to Equation 10-20 with the following parameter values: $D^* = 4 \times 10^{-7}\,cm^2/s$; $R = 0.032\,cm$; $k^* = 1.9 \times 10^{-4}\,s^{-1}$ ($t_{1/2} = 1\,h$). The dashed line represents the steady-state solution (i.e., Equation 10-22) for the same parameters. (b) Solid lines in this plot represent Equation 10-22 with the following parameters: $D^* = 4 \times 10^{-7}\,cm^2/s$; $R = 0.032\,cm$. Each curve represents the steady-state concentration profile for drugs with different elimination half-lifes in the brain, corresponding to different dimensionless moduli, ϕ: $t_{1/2} = 10\,min$ ($\phi = 1.7$); $1\,h$ (0.7); $34\,h$ (0.12); and $190\,h$ (0.016). As the half-life increases, ϕ decreases and the concentration profile is shifted farther to the right indicating better penetration of drug.

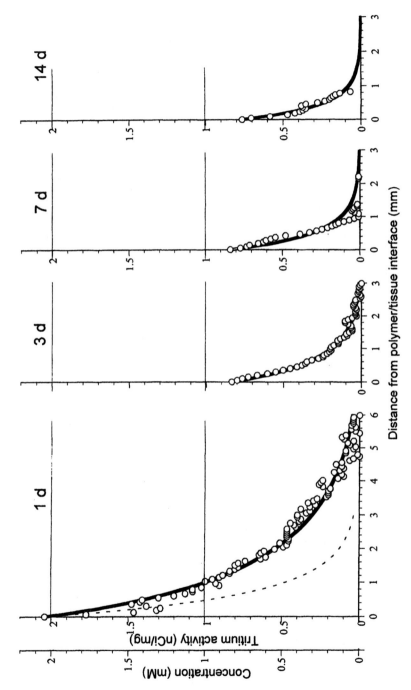

Figure 10.7 Concentration profiles after implantation of a BCNU-releasing implant. Solutions to Equation 10-22 were compared to experiment data obtained by quantitative autoradiographic techniques. The solid lines in the three panels labeled 3, 7, and 14 days were all obtained using the following parameters: $R = 0.15$ cm; $\phi = 2.1$; and $c_i = 0.81$ mM. The solid line in the panel labeled 1 day was obtained using the following parameters; $R = 0.15$ cm; $\phi = 0.7$ and $c_i = 1.9$ mM. Redrawn from [22].

Table 10.2   Implant design applied to chemotherapy compounds

	BCNU	4-HC	Methotrexate
M_w	214	293	454
C_s (mM)	12	100	100
$\log_{10} K$	1.53	0.6	−1.85
k^* (s^{-1})	0.02	0.02	0.0003
D^* (10^{-7} cm^2/s)	14	14	5
Toxic concentration in culture (μM)	25	10	0.04
Max. release rate (mg/day)	1.2	14	17
Implant lifetime at max. rate (days)	0.85	0.07	0.06
Max. concentration in tissue for 1-week-releasing implant (mM)	1.5	1.1	1.8
R_T (mm)	1.3	2.5	5

K is the octanol/water partition coefficient, k is the rate of elimination due to permeation through capillaries, D_b is the diffusion coefficient of the drug in the brain. The following values are assumed, consistent with results from polymer delivery to rats and rabbits: radius of spherical implant, $R = 1.5$ mm; mass of implant, $M = 10$ mg; drug loading in implant $= 10\%$.

$$\left(\frac{dM_t}{dt}\right)_{max} = (\text{Maximum flux}) \times (\text{Surface area}) = -D^*\frac{\partial c_t}{\partial r}\bigg|_R 4\pi R^2 \qquad (10\text{-}23)$$

Evaluating the derivative in Equation 10-23 from the steady-state concentration profile (Equation 10-22) yields:

$$\left(\frac{dM_t}{dt}\right)_{max} = 8\pi D^* c_s R \qquad (10\text{-}24)$$

Regardless of the properties of the implant, it is not possible to release drug into the tissue at a rate faster than determined by Equation 10-24. If the release rate from the implant is less than this maximum rate, c_i (the concentration in the tissue immediately outside the implant) is less than the saturation concentration, c_s. The actual concentration c^* can be determined by balancing the release rate from the implant (dM_t/dt, which can be determined from Equation 10-11 provided that diffusion is the mechanism of release from the implant) with the rate of penetration into the tissue obtained by substituting c_i for c_s in Equation 10-24:

$$c^* = \frac{dM_t}{dt}\left(\frac{1}{8\pi D^* R}\right) \qquad (10\text{-}25)$$

The effective region of therapy can be determined by calculating the distance from the surface of the implant to the point where the concentration drops below the cytotoxic level ($c_{cytotoxic}$, which is usually estimated as the cytotoxic concentration determined from *in vitro* experiments). Using Equation 10-22, and defining the radial distance for effective treatment as R_T, yields:

$$\frac{c_{\text{cytotoxic}}}{c_i} = \frac{R}{R_T} \exp\left\{-R\sqrt{\frac{k^*}{D^*}}\left(\frac{R_T}{R} - 1\right)\right\} \qquad (10\text{-}26)$$

Alternatively, an effective penetration distance, d_p, can be defined as the radial position at which the drug concentration has dropped to 10% of the peak concentration:

$$0.10 = \frac{R}{d_p} \exp\left\{-R\sqrt{\frac{k^*}{D^*}}\left(\frac{d_p}{R} - 1\right)\right\} \qquad (10\text{-}27)$$

Table 10.2 shows the results from some of these calculations, applied to three chemotherapy compounds, where a 10-mg implant containing 10% of drug was used.

In summary, a simple pseudo-steady-state equation (Equation 10-22) yielded simple guidelines (Equation 10-23 through Equation 10-27) for device design. Because the assumptions of the model were satisfied over a substantial fraction of the release period (days 3 to 14, based on the data shown in Figure 10.7), this analysis may be useful for BCNU release from biodegradable implants. The pseudo-steady-state assumptions are reasonable during this period of drug release, presumably because the time required to achieve steady state (on the order of minutes) is much less than the characteristic time associated with changes in the rate of BCNU release from the implant (days).

However, experimental concentration profiles measured 1 day after implantation were noticeably different: the peak concentration was substantially higher and the drug penetration into the surrounding tissue was deeper (see the left-hand panel of Figure 10.7). This behavior cannot be easily explained by the pseudo-steady-state models described above. For example, if the difference observed at day 1 represents transient behavior, the concentration observed at a fixed radial position should increase with time (Figure 10.6); in contrast, the concentration at any radial position on day 1 is higher than the concentration measured at that same position on subsequent days. Alternatively, the observed difference at day 1 might represent variability in the rate of BCNU release from the polymer implant over this period, with transport in the tissue remaining constant. When BCNU-releasing implants are tested *in vitro*, the rate of drug release did decrease over time (Figure 10.5). Equation 10-25 predicts the variation in peak concentration with release rate; the two-fold higher concentration observed at the interface on day 1 (as compared to days 3 through 14) could be explained by a two-fold higher release rate on day 1, but the effective penetration distance, d_p, does not depend on release rate. Experimentally measured penetration distances are $\sim 1.4\,\text{mm}$ on days 3, 7, and 14 and ~ 5 mm on day 1. This observation is shown more clearly in the day-1 panel of Figure 10.7: the dashed line shows the predicted concentration profile if k^* and D^* were assumed equal to the values obtained for days 3, 7, and 14. Therefore, changes in the rate of BCNU release are insufficient to explain the differences observed experimentally.

Penetration of BCNU is enhanced at day 1 relative to penetration at days 3, 7, and 14. For an implant of fixed size, penetration depends only on the ratio of elimination rate to diffusion rate: k^*/D^*. Increased penetration results from a decrease in this ratio (Figure 10.6), which could occur because of a decreased rate of elimination (smaller k^*) or an increased rate of diffusion (larger D^*). Neither of these changes seems probable for BCNU, with its high lipid solubility. BCNU can diffuse readily through brain tissue. In addition, elimination of BCNU from the brain will occur predominantly by partitioning into the circulation; since BCNU can permeate the capillary wall by diffusion, elimination is not a saturable, or concentration dependent, process. Perhaps the enhanced penetration of BCNU is due to the presence of another process for drug dispersion, such as bulk fluid flow, which was neglected in the previous analysis.

10.2.5 Limitations and Extensions of the Diffusion–Elimination Model

While this diffusion/elimination model compares very well with available experimental data, the assumptions used in estimating the concentration profiles in the brain may not be appropriate in all cases. Deviations from the predicted concentration profiles may occur due to (1) ECF flows in the brain, (2) complicated patterns of drug binding to extracellular proteins or other tissue components, or (3) multistep elimination pathways (which could involve phagocytic cells, extracellular enzymes, etc.). The motion of interstitial fluid in the vicinity of the polymer and the tumor periphery may not always be negligible, particularly in the region of a tumor. The interstitial fluid velocity is proportional to the pressure gradient in the interstitium; higher interstitial pressure in tumors—due to tumor cell proliferation, high vascular permeability, and the absence of functioning lymphatic vessels—may lead to steep interstitial pressure gradients at the periphery of the tumor [31]. As a result, interstitial fluid flows within the tumor may influence drug transport, because a drug at the periphery of the tumor must overcome outward convection to penetrate into the tumor [31]. Furthermore, local edema after surgical implantation of the polymer may cause significant fluid movement in the vicinity of the polymer. Improved mathematical models are needed to account for the convective contribution to drug transport.

The metabolism, elimination, and binding of drug are assumed to be first-order processes in our simple analysis. This assumption may not be realistic in all cases, especially for complex agents, like antibodies that target tumor-associated antigens. The metabolism of antibodies in normal and tumor tissues is still poorly understood. In addition, antibody concentration profiles are affected by a number of factors including molecular weight, binding affinity, antigen density, vascular permeability, metabolism, and heterogeneity within the tumor. Other cellular factors (e.g., the heterogeneity of tumor-associated antigen expression and multidrug resistance) that influence the uptake of therapeutic agents may not be accounted for by our simple first-order elimination.

Finally, changes in the brain that occur during the course of therapy are not properly considered in this model. Irradiation can be safely administered when a BCNU-loaded polymer has been implanted in monkey brains, suggesting the feasibility of adjuvant radiotherapy. However, irradiation also causes necrosis in the brain. The necrotic region has a perfusion rate and interstitial pressure lower than that of tumor tissue, thus the convective interstitial flow due to fluid leakage is expected to be smaller. Interstitial diffusion of macromolecules is lower in normal tissue and higher in tumor tissue as the latter has larger interstitial space [32]. The progressive changes in tissue properties—due to changes in tumor size, irradiation, and activity of chemotherapy agent—may be an important determinant of drug transport and effectiveness of therapy in the clinical situation.

When bulk fluid flow is present ($v \neq 0$), concentration profiles can be predicted from Equation 10-17, subject to the same boundary and initial conditions. This set of equations has been used to describe concentration profiles during micro-infusion of drugs into the brain [33]. In addition to Equation 10-17, conservation equations for water are needed to determine the variation of fluid velocity in the radial direction. Relative concentrations are predicted by assuming that the brain behaves as a porous medium (i.e., velocity is related to pressure gradient by Darcy's law, see Equation 6-9). Water introduced into the brain can expand the interstitial space; this effect is balanced by the flow of water in the radial direction away from the infusion source and, to a lesser extent, by the movement of water across the capillary wall.

In the presence of fluid flow, penetration of drug away from the source is enhanced (Figure 10.8). The extent of penetration depends on the velocity of the flow and the rate of elimination of the drug. The calculations used in constructing Figure 10.8 were performed for macromolecular drugs, which have limited permeability across the brain capillary wall [33]. The curves indicate steady-state concentration profiles for three different proteins with metabolic half-lives of 10 min, 1 h, or 33.5 h. In the absence of fluid flow, drugs with longer half-lives penetrate deeper into the tissue (solid lines in Figure 10.8 were obtained from Equation 10-22). This effect is amplified by the presence of flow (dashed lines in Figure 10.8).

During micro-infusion, drug is introduced by pressure-driven fluid flow from a small catheter. Therefore, pressure gradients are produced in the brain interstitial space, which lead to fluid flow through the porous brain microenvironment. Volumetric infusion rates of $3 \, \mu L/min$ were assumed in the calculations reproduced in Figure 10.8. Since loss of water through the brain vasculature is small, the velocity can be determined as a function of radial position:

$$v_r = \frac{q}{4\pi r^2 \varepsilon} \tag{10-28}$$

where q is the volumetric infusion rate and ε is the volume fraction of the interstitial space in the brain (~ 0.20). Fluid velocity decreases with radial distance from the implant (Table 10.3); within the first 20 mm of the implant

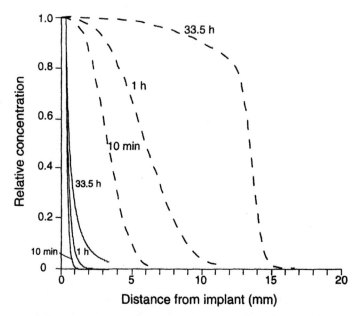

Figure 10.8   Effect of interstitial flow on drug penetration. Redrawn from [33].
Concentration profiles in the presence (dashed lines) and absence (solid lines) of fluid
flow. Solid lines were drawn using Equation 10-22 with $R = 0.032$ cm,
$D = 4 \times 10^{-7}$ cm^2/s; and $k = \ln(2)/t_{1/2}$ where $t_{1/2}$ is either 10 min, 1 h, or 33.5 h as
indicated. Dashed lines indicate steady-state concentration profiles in the presence of
an infusion of fluid of 3 μL/min at $x = 0$. The dashed line for $t_{1/2} = 35.5$ h is not at
steady state, but at 12 h after initiation of flow.

site, the predicted velocity was much greater than the velocities estimated previously during edema or tumor growth in the brain (cf. other fluid velocities in
tissues, compiled in Table 6.1).

The profiles predicted in Figure 10.8 were created by the introduction of
substantial volumes of fluid at the delivery site. Flow-related phenomena are
probably much less important in drug delivery by controlled-release implants.
Still, this transport model provides useful guidelines for predicting the influence

Table 10.3   Effect of radial position on interstitial
velocity during perfusion from a local source

Radial position (mm)	Interstitial velocity (μm/s)
2	5.0
5	0.8
10	0.2
20	0.05

Calculated using Equation 10-28.

of fluid flow on local rates of drug movement. Clearly, the effect of flow velocity on drug distribution is substantial; even relatively low flows, perhaps as small as $0.3\,\mu\text{m/s}$, are large enough to account for the enhancement in BCNU penetration observed at day 1 in Figure 10.7.

10.2.6 New Approaches to Drug Delivery Suggested by the Model

Mathematical models, which describe the transport of drug following controlled delivery, predict the penetration distance of drug and the local concentration of drug as a function of time and location. The calculations indicate that drugs with slow elimination will penetrate deeper into the tissue. The modulus ϕ, which represents the ratio of elimination to diffusion rates in the tissue, provides a quantitative criterion for selecting agents for interstitial delivery. For example, high molecular weight dextrans were retained longer in the brain space, and penetrated a larger region of the brain, than low molecular weight molecules following release from an intracranial implant. This suggests a strategy for modifying molecules to improve their tissue penetration by conjugating active drug molecules with inert polymeric carriers. For conjugated drugs, the extent of penetration should depend on the modulus ϕ for the conjugated compound as well as on the degree of stability of the drug–carrier linkage.

The effects of conjugation and stability of the linkage between drug and carrier on enhancing tissue penetration in the brain have been studied in model systems [34]. Methotrexate (MTX)–dextran conjugates with different dissociation rates were produced by linking MTX to dextran (M_w 70,000) through a short-lived ester bond ($t_{1.2} \sim 3\,\text{days}$) and a longer-lived amide bond ($t_{1/2} > 20\,\text{days}$). The extent of penetration for MTX–dextran conjugates was studied in three-dimensional human brain tumor cell cultures; penetration was significantly enhanced for MTX–dextran conjugates and the increased penetration was correlated with the stability of the linkage. These results suggest that modification of existing drugs may increase their efficacy against brain tumors when delivered directly to the brain interstitium.

10.2.7 Conclusion

Controlled-release polymer implants are useful for delivering drugs directly to the brain interstitium. This approach may improve the therapy of brain tumors or other neurological disorders. The mathematical models described in this section—which are based on methods of analysis developed in earlier chapters—provide a useful framework for analyzing mechanisms of drug distribution after delivery. These models describe the behavior of chemotherapy compounds very well and allow prediction of the effect of changing properties of the implant or the drug. More complex models are needed to describe the behavior of macromolecules, which encounter multiple modes of elimination and metabolism and are subject to the effects of fluid flow.

10.3 TOPICALLY APPLIED DEVICES FOR CONTROLLED RELEASE

Topical controlled release of contraceptives to the female reproductive tract is an alternative to more conventional means of birth control such as oral contraceptives, intra-uterine devices, or condoms. The World Heath Organization and the International Committee for Contraception Research of the Population Control have studied vaginal rings composed of Silastic® that were designed to release hormonal contraceptives for prolonged periods. Clinically, the rings were used over several menstrual cycles (inserted for one month and then removed for one week). These devices provided effective hormone levels in the plasma within several hours after insertion and maintained the agent at levels sufficient to prevent pregnancy for as long as the device was in place [35].

Vaginally applied spermicides have been used since the 1920s, with varying degrees of success. Long-term studies suggest a 40% failure rate with traditional methods of application. Since spermicides, including nonoxynol-9, are highly effective at immobilizing human spermatozoa *in vitro*, this high failure rate in women has been attributed to inadequate mixing of the spermicide with the seminal fluid. Improved methods of delivery may circumvent this problem, but developments in the long-term delivery of spermicides has lagged hormone delivery. Hormones are typically small molecules, while many spermicides are polymers with low permeability through membranes or low solubility in mucus [36]. Interest in the development of spermicide delivery systems has increased recently, primarily because many agents that are spermicidal—such as nonoxynol-9 and antibodies—also prevent transmission of sexually transmitted diseases (STDs) [37–39].

This section considers the design and performance of an intravaginal device for the controlled delivery of an agent that is active at the mucosal surface. The goal is to produce a delivery system that is easily self-administered and provides long-term (i.e., one month) protection against disease or unwanted pregnancy.

10.3.1 Delivery of Agents to the Female Lower Reproductive Tract

The human female reproductive tract consists of the vulva, located outside the pelvic cavity, and the ovaries, fallopian tubes, uterus, and vagina, which are located inside the pelvic cavity (Figure 10.9a). Semen is deposited in the vagina during sexual intercourse, making it the likely site of infection. Therefore, agent must be distributed throughout the vagina, from the base of the pelvic cavity to the surface of the cervix.

The vaginal tract is a hollow, thin-walled tube, between 8 and 11 cm long, consisting of an epithelial layer, and a layer of elastic/collagenous tissue, which binds the epithelium to the underlying smooth muscle. The surface area of the vagina is estimated at $100\,cm^2$. The tissue of the vagina is extremely elastic, with multiple folds in the walls to allow for expansion of the cavity. The epithelium of the vagina varies in thickness, depending on both the age of

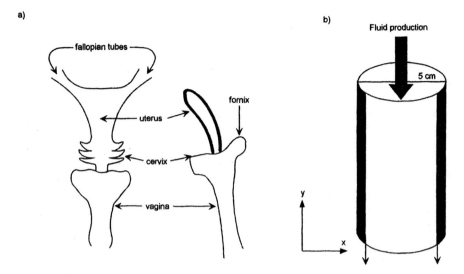

Figure 10.9 Model of the female reproductive tract. (a) Schematic representation of the female lower reproductive tract shown as anterior/posterior projection and mid-sagittal section. (b) Simple model for the lower reproductive tract.

the women and the point in the menstrual cycle. It is twice as thick at the time of ovulation (0.25 mm) as at the onset of menstruation.

The fluid within the lower reproductive tract originates from variety of sources. The vaginal epithelium produces a small amount of fluid, which is bacteriostatic because of its high acidity (pH 3–4). This acidic fluid mixes with mucus, which is produced at the cervix, to create an overall acidic environment (pH 5–7) within the lumen of the vagina. The cervix produces $\sim$ 20–60 mg of mucus per day, with as much as 1 g/day during mid-cycle. The overall volume of vaginal fluid appears to be 2 to 8 mL/day [40].

To examine the dynamics of drug delivery to the vagina, it was modeled as a hollow cylinder (Figure 10.9b). The diameter of the cylinder was estimated at 5 cm (actual diameter of the vagina varies among women). Production of vaginal fluid was assumed to be 2–8 mL/day, creating a constant rate of flow from top to bottom [40]. Steady plug flow (see Section 7.3.1) was assumed, with no mixing along the y-axis. The thickness of the mucus layer at the wall of the cylinder, X, was assumed to be 200 μm [41]. Because the mucus within the lower reproductive tract is under flow, the clearance time (t_c) can be determined from the surface area of the vagina (S, $\sim$ 100 cm^2), the thickness of the mucus layer (X, $\sim$ 200 μm), and the rate of mucus production ($K_m = 2$–8 mL/day):

$$t_c = S^* \frac{X}{K_m} \qquad (10\text{-}29)$$

The clearance time varies between 6 and 17 h.

10.3.2 Concentrations after Bolus Delivery

The simplest mechanism of delivery is insertion of a bolus dose, in which the agent is suspended into a fluid and directly deposited within the vaginal lumen. This method is used currently for spermicide and antifungal drugs. Bolus delivery is a convenient method; however, the duration of efficacy is limited. To utilize bolus delivery effectively, the initial dosing requirements (volume and concentration) and the duration of efficacy must be determined.

The overall system for the bolus dosing is divided into two components: (i) the application of the agent ($t = 0$) and (ii) the removal of the agent through plug flow ($t > 0$). The mucus/agent solution is assumed to be well mixed during application. This assumption simplifies the calculation, but may not be reasonable for all agents. Introduction of fluid into the reproductive tract increases the volume; it is further assumed that any extraneous volume would flow out immediately after application. This assumption is consistent with previous studies in mice; elimination after bolus dosing occurred in two stages, a fast process ($t_{1/2} = 0.7 \, \text{h}$) associated with application of the dose and a slower process ($t_{1/2} = 4.6 \, \text{h}$) associated with the rate of mucus shedding from the tract [42]. In extrapolating to humans, the volume in the reproductive tract before application (V_i) and the volume remaining after application (V_f) were assumed to be equal. The total mass of agent (M_a^*) in the mucus/agent solution volume ($V_i + V_a$), where V_a is the volume applied, was determined before flow of the solution out of the system:

$$M_a^* = C_i V_i + C_a V_a \tag{10-30}$$

This mass was then used to calculate the concentration of agent in the mixture (C_a^m):

$$C_a^m = \frac{M_a^*}{V_i + V_a} \tag{10-31}$$

A volume equal to that applied would leave the system because of overflow. The removal of this fluid, however, did not alter the overall concentration of the mixture although the total mass within the solution was decreased:

$$M_a^m = C_a^m V_f \tag{10-32}$$

The concentration of the agent immediately after application ($t = 0$) was assumed to be uniform throughout the region $0 < y < L$.

The concentration as a function of time was determined from a mass balance:

$$K_m C_{in} - K_m C_{out} = \frac{d(CV_f)}{dt} \tag{10-33}$$

where $C_{in} = 0$ (no agent added after $t = 0$), K_m is the production rate of fluid, C_{out} is the concentration of the fluid leaving the system, and C is the overall concentration within the system. For a perfectly mixed system, Equation 10-33 can be solved by assuming $C_{out} = C$; the concentration decays exponentially with time (see Figure 7.9). For plug flow, the mucus/agent solution moves down the cylinder at a rate determined by fluid production. The y-position of the front between agent- and non-agent-bearing fluid is:

$$Y = \frac{K_m t}{X(SA/L)} \tag{10-34}$$

where L is the length (8–11 cm). Therefore, at time t the fluid within the region $0 < y < Y$ has $C = 0$ while the region $Y < y < L$ has $C = C_{out}$.

10.3.3 Design of a Controlled-Delivery System

Vaginal controlled-delivery systems for IgG have been previously modeled in mice [43, 44]. The device was a flat disk, with IgG diffusing out of the face of the disk. To adapt this model for human use, the disk was replaced by a doughnut-shaped ring, with characteristics based on devices that have been used clinically. EVAc vaginal rings (54 mm) of different cross-sectional thickness (3, 3.5, or 4 mm) were acceptable in the majority of women tested (91%) [45] Similar rings have been used successfully in a number of clinical studies [46, 47].

Vaginal Ring. After placement of the ring within the vaginal lumen, the fluid immediately surrounding the ring was assumed to be well mixed. The ring was approximated as a cylinder with drug released radially by diffusion (Figure 10.10). As in the previous examples, the diffusion equation was used to determine the concentration profile of the drug within the device:

$$\frac{\partial C_p}{\partial t} = D_p \left(\frac{\partial^2 C_p}{\partial r^2} + \frac{1}{r} \frac{\partial C_p}{\partial r} \right) \tag{10-35}$$

where C_p is the concentration of the agent in the ring and D_p is the diffusion coefficient of the agent through the polymer matrix. The initial concentration in the ring is C_{p0}; the concentration at the surface of the ring is $C_1(t)$, which was assumed to be equal to the concentration within the well-mixed region of fluid (a cylindrical segment of the vagina of height a). The solution to this equation is:

$$C_p(t) = 2\frac{C_{p0}}{a} \sum_{n=1}^{\infty} e^{-D_p \alpha_n^2 t} \frac{J_0(r\alpha_n)}{\alpha_n J_1(a\alpha_n)} + 2\frac{D_p}{a} e^{-D_p \alpha_n^2 t} \frac{\alpha_n J_0(r\alpha_n)}{J_1(a\alpha_n)} \int_0^t e^{D_p \alpha_n^2 \lambda} C_1(\lambda) d\lambda \tag{10-36}$$

where α_n are the roots of $J_0(a\alpha_n) = 0$ [5].

The concentration of the agent in the mucus (C_1) was determined by a mass balance around the well-mixed compartment:

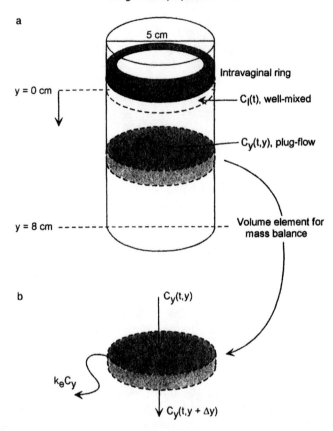

Figure 10.10   Mass balances in model of the lower reproductive tract. (a) A mass balance was obtained for a section along the *y*-axis. (b) When inserted into the reproductive tract, the region immediately surrounding the vaginal ring is considered to be well mixed:: C_I is the concentration in the middle of the tract and on the outside of the ring. The volume (V_I) of the well-mixed region can be determined by subtracting the volume of the ring from the volume of the reproductive tract immediately surrounding the ring (height $= 2a$, radius $= 2.5$ cm).

$$V_I \frac{dC_I}{dt} = -\frac{dM_p}{dt} - K_m C_I(t) \tag{10-37}$$

where:

$$M_p(t) = 2\pi h \int_0^a C_p r \, dr \tag{10-38}$$

and V_I is the volume of the mucus immediately surrounding the ring, K_m is the rate of fluid production, and $h \ (= 2\pi R_{ring})$ is the height of the cylinder. The mass released from the ring was determined by substituting C_p from Equation

10-36 into Equation 10-38. Solving for $C_I(t)$ in Equation 10-37, the concentration within the well-mixed mucus was found to be:

$$C_I(t) = \frac{2\pi h}{V_m} C_{p0} \sum_{n=1}^{\infty} \frac{2}{\alpha_n^2} \left(1 - e^{-D_p \alpha_n^2 t}\right) + \frac{4\pi h}{V_m} D_p \int_0^t e^{-D_p \alpha_n^2 (t-\lambda)} C_I(\lambda) d\lambda$$
$$- \frac{k}{V_m} \int_0^t C_I(t) dt \tag{10-39}$$

The concentration within the well-mixed portion can be determined using numerical methods to approximate the integral.

After the agent is released into the mucus, the fluid production produces a flow that moves the agent towards the base of the pelvic cavity. It was assumed that no elimination of the agent from the reproductive tract took place along the y-axis (this assumption was valid in the mouse [42, 44]). The concentration in the well-mixed portion of the tract was related to the concentration emerging from the vagina:

$$C_I(t) = C_{out}(t - \tau) \tag{10-40}$$

where τ is equal to the length of the tract divided by the velocity of the fluid in the y-direction. The concentration at any axial location y is:

$$C_y(y, t) = C_I(t - z)u(t - z) \tag{10-41}$$

where $z = y/v_y$ and $u(t - z)$ equals 1 if $t > z$ and 0 if $t \leq z$.

To investigate the difference between C_y with or without elimination through the wall of the reproductive tract, elimination was included in a mass balance to produce:

$$\frac{\partial C_y}{\partial t} = -v_y \frac{\partial C_y}{\partial y} - \frac{2\pi R^2}{S} k_e C_y \tag{10-42}$$

where C_y is the concentration along the y-axis of the cylinder at time t, R is the radius of the cylinder, v_y is the velocity along the y-axis and is assumed to be constant, S is the surface area and k_e is the elimination of the agent through the wall of the cylinder. Using Laplace transforms and the boundary conditions as above, the solution of this equation is:

$$C_y(y, t) = C_I\left(t - \frac{y}{v_y}\right) e^{-\left(\frac{2\pi R^2 k_e}{SA}\right) y} u\left(t - \frac{y}{v_y}\right) \tag{10-43}$$

where $u(\cdot)$ is a unit step function (equal to 1 at $t > y/v_y$ and 0 at $t < y/v_y$).

Results. To study the effect of various parameters—such as D_p, C_{p0}, or K_m—on the effectiveness of an agent, a hypothetical minimum effective concentration was selected ($C_{eff} = 0.1$ mg/mL). Low concentrations of agent were required to attain the effective concentration with bolus delivery (Table 10.4). However, the effective dose persists for a short time, leaving half the reproductive tract unprotected within 4–16 h. Plug flow assumes no

Table 10.4 Concentrations in the model system after different bolus
dosing schedules

Volume of dose (mL)	Concentration of agent in dose (g/mL)	Final concentration immediately after bolus (g/mL)	System volume (mL)
0.5	10	2	2
1	10	3.3	2
2	10	5	2
0.5	1	0.2	2
1	1	0.33	2
2	1	0.5	2
0.5	0.1	0.02	2
1	0.1	0.033	2
2	0.1	0.05	2
0.5	0.01	0.002	2
1	0.01	0.0033	2
2	0.01	0.005	2
0.5	0.001	0.0002	2
1	0.001	0.00033	2
2	0.001	0.0005	2

mixing along the y-axis and therefore produces a concentration profile
where along the tract there is either no agent ($C = 0$), or a constant
concentration of agent ($C = C_{dose}$). Using the low volume dose (0.5 mL,
1 mg/mL), both the well-mixed (Figure 10.11a) and the plug flow (Figure
10.11b) profiles were determined. While the plug flow model predicted
regions that were completely devoid of agent, the remainder of the fluid
had a concentration of C_{dose}. In the well-mixed system, the concentration
decreased over time until, after ~ 6.5 h, the dose in the fluid was below the
effective concentration. The difference between the well-mixed and the plug
flow models was not the mass remaining in the fluid (which is determined
by the elimination rate) but the distribution of agent within the tract.
Bolus dosing could be an effective option for short-term delivery,
depending on the distribution of the material within the tract. One method
of improving this type of delivery would be to increase the viscosity of the
agent solution so that more mass would remain in the reproductive tract
by changing the rate of mucus flow and therefore elimination, although
there are significant limits to this approach.

A vaginal ring provides long-term delivery. Figure 10.12a shows the
kinetics of agent release from the vaginal ring as a function of the rate of
agent diffusion within the polymer (D_p). If the agent accumulated within the
vaginal lumen, and was not eliminated, these curves would represent the fraction of dose within the vagina as a function of time. In practice, the agent is
eliminated from the body, with an overall rate that is determined by the rate of
fluid flow (K_m). Figure 10.12b shows the amount of agent within the lumen as a

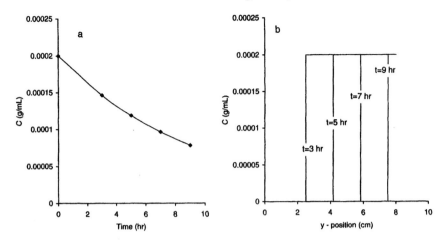

Figure 10.11 Agent concentrations after bolus delivery. The concentration of agent as a function of time and position within the vaginal tract was determined using a well-stirred (a) or plug-flow (b) model of mixing within the vagina.

function of time for a ring in which D_p is 10^{-9} cm^2/s and the rate constant for elimination (K_m/V) varies. Reasonable values of these two parameters were used, suggesting that vaginal rings can provide high levels of active agent within the vagina for prolonged periods.

Using the model described above, it was possible to investigate delivery to the well-mixed portion of the mucus as a function of characteristics of the vaginal ring, the agent, and the reproductive tract. The initial concentration required for an effective vaginal ring was greater than that required for bolus dosing (0.1 g/mL for vaginal ring versus 0.001 g/mL for bolus). However, the effective dose lasted for much longer periods when applied by vaginal ring. With no elimination from the system other than fluid flow ($K_m = 7$ mL/day), the concentration within the well-mixed compartment was greater than the C_{eff} for up to 6 months.

The duration of efficacy also depends on the diffusion coefficient of the agent in the ring (D_p). D_p was varied while keeping other parameters constant ($C_{p0} = 1$ g/mL, $K_m = 8.1 \times 10^{-5}$ mL/s, $a = 0.35$ cm) (Figure 10.12). D_p determined the release rate from the ring and, therefore, the time required to reach an effective concentration. For example, when $D_p = 5 \times 10^{-10}$ cm^2/s, the concentration of the system was greater than $C_{effective}$ within 1 h; when $D_p = 1 \times 10^{-10}$ cm^2/s it took ~ 3 h to reach $C_{effective}$.

To investigate the distribution of the released agent within the reproductive tract, Equation 10-41 (without elimination) and Equation 10-43 (with elimination) were solved using reasonable parameters ($D_p = 1 \times 10^{-9}$ cm^2/s, $C_{p0} = 1$ g/mL, $a = 0.35$ cm, $K_m = 8.1 \times 10^{-5}$ mL/s, $v_y = 3.3 \times 10^{-4}$ cm/s). The agent is distributed throughout the lower reproductive tract within several hours of insertion (Figure 10.13), providing a concentration greater than

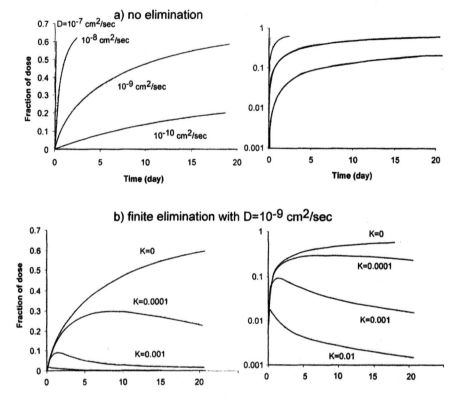

Figure 10.12   Concentration within the lower reproductive tract after insertion of controlled-release device. (a) Mass released into the body as a function of time with changing D_p (cm^2/s) when the initial concentration of the ring is 1 g/mL. (b) Mass of agent within the reproductive tract as a function of time for a ring with fixed D_p (10^{-9} cm^2/s) but varying rates of elimination.

$C_{\text{effective}}$. Once the agent has distributed throughout the tract, the variations in concentration with axial location are minimal.

10.3.4 Conclusions

Both bolus dosing and controlled-release delivery have advantages. Bolus delivery, while lasting only several hours, provides immediate protection with low concentrations and volumes of the agent. This type of delivery may be advantageous for short-term protection, especially if the agent is expensive. However, controlled release is a better option for long-term protection. After insertion, protective concentrations were achieved within several hours and the dosing lasted over several months. The concentration of the agent required within the ring was higher than for bolus delivery which may make the ring more expensive. However, because of the possibility for protection for up to 6 months, the ring should have a cost advantage over the bolus delivery.

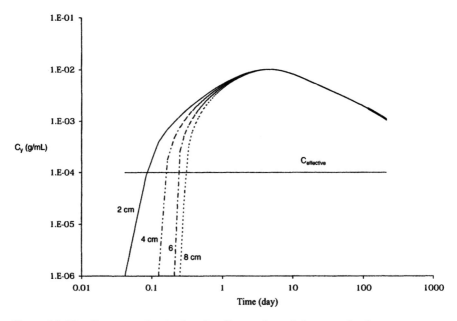

Figure 10.13 Concentration in the plug-flow region of the reproductive tract. Concentration in the plug flow region of the reproductive tract over time for four different y-positions (2, 4, 6, and 8 cm).

REFERENCES

1. Saltzman, W.M., Interstitial transport in the brain: principles for local drug delivery, in *Handbook of Biomedical Engineering*. Boca Raton, FL: CRC Press, in press.
2. Flexner, C., *et al.*, Pharmacokinetics, toxicity, and activity of intravenous dextran sulfate in HIV infection. *Antimicrob. Agents Chemother.*, 1991, **35**, 2544–2550.
3. Shacker, T., *et al.*, Phase I study of high-dose, intravenous rsCD4 in subjects with advanced HIV-1 infection. *J. Acquir. Immune Defic. Syndr. Hum. Retrovirol.*, 1995, **9**, 145–152.
4. Kilby, J., *et al.*, Potent suppresion of HIV-1 replication in humans by T-20, a peptide inhibitor of gp41-mediated virus entry. *Nature Medicine*, 1998, **4**(11), 1302–1307.
5. Carslaw, H.S. and J.C. Jaeger, *Conduction of Heat in Solids*. 2nd ed. Oxford: Oxford University Press, 1959, 510 pp.
6. Pardridge, W.M., *Peptide Drug Delivery to the Brain*. New York: Raven Press, 1991.
7. Lieb, W. and W. Stein, Biological membranes behave as non-porous polymeric sheets with respect to the diffusion of non-electrolytes. *Nature*, 1969, **224**, 240–249.
8. Stein, W.D., *The Movement of Molecules across Cell Membranes*. New York: Academic Press, 1967.
9. Simpkins, J., N. Bodor, and A. Enz, Direct evidence for brain-specific release of dopamine from a redox delivery system. *Journal of Pharmaceutical Sciences*, 1985, **74**, 1033–1036.
10. Gregoriadis, G., The carrier potential of liposomes in biology and medicine. *New England Journal of Medicine*, 1976, **295**, 704–710.

11. Neuwelt, E., *et al.*, Delivery of melanoma-associated immunoglobulin monoclonal antibody and Fab fragments to normal brain utilizing osmotic blood–brain barrier disruption. *Cancer Research*, 1988, **48**, 4725–4729.

12. Triguero, D., *et al.*, Blood–brain barrier transport of cationized immunoglobulin G: enhanced delivery compared to native protein. *Proceedings of the National Academy of Sciences USA*, 1989, **86**, 4761–4765.

13. Tokuda, H., Y. Takakura, and M. Hashida, Targeted delivery of polyanions to the brain. *Proceed. Intern. Symp. Control. Rel. Bioact. Mat.*, 1993, **20**, 270–271.

14. Friden, P., *et al.*, Anti-transferrin receptor antibody and antibody–drug conjugates cross the blood–brain barrier. *Proceedings of the National Academy of Sciences USA*, 1991, **88** 4771–4775.

15. Friden, P.M., *et al.*, Blood-brain barrier penetration and *in vivo* activity of an NGF conjugate. *Science*, 1993, **259**, 373–377.

16. Chien, Y.W. and S.-F. Chang, Intranasal drug delivery for systemic medications. *Critical Reviews of Therapeutic Drug Carrier Systems*, 1987, **4** 67–194.

17. Blasberg, R., C. Patlak, and J. Fenstermacher, Intrathecal chemotherapy: brain tissue profiles after ventriculocisternal perfusion. *Journal of Pharmacology and Experimental Therapeutics*, 1975, **195**, 73–83.

18. Dang, W. and W.M. Saltzman, Controlled release of macromolecules from a biodegradable polyanhydride matrix. *Journal of Biomaterials Science, Polymer Edition*, 1994, **6**(3), 291–311.

19. Dang, W., *Engineering Drugs and Delivery Systems for Brain Tumor Therapy*. Baltimore, MD: The Johns Hopkins University Press, 1993.

20. Mahoney, M.J. and W.M. Saltzman, Controlled release of proteins to tissue transplants for the treatment of neurodegenerative disorders. *Journal of Pharmaceutical Sciences*, 1996, **85**(12), 1276–1281.

21. Bird, R.B., W.E. Stewart, and E.N. Lightfoot, *Transport Phenomena*. New York: John Wiley, 1960, 780 pp.

22. Fung, L., *et al.*, Chemotherapeutic drugs released from polymers: distribution of 1,3-bis(2-chloroethyl)-1-nitrosourea in the rat brain. *Pharmaceutical Research*, 1996, **13**, 671–682.

23. Nicholson, C., Diffusion from an injected volume of a substance in brain tissue with arbitrary volume fraction and tortuosity. *Brain Research*, 1985, **333**, 325–329.

24. Morrison, P. and R.L. Dedrick, Transport of cisplatin in rat brain following microinfusion: an analysis. *Journal of Pharmaceutical Sciences*, 1986, **75**, 120–128.

25. Saltzman, W.M. and M.L. Radomsky, Drugs released from polymers: diffusion and elimination in brain tissue. *Chemical Engineering Science*, 1991, **46**, 2429–2444.

26. Reinhard, C., *et al.*, Polymeric controlled release of dexamethasone in normal rat brain. *Journal of Controlled Release*, 1991, **16**, 331–340.

27. Dang, W. and W.M. Saltzman, Dextran retention in the rat brain following controlled release from a polymer. *Biotechnology Progress*, 1992, **8**, 527–532.

28. Krewson, C.E., M. Klarman, and W.M. Saltzman, Distribution of nerve growth factor following direct delivery to brain interstitium. *Brain Research*, 1995, **680**, 196–206.

29. Strasser, J.F., *et al.*, Distribution of 1,3-bis(2-chloroethyl)-1-nitrosourea (BCNU) and tracers in the rabbit brain following interstitial delivery by biodegradable polymer implants. *Journal of Pharmacology and Experimental Therapeutics*, 1995, **275**(3), 1647–1655.

30. Fung, L.K., *et al.*, Pharmacokinetics of interstitial delivery of carmustine, 4-hydro-peroxycyclophosphamide, and paclitaxel from a biodegradable polymer implant in the monkey brain. *Cancer Research*, 1998, **58**, 672–684.
31. Jain, R.K., Barriers to drug delivery in solid tumors. *Scientific American*, 1994, **271**(1), 58–65.
32. Clauss, M.A. and R.K. Jain, Interstitial transport of rabbit and sheep antibodies in normal and neoplastic tissues. *Cancer Research*, 1990, **50**, 3487–3492.
33. Morrison, P.F., *et al.*, High-flow microinfusion: tissue penetration and pharmaco-dynamics. *American Journal of Physiology*, 1994, **266**, R292–R305.
34. Dang, W.B., *et al.*, Covalent coupling of methotrexate to dextran enhances the penetration of cytotoxicity into a tissue-like matrix. *Cancer Research*, 1994, **54**, 1729–1735.
35. Landgren, B., *et al.*, Pharmacokinetic studies with a vaginal delivery system releasing levonorgestrel at a near zero order rate for one year. *Contraception*, 1986, **33**, 473–485.
36. Saltzman, W.M. and L.B. Tena, Spermicide permeation through biocompatible polymers. *Contraception*, 1991, **43**, 497–505.
37. Whaley, K.J., *et al.*, Nonoxynol-9 protects mice against vaginal transmission of genital herpes infections. *Journal of Infectious Disease*, 1993, **168**, 1009–1011.
38. Moench, T.R., *et al.*, The cat-feline immunodeficiency virus model for transmucosal transmission of AIDS: nonoxynol-9 contraceptive jelly blocks transmission by an infected cell inoculum. *AIDS*, 1993, **7**, 797–802.
39. Cone, R.A. and K.J. Whaley, Monoclonal antibodies for reproductive health: Part I. Preventing sexual transmission of disease and pregnancy with topically applied antibodies. *American Journal of Reproductive Immunology*, 1994, **31**, 1–18.
40. Owen, D. and D. Katz, A vaginal fluid simulant. *Contraception*, 1999, **59**(2), 91–95.
41. Parkhurst, M.R. and W.M. Saltzman, Leukocyte migration in three-dimensional gels of midcycle cervical mucus. *Cellular Immunology*, 1994, **156**, 77–94.
42. Sherwood, J.K., *et al.*, Residence half-time of IgG administered topically to the mouse vagina. *Biology of Reproduction*, 1996, **54**, 264–269.
43. Sherwood, J.K., *et al.*, Controlled release of antibodies for sustained topical passive immunoprotection of female mice against genital herpes. *Nature Biotechnology*, 1996, **14**, 468–471.
44. Saltzman, W., *et al.*, Long-term vaginal antibody delivery: delivery systems and biodistribution. *Biotechnology and Bioengineering*, 2000, **67**, 253–264.
45. Roumen, F.J.M.E. and T.O.M. Dieben, Clinical acceptability of an ethylene-vinyl-acetate nonmedicated vaginal ring. *Contraception*, 1999, **59**(1), 59–62.
46. Eriksen, B.C., A randomized, open, parallel-group study on the preventive effect of an estradiol-releasing vaginal ring (Estring) on recurrent urinary tract infections in postmenopausal women. *American Journal of Obstetrics and Gynecology*, 1999, **180**(5), 1072–1079.
47. Bachmann, G., Estradiol-releasing vaginal ring delivery system for urogenital atro-phy—experience over the past decade. *Journal of Reproductive Medicine*, 1998, **43**(11), 991–998.

Postscript

But why did I admire nature so?
Was it that I liked

the absence of a Master
neuron in the brain—

the absence of a Master
cell in embryos—

the nothing in the way of
center that would hold?

Alice Fulton, *Wonder Bread* (1995)

Drug delivery—like other technologies—is a result of scientific inquiry, a natural collaboration between discovery and invention. This spirit is obvious in drug therapy: it links Harvey's inquiry into the circulatory system with intravenous injections, the molecular biology of receptors to targeted drug delivery systems. We must invent to discover, and discover to invent.

Trail-marking has great benefits, but also perils; being shown a direction can make it more difficult to leave the pathway. It is my hope that you will have found something here that engages you, a flash that persuades you off the path and into the forest.

OVERVIEW OF POLYMERIC BIOMATERIALS

And what if all of animated nature
Be but organic harps diversely fram'd,
That tremble into thought, as o'er them sweeps
Plastic and vast, one intellectual breeze,
At once the Soul of each, and God of all?
Samuel Taylor Coleridge, *The Eolian Harp* (1795)

Drug delivery systems often require synthetic components or biomaterials. Frequently these materials are polymers (see Table A.1).

Much of the initial progress in biomaterials was related to repair in the vascular system, primarily in an effort to reapproximate severed vessels. From ancient times until the early 1970s, only collagenous materials like catgut found general acceptance as absorbable sutures, although a variety of innovative approaches were used. Absorbable sutures derived from animal sinews were described by Sushruta, an Indian surgeon, in 600 B.C. [1]. Hallowell used a wooden peg and thread to repair a brachial artery in 1759; Gluck used ivory clamps and Jassinowsky used silk and needles to repair vessels in the 1880s [2]. The first absorbable synthetic material, poly(glycolic acid) (pGA), was developed by American Cyanamid in the 1960s (see [3] for a historical review of this development).

Polymers have been used as biomaterials in dental applications since the last century (see [4] for review); gutta-percha was used for dental impressions beginning in 1848. Vulcanized caoutchouc was used in 1854 for dental bases, and celluloid was used in 1868 for dental prostheses. In 1909, bakelite was first used and poly(methyl methacrylate) (PMMA) has been used since 1930 for denture bases, artificial teeth, removable orthodontics, surgical splinting, and fillings. Many new polymers appeared during the 1930s, including polyamides, polyesters, and polyethylene

Catheters—thin, hollow tubes formed of polymer and used medically to introduce or remove fluid within vessels far from the body surface—are an important tool in diagnosis and disease management. Fritz Bleichroeder was the first individual to perform catheterization, when he inserted a catheter into his own femoral artery. The first cardiac catheterization was performed by another brave individual, Werner Forssman, in 1929. Forssman, a 23-year-old urology student, inserted a urethral catheter via the antecubital vein into his heart. With the catheter in place, Forssman reportedly ascended a flight of stairs to the X-ray room, where he documented this experiment, which eventually earned him a Nobel Prize.

Synthetic materials are critical components in extracorporeal systems for blood purification or treatment. Willem Kolff, a Dutch physician, developed the first successful kidney dialysis unit in 1943, using cellophane to remove urea from the blood of diabetics [5].

The first implanted synthetic polymeric biomaterial appears to be PMMA, which was used as a hip prosthesis in 1947 (see USP XVIII, The Pharmacopia of the USA, (18th Revision), US Pharmacopoeial Convention, Inc., Rockville, MD, 1 September 1980). Polyethylene, and then other polymers, were used as implants in the middle ear in the early 1950s, yielding good initial results, but local inflammation limited the use of these materials.

An artificial heart valve, composed of a plastic ball in a metal socket, was implanted by Charles A. Hufnagel in 1952. Implantable cardiac pacemakers were also developed in the 1950s. Research on artificial hearts began during this same period. An artificial ventricle, the Liotta-DeBakey tube pump composed of (polyethylene terephthalate) (PET)-reinforced silicone, was implanted into a patient in 1963 [6]. A large, bed-side artificial heart machine was first used on a human in Houston in 1969 [7]; it was designed by Domingo Liotta and Michael DeBakey, and used to keep Haskell Karp alive, without a heart, and awaiting a donor for 63 h. The first permanent total artificial heart (the Jarvik 7-100, composed of PET with a polyurethane diaphragm) was implanted into a human in 1982 by DeVries *et al.* [8]; the longest duration of survival to date is 620 days.

Vascular surgeons have been traumatizing arteries for wound treatment since first demonstrated by Sushruta in India and Galien in Rome (see historical review in [9]). Hufnagel used smooth, tubes as vascular grafts in the late 1940s [10]. In the early 1950s, textile grafts were first introduced, primarily composed of silk fibers initially and then poly(vinyl chloride-*co*-acrylonitrile). Currently, PET and poly(tetrafluoroethylene) (PTFE) grafts are the most widely used.

Biomaterials have been used frequently in the nervous system, particularly as shunts to divert excess cerebrospinal fluid from the ventricular system in patients with hydrocephalus (see [11] for review of early work). Miculicz reported the use of a glass wool "nail" for this purpose in 1890. Autologous vessels and rubber tubes were used in the period 1900–1930, with some success (see [12]). Poly(vinyl chloride) and silicone tubing was first used in the 1950s, yielding better results. Implanted polymeric materials are now being used in a variety of neurological settings, including regeneration of damaged peripheral nerves and drug delivery to the brain.

Today, many medical devices involve polymeric biomaterials (see Table A.1). Polymers of a variety of chemistries, in many shapes and forms, are introduced into patients around the world each day. Many of these materials are now common: catheters, coatings for pacemaker leads, and contact lenses. In some cases, appropriate synthetic polymer materials have allowed dramatic and heroic technologies, like the artificial hip and total artificial heart. Still, there is considerable work remaining: even the most modern artificial hips are a pale replacement of the natural material and no patient has survived for long on a total artificial heart. Clearly, the development of better materials and smarter ways for using existing materials will improve human health care.

The sections that follow provide a schematic overview of the polymeric materials most used in drug delivery systems. For more details, readers are referred to other recent books for aspects of biomaterials that are not covered in detail here. For example, biocompatibility and interactions with implanted polymers are reviewed in several edited volumes [2, 13, 14].

Non-degradable Polymers

Ben—I want to say one word to you—just one word—*plastics*.

Buck Henry and Calder Willingham, *The*
Graduate (screenplay) (1967), from the novel
(1962) by Charles Webb

Among the many classes of polymeric materials now available for use as bio-materials, non-degradable, hydrophobic polymers are the most widely used. Silicone, polyethylene, polyurethanes, PMMA, and EVAc account for the majority of polymeric materials currently used in clinical applications. Consider, for example, the medical applications listed in Table A.1; most of these applications require a polymer that does not change substantially during the period of use. This chapter describes some of the most commonly used non-degradable polymers that are used as biomaterials, with an emphasis on their use in drug delivery systems.

A.1.1 SILICONE ELASTOMERS

Background and Nomenclature

Elastomers of silicone are widely used as biomaterials. In general, silicone elastomers have excellent biocompatibility, inducing only a limited inflammatory response following implantation. In fact, until very recently, it was assumed that silicones were almost completely inert in biological systems. It is now known, however, that certain silicone polymers can provoke inflammatory and immune responses. The biological response to implanted silicone, and the variability of that response among individuals, is the subject of considerable debate and interest.

Table A.1 Some of the polymers that might be useful in drug delivery, based on past use in biomedical devices

Polymer	Typical applications
Poly(dimethylsiloxane), Silicone elastomers (PDMS)	Breast, penile, and testicular prostheses Catheters Drug delivery devices Heart valves Hydrocephalus shunts Membrane oxygenators
Polyurethane (PEU)	Artificial hearts and ventricular assist devices Catheters Pacemaker leads
Poly(tetrafluoroethylene) (PTFE)	Heart valves Vascular grafts Facial prostheses Hydrocephalus shunts Membrane oxygenators Catheters Sutures
Polyethylene (PE)	Hip prostheses Catheters
Polysulfone (PSu)	Heart valves Penile prostheses
Poly(methyl methacrylate) (pMMA)	Fracture fixation Intraocular lenses Dentures
Poly(2-hydroxyethyl methacrylate) (pHEMA)	Contact lenses Catheters
Polyacrylonitrile (PAN)	Dialysis membranes
Polyamides	Dialysis membranes Sutures
Polypropylene (PP)	Plasmapheresis membranes Sutures
Poly(vinyl chloride) (PVC)	Plasmapheresis membranes Blood bags
Poly[ethylene-*co*-(vinyl acetate)]	Drug delivery devices
Poly (L-lactic acid), poly(glycolic acid), and poly(lactide-*co*-glycolide) (PLA, PGA, and PLGA)	Drug delivery devices Sutures
Polystyrene (PS)	Tissue culture flasks
Poly(vinyl pyrrolidone) (PVP)	Blood substitutes

Collected from reviews by Peppas and Langer [52] and Marchant *et al.* [53]

Synthesis and Chemical Variations

Silicones, more correctly referred to as polysiloxanes, are partially inorganic polymers with the general structure:

$$\left[\begin{array}{c} R_1 \\ | \\ -Si-O- \\ | \\ R_2 \end{array} \right]_N \qquad\qquad (A.1\text{-}1)$$

They are usually prepared by hydrolysis of alkylsilicon or arylsilicon halides. Because of their chemical inertness, stability, and excellent mechanical properties, polysiloxanes are used in industrial applications as elastomers, sealants, coatings, lubricating oils, and hydraulic fluids. These polymers are characterized by high chain flexibility and unusually high oxygen permeability. Polysiloxanes are very stable toward hydrolysis, probably as a result of their hydrophobicity. The physical characteristics of polysiloxanes can be modified by varying the polymer molecular weight, degree of crosslinking, and chemical modification. Chemical modification commonly involves introducing substituents in place of one or both of the pendant methyl groups in the structure shown in Equation A.1-1.

Many of the silicone elastomers that are used in biomedical applications are produced by Dow Chemical Corp., under the trade name SILASTIC®. For example, a typical medical-grade silicone (like SILASTIC® MDX4-4210 Medical grade elastomer) contains, after curing, cross-linked dimethylsiloxane polymer and silica for reinforcement. Silcones are also reinforced with PET (Dacron) fiber meshes for certain biomedical applications. For implantable medical devices, it is important to realize that the cured polymer contains residual catalysts and silicone cross-linkers, which are necessary for the polymerization.

Interesting polymers have been produced by copolymerization of siloxanes with other polymers. For example, dimethylsiloxane was copolymerized with ethylene oxide and methyl methacrylate to create a series of polymers with controlled permeability to hydrophilic or hydrophobic steroids, enhanced mechanical properties, and improved adhesion to tissues [15–17].

Silicone matrix systems have good biocompatibility when implanted subcutaneously [18]. Silicone has received FDA approval for many biomedical applications, including breast prostheses and heart-valve prostheses.

Drug Delivery Applications

Silicone elastomers were first used as the basis for controlled drug delivery systems in 1962, when Folkman and Long discovered that low molecular weight compounds could diffuse, at a controlled rate, through the walls of silicone tubing [18–20]. That observation led to the development of several clinical products for controlled delivery of pharmaceutical agents. Most notably, the Population Council developed tubes of SILASTIC® (in this case, copolymers of dimethylsiloxane and methylvinylsiloxane) for the controlled release of the contraceptive hormone levonorgestrel [21, 22]. Norplant devices, which provide reliable contraception for 5 years after implantion into the forearm, have been extensively tested in women throughout the world and were approved for use in the United States in late 1990. By December of 1992, over 600,000 Norplant systems had been implanted in women in the United States [23]. Early surveys of a fraction of these users suggest enthusiasm for the product [24], although removal of the devices at the end of the delivery period is sometimes difficult due to the formation of a fibrous capsule.

Silicone can also be formulated into a matrix with dispersed drug. These matrices can provide a controlled release of small organic molecules, such as steroids .

Other Relevant Information

Silicone elastomers are an important element in breast prostheses. For example, a common type of breast implant is a silicone device filled with a silicone gel. A possible problem with these devices is the rupture of the outer layer of silicone, with subsequent leakage of the gel; this gel has been associated with a variety of diseases. Breast implants based on silicone were approved for use in humans over 30 years ago, but in a period before the FDA had clear jurisdication over the approval of medical devices (that jursidication was provided by the 1976 Medical Device Amendments to the Food, Drug, and Cosmetic Act). Since that time, and following implantation in over 1 million women, there are no definitive studies regarding the safety of these devices. Based on the best information regarding risks and benefits, the FDA announced in April of 1992 that breast implants filled with silicone gel would be only available to women who agreed to participate in clinical studies to evaluate the safety of the product [25]. The best scientific evidence provided to date provides no link between breast implants and the development of connective-tissue diseases (like rheumatoid arthritis and systemic lupus erythematosus) [26], and the most recent legal decisions agree that the scientific evidence of silicone implants being harmful is lacking [27], but questions of linkage between disease and long-term exposure to implanted silicone (or any material, for that matter) are extremely difficult to answer.

A.1.2 POLY[ETHYLENE-*CO*-(VINYL ACETATE)]

Background and Nomenclature

Certain copolymers of ethylene and vinyl acetate, poly[ethylene-*co*-(vinyl acetate)] (EVAc, ELVAX®, Dupont Corp.), have exceptionally good biocompatibility and are therefore widely used in implanted and topical devices.

Synthesis and Chemical Variations

The most commonly used copolymer contains 40% vinyl acetate and has the general structure:

$$\left[\begin{array}{c} C-C \\ | \ | \\ H_2 \ H_2 \end{array}\right]_x \left[\begin{array}{c} C-C- \\ | \ | \\ H_2 \ O \\ \ \ | \\ \ \ C=O \\ \ \ | \\ \ \ CH_3 \end{array}\right]_y$$

(A.1-2)

The copolymer is synthesized by free-radical polymerization from ethylene and vinyl acetate. The most commonly used EVAc (ELVAX-40, DuPont) consists of ~ 40% vinyl acetate, with a low degree of crystallinity (5–20%). EVAc is hydrophobic; it swells less than 0.8% in water [28].

Drug Delivery Applications

Matrices of EVAc are among the most well-studied drug delivery systems for low molecular weight substances, as well as macromolecules. Matrices composed of EVAc and protein can be fabricated by solvent evaporation or compression molding [29]. In solvent evaporation, EVAc, which has been extensively washed to remove low molecular weight oligomers and impurities, is dissolved in methylene choride. The protein of interest is dried or lyophilized, ground, and sieved to a desired particle size range, and suspended in the polymer solution. The suspension is poured into a chilled mold and allowed to solidify. The matrix is then removed from the mold and dried at atmospheric pressure and −4 °C for 48 h and then dried under vacuum at 20 °C for 48 h.

EVAc has been used in the fabrication of a variety of devices for drug delivery. For example, EVAc was used by Alza in devices to deliver pilocarpine to the surface of the eye for glaucoma treatment (Ocusert). Currently, EVAc is used in the Progestasert intra-uterine device for the delivery of contraceptive hormones to the female reproductive tract and as a rate-controlling membrane in a number of transdermal devices. Since EVAc is one of the most biocompatible of the polymers that have been tested as implant materials [30], it has been widely studied as a matrix for controlled drug delivery (see [31, 32] for reviews).

Other Relevant Information

The biocompatibility of EVAc matrices has been studied quite extensively. When implanted in the cornea of rabbits—which is sensitive to edema, white-cell infiltration, and neovascularization associated with inflammation— purified EVAc caused no inflammation, while unpurified EVAc caused mild inflammation [33]. After seven months of subcutaneous implantion, only a thin capsule of connective tissue surrounded EVAc implants; no inflammation was present and the adjacent loose connective tissue was normal [34]. When implanted in the brains of rats, EVAc matrices produced only mild gliosis

[35]. EVAc has shown good biocompatibility in humans over the years and has been approved by the FDA for use in a variety of implanted and topically applied devices.

A.1.3 POLYURETHANES

Background and Nomenclature

Polyurethanes were first suggested for use as biomaterials in 1967 [36]. Polyurethane materials have excellent mechanical properties, making them suitable for many different biomedical applications. Currently, a variety of polyurethanes are used in biomedical devices like coatings for catheters and pacemaker leads (Table A.2). The biocompatibility of biomedical polyurethanes appears to be determined by their purity: i.e., the effectiveness of the removal from the polymer of catalyst residues and low molecular weight oligomers [37]. The surface properties of commercially available polyurethanes, which are critically important in determining biocompatibility, can vary considerably, even among lots of the same commercially available preparation [38].

Synthesis and Chemical Variations

Polyurethanes can be formed by reacting a bischloroformate with a diamine, for example:

$$(A.1-3)$$

or by reacting a diisocyanate with a dihydroxy compound. For example, ethanediol and hexanediisocyanate react as shown:

$$(A.1-4)$$

One of the most interesting and useful characteristics of polyurethanes involves block copolymers containing "hard" and "soft" segments to produce an elastomeric material. The production of these polyurethanes is a two-step process, where first an aromatic isocyanate-terminated polymer, in large excess,

Table A.2 Commercially available polyurethanes of known chemical structure

Name	Supplier	Description	Structure	Advantage	Disadvantage
Angioflex	Abiomed (Danvers, MA)	Silicone–urethane copolymer	MDI-PTMEG BD-Sil	Good blood compatibility	Difficult to make
Biomer[a]	Ethicon (Somerville, NJ)	Aromatic co(polyether urea)	MDI-PTMEG EDA	Outstanding flex endurance	Research use only
Cardiothane	Kontron, Inc. (Everett, MA)	Silicone–urethane copolymer	MDI-PTMEG BD-Sil	Good blood compatibility	Difficult to fabricate
Chronoflex	PolyMedica (Woburn, MA)	Aliphatic monoether	HMDI-???-BDI	Biostable	
Hemothane	Sarms, Division of 3M (Ann Arbor, MI)	Similar to Biomer	Similar to Biomer	Similar to Biomer	Internal use only
Mitrathane	PolyMedica	Similar to Biomer	Similar to Biomer	Similar to Biomer	Internal use only
Pellethane[a]	Dow Chemical (La Porte, TX)	Aromatic ether based	MDI-PTMEG BD	Thermoplastic elastomer	Microcracks
Surethane	Cardiac Control (Palm Coast, FL)	Purified Lycra	MDI-PTMEG EDA	Similar to Biomer	Microcracks
Tecflex	Thermedics (Woburn, MA)	Aliphatic ether based	HMDI-PTMEG BD	Thermoplastic elastomer	Microcracks

From [39].

[a] Withdrawn from market. MDI (methylenebisphenyldiisocyanate (aromatic)); PTMEG (poly(oxyetramethylenether glycol)); BD (butanediol); HMD (hydrogenated MDI (cycloaliphatic)); EDA (ethylenediamine). Other commercially available polymers have been investigated for possible use: Lykra Spandex and Tecoflex. Structure and composition of these polymers are discussed in [54]

is reacted with a polyether or polyester containing terminal hydroxyl groups. The product of this reaction is chain extended with a diamine, producing a polymer with urea bonds, $-N-(C=O)-N-$, in addition to the urethane linkages, $-N-(C=O)-O-$. For example:

(A.1-5)

For chain-extended polyurethane elastomers like the one shown in Equation (A.1-5), the "hard" segments tend to associate into crystalline domains, while the soft segments, which form a continuous phase surrounding the discrete crystalline regions, remain amorphous. The utility of segmented polyurethanes, such as Lycra, for biomedical applications including the total artificial heart, catheters, heart valves, and pacemaker leads was first proposed by Boretos and Pierce [36]. Lycra, a linear segmented polyurethane, was developed by DuPont and is used to make the fiber Spandex. Ethicon manufactures a modified version of this polymer as Biomer. Biomer was used as the basis for the Jarvik-7 artificial heart, but Ethicon removed Biomer from the general market because of concerns for legal liability [39].

Of the commonly used diisocyanates, 4,4′-diphenylmethane diisocyanate (MDI) forms a semicrystalline hard-segment domain which apparently contributes to the excellent mechanical properties of the resulting polyurethane. Other diisocyanates, such as 4,4′-dicyclohexylmethane diisocyanate, do not crystallize, but still produce polyurethanes with good mechanical properties, probably because these hard segments produce more numerous and smaller microdomains than an MDI polyurethane with the same composition, leading to a more stabilized microstructure [40]. The chemistry of the chain extender can also affect the properties of the polymer. Diamines produce stronger polyurethanes than diols, because the resulting urea linkages within the hard segment participate in hydrogen bonding more readily than the urethane linkages (see Equation A.1-5). Unfortunately, polyurethanes chain-extended with diamines are more difficult to prepare because they do not dissolve easily in common solvents and, when the temperature is increased, they tend to decompose before they melt. The urea linkages in the diamine-extended polymers may be more susceptible to enzymatic degradation than urethane linkages, and this would lead to unwanted degradation in the body [40]. When polyurethane polymer films were implanted subcutaneously in rats, polymers that were chain-extended with butanediol retained mechanical properties better than polymers chain-extended with ethylenediamine. In addition, polymers with MDI in the hard segment were more resistant to cracking and molecular weight changes than polymers with 4, 4'-dicyclohexylmethane diisocyanate.

Polyurethanes can be surface modified to produce materials that are resistant to thrombosis or that interact with cells and tissues in specific ways. Goodman *et al.* [41] used several different polyurethaneureas composed of hard segments of 4, 4'-methylenebis(phenyl diisocyanate) and ethylenediamine (ED) with one of four different soft segments ($M_w \sim 1,000$ except PDMS $\sim 2,000$): hydroxybutyl terminated poly(dimethyl siloxane) (PDMS), poly(-ethylene oxide) (PEO), poly(tetramethylene oxide) (PTMO), poly(propylene oxide) (PPO). The ratio of MDI/ED/soft segment was $2:1:1$ for all of the polymers. Polyurethanes such as these can be modified (i) to add sulfonate groups by treating dissolved polymer (usually in dimethylformamide) with sodium hydride (NaH) to remove urethane hydrogen and propanesulfone producing $(CH_2)_3SO_3^-$ (this change makes the polymer more polar and it now swells considerably in water); (ii) to add C_{18} alkyl groups by treating with NaH and octadecyl iodide $C_{18}H_{37}I$) (this change makes the surface more hydrophobic [42, 43]); and (iii) to add covalently bound Arg-Gly-Asp (RGD)-containing peptides to the surface by treating with NaH and β-propiolactone to replace urethane hydrogen with ethyl carboxylate groups followed by coupling of a protected peptide by the N-terminus using a coupling reagent [41, 44–46].

Other Relevant Information

Many polyurethenes for biomedical applications are available from commercial manufacturers (see Table A.2): Biomer (Ethicon) is composed primarily of methylenediphenylenediisocyanate, ED, and PTMO $(2:1:1)$, see [42]; Biomer and Pellethane were used in the artificial-heart program at the University of Utah. Medtronics produces polyurethanes Biostable C-19 and Biostable C-36 using cyclohexane-1,4-diisocyanate, 1,6-hexanediol, and soft segments with 18 or 36 carbons.

Polyurethanes are considered non-degradable, but a well-documented failure of a commonly used breast implant material was due to degradation of the polyurethane coating on the silicone prosthesis. The polyurethane coating on the Meme® implant (Surgitel, Racine, WI) degrades after several years of implantation, releasing small quantities of carcinogenic material (toluenediamine) (TDA) [47]. In fact, reasonably high concentrations of TDA have been found in the urine and tissues of women undergoing removal of failing implants [48]. The ester bonds within the polyurethane are subject to hydrolysis within the body, eventually leading to breakdown of the polyurethane chains and physical degradation of the material.

The microscopic events that eventually lead to breakdown and degradation of implanted polyurethanes have been studied . Following implantation, the polymer surface becomes coated with a layer of protein, which enhances the adhesion of macrophages. The activated macrophages release oxidative factors, such as peroxide and superoxide anion, which accelerate chemical degradation of the polymer. The complement protein C3bi appears to be critical in the adhesion and activation of phagocytic cells [50].

A.1.4 OTHER IMPORTANT POLYMERS

Polyethylene, Polypropylene, and Polystyrene

Several other common industrial polymers are also used in biomedical applications [51]. Because of its low cost and easy processibility, polyethylene is frequently used in the production of catheters. High-density polyethylene is used to produce hip prostheses, where durability of the polymer is critical. Polypropylene, which has a low density and high chemical resistance, is frequently employed in syringe bodies, external prostheses, and other non-implanted medical applications. Polystyrene is used routinely in the production of tissue culture dishes, where dimensional stability and transparency are important. Styrene–butadiene copolymers or acrylonitrile–butadiene–styrene copolymers are used to produce opaque, molded items for perfusion, dialysis, syringe connections, and catheters.

Poly(tetrafluoroethylene)

Poly(tetrafluoroethylene) (PTFE), more commonly known as Teflon (DuPont) and Goretex (expanded PTFE), is found in vascular grafts. The polymer has exceptionally good resistance to chemicals.

$$\left[\begin{matrix} \underset{|}{\overset{|}{C}} & \underset{|}{\overset{|}{C}} \\ F & F \\ F & F \end{matrix}\right]_n \qquad \text{(A.1-6)}$$

Poly(vinyl chloride) and Poly[acrylonitrile-*co*-(vinyl chloride)]

Poly(vinyl chloride) has excellent resistance to abrasion, good dimensional stability, and chemical resistance; therefore, it is often used in medical tubing and catheter tubes. Poly[acrylonitrile-*co*-(vinyl chloride)]:

$$
\left[\begin{array}{c} \text{H} \ \text{H} \\ | \ | \\ -\text{C}-\text{C}- \\ | \ | \\ \text{H} \ \text{Cl} \end{array}\right]_x \left[\begin{array}{c} \text{H} \ \text{H} \\ | \ | \\ -\text{C}-\text{C}- \\ | \ | \\ \text{H} \ \text{C}{\equiv}\text{N} \end{array}\right]_y \tag{A.1-7}
$$

has been used to make semipermeable membranes, and is often used in the construction of hollow fibers.

Poly(ethylene Terephthalate)

A commonly used linear polyester is poly(ethylene terephthlate) (PET), which is synthesized by condensation of terephthalic acid and ethylene glycol, and produced under the trade name Dacron.

$$
\left[\begin{array}{c} \text{O} \\ || \\ -\text{C}- \end{array} \bigcirc \begin{array}{c} \text{O} \\ || \\ -\text{C}-\text{O}- \end{array} \begin{array}{c} \text{H} \ \text{H} \\ | \ | \\ \text{C}-\text{C}-\text{O}- \\ | \ | \\ \text{H} \ \text{H} \end{array}\right] \tag{A.1-8}
$$

Woven PET fibers are used as vascular grafts and as a reinforcement for a variety of other materials, including laryngeal and esophageal prostheses.

Poly(methyl methacrylate)

Poly(methyl methacrylate) (PMMA) is another vinyl polymer, which is prepared in large quantities commercially using free-radical polymerization:

$$
\left[\begin{array}{c} \text{H} \ \ \text{CH}_3 \\ | \ \ | \\ -\text{C}-\text{C}- \\ | \ \ | \\ \text{H} \ \ \text{C}{=}\text{O} \\ \quad | \\ \quad \text{O} \\ \quad | \\ \quad \text{CH}_3 \end{array}\right]^N \tag{A.1-9}
$$

It has exceptionally good optical properties; its transparency has made it a popular substitute for glass in applications where breakage must be avoided (plexiglass). It has a variety of industrial uses including automotive parts and glazings. PMMA was the first implanted synthetic polymeric biomaterial; it was used as a hip prosthesis in 1947 (see USP XVIII, The Pharmacopia of the USA, (18th Revision), US Pharmacopoeial Convention, Inc., Rockville, MD, 1 September 1980). PMMA is currently used in orthopedic applications, as bone cement, and in intraocular lenses.

References

1. Goldenberg, I.S., Catgut, silk, and silver—the story of surgical sutures. *Surgery*, 1959, **46**, 908–912.
2. Greco, R.S., ed. *Implantation Biology: The Host Response and Biomedical Devices.* Boca Raton, FL: CRC Press, 1994.
3. Frazza, E.J. and E.E. Schmitt, A new absorbable suture. *Journal of Biomedical Materials Research*, 1971, **1**, 43–58.

The

4. Bascones, A., *et al.*, Polymers for dental and maxillofacial surgery, in S. Dimitiriu, *Polymeric Biomaterials*. New York: Marcel Dekker, 1994, pp. 277–311.
5. Deaton, J.G., *New Parts for Old: the Age of Organ Transplants*. Palisade, NJ: Franklin Publishing Co., 1974.
6. Liotta, D., *et al.*, Prolonged assisted circulation during and after cardiac or aortic surgery. American *Journal of Cardiology*, 1963, **18**, 399.
7. Cooley, D., *et al.*, Orthotopic cardiac prosthesis for two-staged cardiac replacement. *American Journal of Cardiology*, 1969, **24**, 723–730.
8. DeVries, W., *et al.*, Clinical use of the total artificial heart. *New England Journal of Medicine*, 1984, **310**, 273–278.
9. Paris, E., *et al.*, Innovations and deviations in therapeutic vascular devices, in S. Dumitriu, *Polymeric Biomaterials*. New York: Marcel Dekker. 1994, pp. 245–275.
10. Hufnagel, C.A., Permanent intubation of the thoracic aorta. *Archives of Surgery*, 1947, **54**, 382.
11. Davidoff, L.M., Treatment of hydrocephalus. Historical review and description of a new method. *Archives of Surgery*, 1927, **18**, 1737.
12. Nosko, M.G. and L.D. Frenkel, Biomaterials used in neurosurgery, in R.S. Greco, *Implantation Biology*. Boca Raton, FL: CRC Press, 1994.
13. Hastings, G.W., ed. *Cardiovascular Materials*. London: Springer-Verlag, 1991.
14. Dumitriu, S., ed. *Polymeric Biomaterials*. New York: Marcel Dekker, 1994.
15. Ulman, K., *et al.*, Drug permeability of modified silicone polymers. I. Silicone–organic block copolymers. *Journal of Controlled Release*, 1989, **10**, 251–260.
16. Ulman, K. and C. Lee, Drug permeability of modified silicone polymers. III. Hydrophilic pressure-sensitive adhesives for transdermal controlled drug release applications. *Journal of Controlled Release*, 1989, **10**, 273–281.
17. Ulman, K., *et al.*, Drug permeability of modified silicone polymers. II. Silicone–organic graft copolymers. *Journal of Controlled Release*, 1989, **10**, 261–272.
18. Folkman, J. and D. Long, The use of silicone rubber as a carrier for prolonged drug therapy. *Journal of Surgical Research*, 1964, **4**, 139–142.
19. Folkman, J. and D. Long, Drug pacemakers in the treatment of heart block. *Annals of the New York Academy of Sciences*, 1964, **11**, 857–868.
20. Folkman, J., D.M. Long, and R. Rosenbaum, Silicone rubber: a new diffusion property useful for general anesthesia. *Science*, 1966, **154**, 148–149.
21. Segal, S., A new delivery system for contraceptive steroids. *American Journal of Obstetrics and Gynecology*, 1987, **157**, 1090–1092.
22. Segal, S.J., The development of Norplant implants. *Studies in Family Planning*, 1983, **14**(6/7), 159.
23. Huggins, G.R. and A.C. Wentz, Obstetrics and gynecology contempo. *JAMA*, 1993, **270**(2), 234–236.
24. Frank, M.L., *et al.*, Characteristics and attitudes of early contraceptive implant acceptors. *Family Planning Perspectives*, 1992, **24**, 208.
25. Kessler, D.A., The basis of the FDA's decision on breast implants. *New England Journal of Medicine*, 1992, **326**(25), 1713–1715.
26. Gabriel, S.E., *et al.*, Risk of connective-tissue disease and other disorders after breast implantation. *New England Journal of Medicine*, 1994, **330**, 1697–1702.
27. Kaiser, J., Scientific panel clears breast implants. *Science*, 1998, **282**, 1963–1964.
28. Hsu, T. and R. Langer, Polymers for the controlled release of macromolecules: Effect of molecular weight of ethylene-vinyl acetate copolymer. *Journal of Biomedical Materials Research*, 1985, **19**, 445–460.

29. Siegel, R. and R. Langer, Controlled release of polypeptides and other macro-molecules. *Pharmaceutical Research*, 1984, 2–10.
30. Langer, R., H. Brem, and D. Tapper, Biocompatibility of polymeric delivery systems for macromolecules. *Journal of Biomedical Materials Research*, 1981, **15**, 267–277.
31. Langer, R., New methods of drug delivery. *Science*, 1990, **249**, 1527–1533.
32. Saltzman, W.M., Antibodies for treating and preventing disease: the potential role of polymeric controlled release. *Critical Reviews in Therapeutic Drug Carrier Systems*, 1993, **10**(2), 111–142.
33. Langer, R. and J. Folkman, Polymers for the sustained release of proteins and other macromolecules. *Nature*, 1976, **263**, 797–800.
34. Brown, L.R., C.L. Wei, and R. Langer, *In vivo* and *in vitro* release of macromolecules from polymeric drug delivery system. *Journal of Pharmaceutical Research*, 1983, **72**, 1181–1185.
35. During, M.J., *et al.*, Controlled release of dopamine from a polymeric brain implant: *in vivo* characterization. *Annals of Neurology*, 1989, **25**, 351–356.
36. Boretos, J.W. and W.S. Pierce, Segmented polyurethane: a new elastomer for biomedical applications. *Science*, 1967, **158**, 1481–1482.
37. Gogolewski, S., Selected topics in biomedical polyurethanes: a review. *Colloid Polymer Science*, 1989, **267**, 757–785.
38. Tyler, B.J., B.D. Ratner, and D.G. Castner, Variations between Biomer lots. I. Significant differences in the surface chemistry of two lots of a commercial poly (ether urethane). *Journal of Biomedical Materials Research*, 1992, **26**, 273–289.
39. Szycher, M., A.A. Siciliano, and A.M. Reed, Polyurethane elastomers in medicine, in S, Dimitriu, *Polymeric Biomaterials*. New York: Marcel Dekker, 1994, pp. 233–244.
40. Hergenrother, R.W., H.D. Wabers, and S.L. Cooper, Effect of hard segment chemistry and strain on the stability of polyurethanes: *in vivo* biostability. *Biomaterials*, 1993, **14**, 449–458.
41. Goodman, S.L., S.L. Cooper, and R.M. Albrecht, Integrin receptors and platelet adhesion to synthetic surfaces. *Journal of Biomedical Materials Research*, 1993, **27**, 683–695.
42. Wabers, H.D., *et al.*, Biostability and blood-contacting properties of sulfonate grafted polyurethane and Biomer. *Journal of Biomaterial Science Polymer Edition*, 1992, **4**, 107–133.
43. Pitt, W.G. and S.L. Cooper, Albumin adsorption on alkyl chain derivatized polyurethanes: I. The effect of C-18 alkylation. *Journal of Biomedical Materials Research*, 1988, **22**, 359–382.
44. Lin, H.-B., *et al.*, Synthesis of a novel polyurethane co-polymer containing covalently attached RGD peptide. *Journal of Biomaterial Science Polymer Edition*, 1992, **3**, 217–227.
45. Lin, H.-B., *et al.*, Surface properties of RGD-peptide grafted polyurethane block copolymers: Variable take-off angle and cold-stage ESCA studies. *Journal of Biomaterial Science Polymer Edition*, 1993, **4**(3), 183–198.
46. Lin, H., *et al.*, Synthesis, surface, and cell-adhesion properties of polyurethanes containing covalently grafted RGD-peptides. *Journal of Biomedical Materials Research*, 1994, **28**, 329–342.
47. Batich, C. and J. Williams, Toxic hydrolysis product from a biodegradable foam implant. *Journal of Biomedical Materials Research: Applied Biomaterials*, 1989, **23**, 311–319.

48. Chan, S.C., D.C. Birdsell, and C.Y. Gradeen, Detection of toluenediamines in the urine of a patient with polyurethane-covered breast implants. *Clinical Chemostry*, 1991, **37**, 756–758.
49. Anderson, J.M., *et al.*, Cell/polymer interactions in the biodegradation of polyurethanes. *Biodegradable Polymers and Plastics*, 1992, 122–136.
50. McNally, A.K. and J.M. Anderson, Complement C3 participation in monocyte adhesion to different surfaces. *Proceedings of the National Academy of Sciences USA*, 1994, **91**, 10119–10123..
51. Dumitriu, S. and C. Dumitriu-Medvichi, Hydrogel and general properties of biomaterials, in S. Dumitriu, *Polymeric Biomaterials*. New York: Marcel Dekker, 1994, pp. 3–97.
52. Peppas, N.A. and R. Langer, New challenges in biomaterials. *Science*, 1994, **263**, 1715–1720.
53. Marchant, R.E. and I. Wang, Physical and chemical aspects of biomaterials used in humans, in R. S. Greco, *Implantation Biology*. Boca Raton, FL: CRC Press, 1994, pp. 13–53.
54. Richards, J.M., *et al.*, Determination of the structure and composition of clinically important polyurethanes by mass spectrometric techniques. *Journal of Applied Polymer Science*, 1987, **34**, 1967–1975.

A.2

Biodegradable Polymers

The saddest thing I ever did see
Was a woodpecker peckin' at a plastic tree.
                              Shel Silverstein, *Peckin'* (1981)

Polymers that slowly dissolve following implantation into the body have many potential medical uses. A wide variety of biodegradable polymers have been synthesized and characterized. This section reviews the characteristics of several families of biodegradable polymers. Readers interested in more details on the synthesis and chemistry of these classes of materials should consult the references provided, or general texts on biodegradable polymers [1].

A.2.1 POLYESTERS

Background and Nomenclature

Polylactide, polyglycolide, and copolymers of lactide/glycolide, (pLGA), are among the most commonly used biomaterials for drug delivery and tissue engineering. The interest in these materials has resulted from several characteristics: (i) they break down to naturally occurring metabolites, (ii) materials with a variety of useful properties can be obtained by copolymerization of the two monomers, (iii) degradation requires only water, and (iv) early development of successful suture materials based on these polymers has led to a great deal of experience with these materials in humans, so that their safety is now well documented.

The first synthetic absorbable suture was made from a homopolymer of glycolic acid by Davis & Geek Co. and manufactured with the tradename Dexon (1970). This suture was followed by a second material produced by Ethicon, Inc., in 1974, a copolymer of lactide and glycolide known as polyglactine 910 or Vicryl. Both Vicryl and Dexon are made from polymer fibers, which are braided to produce sutures. In addition, Vicryl is Teflon coated for increased smoothness. When used as sutures, Vicryl maintains 50% of its tensile strength for 30 days, compared to 25 days for Dexon, 5–6 days for catgut, and 20 days for chromic catgut. Since their introduction as suture materials, these polymers have been used extensively in biomaterials, particularly as drug delivery devices.

Synthesis and Chemical Variations

The polyesters of lactic acid and glycolic acid have the chemical structures:

$$(A.2-1)$$

Poly(glycolic acid)   Poly(lactic acid)

These biodegradable polyesters can be synthesized by direct polycondensation of lactic and glycolic acid, but only low molecular weight oligomers—poly-(lactic acid) (pLA) and poly(glycolic acid) (pGA)—are produced. Higher molecular weight polymers can be obtained by ring-opening melt polymerization of cyclic diesters:

$$(A.2-2)$$

In that case, the following homopolymers are produced:

Poly(L-lactide)   Poly(D,L-lactide)   Poly(glycolide)

$$(A.2-3)$$

in addition to the the copolymer indicated in Equation A.2-2. Properties of pLGA polymers and copolymers are the subject of several recent reviews [2, 3]. Several groups have studied the synthesis and physical characteristics of this family of polymers in detail [4].

Polymers of ϵ-caprolactone, a biodegradable polyester with many similarities to the pLA/pGA family, are also useful in drug delivery [5]. The general structure of poly(ϵ-caprolactone) is:

$$\left[\begin{array}{c} \overset{O}{\overset{\|}{C}}-\underset{H_2}{C}-\underset{H_2}{C}-\underset{H_2}{C}-\underset{H_2}{C}-\underset{H_2}{C}-O \end{array} \right]_n \tag{A.2-4}$$

This polymer can be produced by a variety of mechanisms, including anionic, cationic, coordination, and radical polymerization [5].

While the copolymers of lactide and glycolide are extremely useful, they are difficult to derivatize. To overcome this limitation, copolymers of L-lactide and L-lysine have also been produced [6]. These polymers have the structure:

$$\left[\begin{array}{c} H \quad O \qquad H \quad O \\ | \quad \| \qquad | \quad \| \\ O-C-C-O-C-C \\ | \qquad\quad | \\ CH_3 \qquad CH_3 \end{array} \right]_x \left[\begin{array}{c} H \quad O \quad H \quad H \quad O \\ | \quad \| \quad | \quad | \quad \| \\ O-C-C-N-C-C \\ | \qquad\quad (CH_2)_4 \\ CH_3 \qquad\quad | \\ \qquad\qquad NH_2 \end{array} \right]_y \tag{A.2-5}$$

and might be useful because other functional groups, like peptides, can be coupled with the polymer through the primary amine on the lysine.

When implanted into the bones or soft tissues of rats, pellets of pLA, pLGA, and copolymers degrade at different rates [7]. The homopolymers, pLA and pGA, degrade slowly in tissue, taking many months to disappear (Table A.3). Copolymers degrade more rapidly, probably due to their decreased crystallinity. This characteristic of LA/GA copolymers, that the physical properties can be adjusted by altering the copolymerization ratio, is important in the design of biomaterials. Amorphous polymers may be more suitable for applications where more rapid mass loss is important, or where ease of dispersion and diffusion through the polymer is essential, as in drug delivery. More crystalline polymers, on the other hand, may be more suitable for applications where the polymer must have increased physical strength.

In general, polymers of ϵ-caprolactone degrade more slowly than comparable lactide or glycolide polymers, making them particularly well suited for long-term drug delivery applications. Alternatively, copolymerization of ϵ-caprolactone with other monomers or polymers, including lactide and glycolide, leads to more rapidly degrading materials.

Drug Delivery Applications

The first reported biomedical applications of pLA polymers involved use as sutures and prosthetics [8]. American Cyanamid developed synthetic, degradable sutures composed of pGA in the 1960s [9], while Ethicon developed similar materials involving pGA and pLA [10, 11].

In the early 1970s, patents for polylactide/drug mixtures were awarded to DuPont [12, 13]. The first applications for controlled drug delivery involved the release of narcotic antagonists from pLA films [14, 15]. Poly(L(+)-lactic acid)

Table A.3   Characteristics of biodegradable polyesters

	Molecular weight	T_g (°C)	T_m (°C)	Crystallinity (%)	Approximate degradation time (month)[c]
Poly(L-lactic acid)	2,000	40	140		
Poly(L-lactide)	50,000 or 100,000	60–67	170–180	37	18–24
Poly(L-lactide-co-glycolide) (90:10)	N.R.[a]	~65	~170	22	[6.5]
Poly(L-lactide-co-glycolide) (80:20)	N.R.	~65	~120	8	
Poly(L-lactide-co–glycolide) (70:30)	N.R.	58	None	0	[0.6]
Poly(L-lactide-co-glycolide) (50:50)	N.R.	~65	None	0	[0.2]
Poly(L-lactide-co-glycolide (30:70)	N.R.	~65	None	0	0[0.6]
Poly(L-lactide-co-glycolide) (20:80)	N.R.	~60	~170	20	
Poly(L-lactide-co-glycolide) (10:90)	N.R.	~60	~205	40	
Polyglycolide	50,000	36	210–220	50	2–4 [5]
Poly(DL-lactide)	N.R.	52–59	None		12–16
Poly(DL-lactide-co-glyco-lide) (85:15)	232,000	45–49	None		5
Poly(DL-lactide-co-glyco-lide) (75:25)	63,000	48	None		
Poly(DL-lactide-co-glyco-lide) (50:50)	12,000–98,000	40–47	None		2
Poly(DL-lactide-co-glyco-lide) (25:75)	N.R.	60	None		
Polyglycolide	50,000	36	210–220	50	
Poly(ε-caprolactone)	5,000–100,000	−60	59–64	40–80[b]	30

Data collected from recent reviews [2, 3, 5].

[a] N.R. = not reported.

[b] Crystallinity varies with molecular weight with higher molecular weights exhibiting the lowest crystallinity.

[c] Bracketed values indicate the half-life for degradation, from [7].

was used to deliver contraceptive steroids [16] and particles of poly(DL-lactide-co-glycolide) (25:75) were used to deliver an antimalarial drug to mice, providing 14 weeks of protection against malarial challenge [17]. Microspheres of poly(DL-lactic acid) were used to deliver the contraceptive steroid norethisterone to baboons, inhibiting ovulation for six months [18]. In the last 15 years, many different groups have evaluated the use of copolymers of lactide and glycolide for the release of small and large molecules. Products based on this technology are currently available in the United States, including Lupron

Depot for endometriosis and prostate cancer. Many other products are under development.

Other Relevant Information

Since pLA and pGA degrade to molecules that are environmentally safe, they have been proposed for agricultural applications, like the sustained release of pesticides [19].

A.2.2 COPOLYMERS OF METHYL VINYL ETHER AND MALEIC ANHYDRIDE

To control the release of drugs from an eroding or degrading matrix, the erosion or degradation process must occur in an orderly, reproducible manner. Most materials, such as the biodegradable polyesters discussed in the previous section, erode in a disorderly pattern: defects, cracks, and holes that initially appear throughout the material increase in size with time, permitting water penetration into the matrix and, eventually, loss of mechanical integrity.

To provide better control of polymer matrix erosion, a more orderly degradation and erosion process is needed. In an effort to achieve this goal, materials that erode heterogeneously have been produced. In particular, for materials that erode from the surface only, the kinetics of dissolution and the release of incorporated drugs can be precisely controlled. The first surface-eroding bioerodible polymer formulation was produced at Alza Corp. in the 1970s; it was a copolymer of methyl vinyl ether and maleic anhydride [20]:

$$\text{(A.2-6)}$$

where the reaction product on the right was obtained after partial esterification. When placed in an aqueous environment, the carboxylic acid groups on the polymer become ionized and the polymer becomes water soluble. When a matrix of this polymer is placed in water, only the carboxylic acids at the polymer–water interface become ionized. Therefore, erosion of the polymer is confined to the polymer surface. The rate of erosion of devices fabricated from these copolymers depends strongly on pH, with the rate increasing as the pH drops. Unfortunately, the erosion products from devices composed of this polymer are macromolecules and, therefore, not easily metabolized or excreted by the body.

A.2.3 POLY(ORTHO ESTERS)

The first poly(ortho ester) for biomedical applications was produced by Choi and Heller at Alza Corporation in the 1970s, and was given the trade name Alzamer®. When placed in an aqueous environment, the polymer degrades to produce a diol and a lactone, which is rapidly converted into γ-hydroxybutyric acid:

$$(A.2\text{-}7)$$

This degradation process is autocatalytic, since the g-hydroxybutyric acid that is produced catalyzes the hydrolysis reaction. To prevent abrupt degradation and erosion, a basic compound must be incorporated into the polymer. For example, sodium bicarbonate can be incorporated into a polymeric device composed of Alzamer to control the rate of polymer degradation and erosion. Although the polymer has been used for a number of drug delivery applications, it is difficult to produce and requires addition of significant amounts of a basic chemical to prevent uncontrolled degradation [21].

Several improvements in the design and synthesis of poly(ortho esters) have been reported over the last 20 years. One of the most useful of these polymers is shown in Equation A.2-8:

$$(A.2\text{-}8)$$

The poly(ortho ester), shown on the left-hand side of Equation A.2-8, can be synthesized under milder conditions those required for Alzamer. Since cross-linked polymeric drug delivery devices are usually fabricated by mixing a pre-polymer and a diol with mild curing ($\sim 40\ ^\circ\text{C}$), care must be taken when using drugs with reactive hydroxyl groups. Most recently, another group of poly-(ortho ester) materials was synthesized [22]:

$$R-\overset{\overset{\displaystyle O}{\|}}{C}-O-\overset{\overset{\displaystyle H_2}{|}}{C}-\overset{\overset{\displaystyle OH}{|}}{C}-(CH_2)_3-\overset{\overset{\displaystyle OH}{|}}{CH_2}$$

$$+$$

$$\overset{\overset{\displaystyle OH}{|}}{H_2C}-\overset{\overset{\displaystyle OH}{|}}{C}-(CH_2)_3-\overset{\overset{\displaystyle OH}{|}}{CH_2} \quad + \quad R-\overset{\overset{\displaystyle O}{\|}}{C}-OH$$

$$+$$

$$\overset{\overset{\displaystyle OH}{|}}{H_2C}-\overset{\overset{\displaystyle OH}{|}}{C}-(CH_2)_3-\overset{\overset{\displaystyle O-\overset{\overset{\displaystyle O}{\|}}{C}-R}{|}}{CH_2}$$

$$(A.2\text{-}9)$$

This polymer is an ointment at room temperature, which may make it appropriate for a variety of topical and peridontal applications. Since the polymer is a viscous liquid at this temperature, proteins and other labile molecules can be mixed into the polymer without using solvents or high temperatures.

A.2.4 POLYANHYDRIDES

Hydrolysis is the most common mechanism of polymer degradation in biomaterials. Of the linkages that commonly occur in polymers, the anhydride linkage is one of the least stable in the presence of water. In fact, polyanhydride polymers are so sensitive to water that they are unsuitable for many potential applications. However, the potential for rapid hydrolysis of the polymer backbone makes anhydride-based polymers attractive candidates as biodegradable materials.

Because of the instability of the anhydride bond in the presence of water, special properties are required for stable polyanhydride devices. A critical element in the development of polyanhydride biomaterials is controlling hydrolysis within a polymeric device. To obtain implants where hydrolysis is confined to the surface of the polymer, hydrophobic monomers can be polymerized via anhydride linkages to produce a polymer that resists water penetration, yet degrades into low molecular weight oligomers at the polymer/water interface. By modulating the relative hydrophobicity of the matrix, which can be achieved by appropriate selection of monomers, the rate of degradation can then be adjusted. For example, copolymers of sebacic acid, a hydrophilic monomer, with carboxyphenoxypropane, a hydrophobic monomer, yield:

$$(A.2\text{-}10)$$

Polymers prepared with different copolymerization ratios can be used to produce implants that degrade in controlled fashion. For more information, recent reviews describe the development, characterization, and applications of poly-anhydride polymers [23, 24].

A.2.5 POLY(AMINO ACIDS)

A common strategy in the design of biodegradable polymers for medical applications has been to use naturally occurring monomers, with the hope that these polymers will degrade into non-toxic components. For example, poly(lactide-*co*-glycolide) degrades into lactic and glycolic acid, commonly occurring metabolites. Amino acids are an obvious choice as monomers for the production of new polymeric biomaterials; the large number of structurally related amino acids should lead to a correspondingly wide variety of new materials. Unfortunately, amino acids polymerized by conventional methods usually yield materials that are extremely antigenic and exhibit poor mechanical properties, making it difficult to engineer suitable medical devices.

Several approaches have been developed for producing biodegradable materials using amino acids as starting materials. A few amino acids, like glutamic acid and lysine, can be modified through their side chains to produce polymers with different mechanical properties. Copolymers of L-glutamic acid and γ-ethyl L-glutamate have been used to release a variety of drugs; variation of the ratio of monomers in the polypeptide influences the rate of degradation of the resulting polymer [25]. Because of the stability of the peptide bond in water, the biodegradation of these implants occurs by dissolution of intact polymer chains and subsequent enzymatic hydrolysis in the liver.

Alternatively, amino acids can be polymerized by linkages other than the conventional peptide bond, yielding pseudopoly(amino acids) [26]. For example, the amino acid serine can be used to produce poly(serine ester), poly(serine imine), or conventional polyserine:

$$(A.2\text{-}11)$$

A.2.6 PHOSPHORUS-CONTAINING POLYMERS

Polyphosphazenes have the general structure:

$$\left[-N = \underset{\underset{R}{|}}{\overset{\overset{R}{|}}{P}} - \right]_n \tag{A.2-12}$$

Polymers with a variety of chemical, physical, and biological properties can be produced by varying the side groups attached to the phosphorus in the polymer backbone. Different polymers are usually produced by performing substitution reactions on the base polymer, poly(dichlorophosphazene), for example:

$$\tag{A.2-13}$$

This basic structure provides for considerable flexibility in the design of biomaterials, as described in a recent review [27]. By selection of the side groups on the polymer chain, both hydrophobic and hydrophilic polymers can be produced. Hydrophobic polyphosphazenes may be useful as the basis of implantable biomaterials, such as heart valves. The hydrophilic polymers can be used to produce materials with a hydrophilic surface or, when the polymer is so hydrophilic that it dissolves in water, cross-linked to produce hydrogels or solid implants. In addition, a variety of bioactive compounds can be linked to polyphosphazene molecules allowing the creation of bioactive water-soluble macromolecules or polymer surfaces with biological activity.

A.2.7 POLYCARBONATES AND POLY(IMINOCARBONATES)

Many of the materials with good biocompatibility have poor processibility or poor mechanical strength. In an effort to design new classes of biodegradable materials with improved physical properties, it was observed that polycarbonates have excellent mechanical strength and are easily processed, but do not degrade under physiological conditions [28]. To prepare polymers with the good mechanical properties of polycarbonates, investigators produced poly-(iminocarbonates) based on bisphenol A:

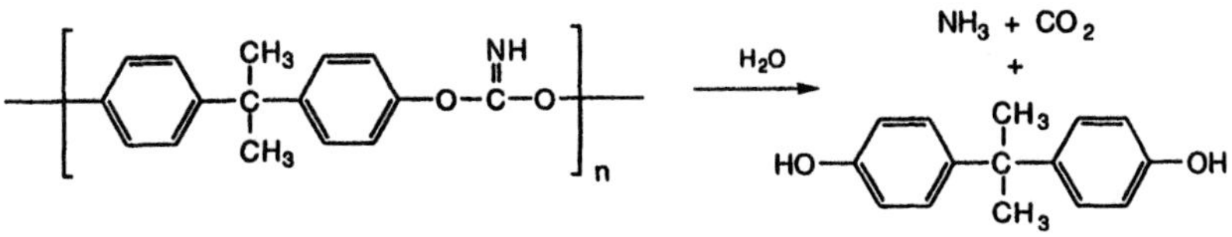

(A.2-14)

which differ from the comparable bisphenol A based polycarbonate only in the replacement of the carbonyl oxygen in the polycarbonate by an imino group. The presence of this imino group decreases the stability of the polymer in water, with hydrolytic biodegradation producing ammonia, carbon dioxide, and the biphenol.

Several other structurally similar polymers have been prepared from biphenols with less toxicity than bisphenol A. For example, derivatives of the amino acid tyrosine have been used to produce poly(iminocarbonates) that have good mechanical properties, such as poly(desaminotyrosyltyrosine hexyl ester iminocarbonate) [29]:

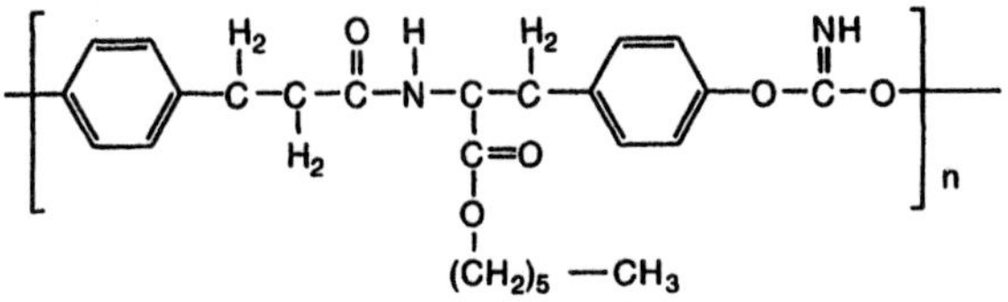

(A.2-15)

A similar approach can also be used to produce a polycarbonate from derivatives of the dipeptide of tyrosine [30]:

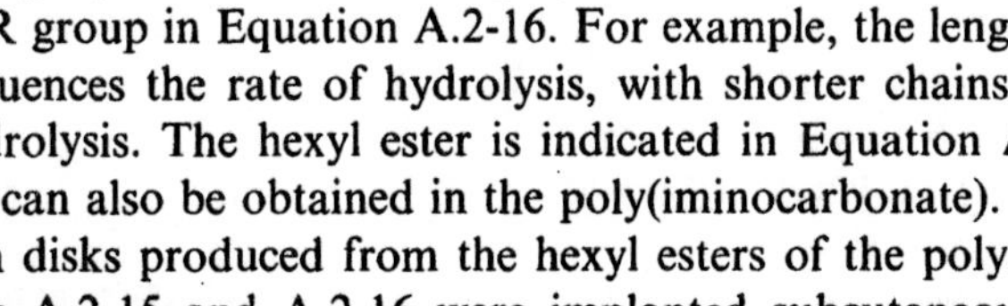

(A.2-16)

where the properties of the final polymer can be modified by changing the pendant R group in Equation A.2-16. For example, the length of the pendant chain influences the rate of hydrolysis, with shorter chains permitting more rapid hydrolysis. The hexyl ester is indicated in Equation A.2-15, but other polymers can also be obtained in the poly(iminocarbonate).

When disks produced from the hexyl esters of the polymers indicated in Equations A.2-15 and A.2-16 were implanted subcutaneously in rats, mild tissue responses were observed, comparable to the response produced by implantation of pLA or polyethylene disks [31]. These materials are being evaluated for orthopedic applications [32].

A.2.8 POLYDIOXANONE

Polydioxanone (PDS) was produced in the early 1980s as an absorbable suture material [33]:

$$\left[O-\overset{\overset{\displaystyle H}{|}}{\underset{\underset{\displaystyle H}{|}}{C}}-\overset{\overset{\displaystyle H}{|}}{\underset{\underset{\displaystyle H}{|}}{C}}-O-\overset{\overset{\displaystyle H}{|}}{\underset{\underset{\displaystyle H}{|}}{C}}-\overset{\overset{\displaystyle O}{\|}}{C} \right]_n \qquad (A.2\text{-}17)$$

PDS is produced by polymerization of *p*-dioxanone. The polymer has unusually high flexibility and, unlike copolymers of lactic and glycolic acid, can be used to produce a variety of monofilament sutures. Since PDS is a polyester, like pLA and pGA, the polymer chains break down by hydrolysis. Currently, PDS is also used in orthopedic applications (Orthosorb®), as a fixation element for bone repair.

References

1. Chasin, M. and R. Langer, eds. *Biodegradable Polymers as Drug Delivery Systems*. New York: Marcel Dekker, 1990.
2. DeLuca, P.P., *et al.*, Biodegradable polyesters for drug and polypeptide delivery, in M.A. El-Nokaly, D.M. Piatt, and B.A. Charpentier. *Polymeric Delivery Systems: Properties and Applications*. Washington, DC: American Chemical Society, 1993, pp 53–79.
3. Lewis, D.H., Controlled release of bioactive agents from lactide/glycolide polymers, in *Biodegradable Polymers as Drug Delivery Systems*. M. Chasin and R. Langer. New York: Marcel Dekker, 1990, pp 1–42.
4. Gilding, D.K. and A.M. Reed, Biodegradable polymers for use in surgery—polyglycolic/poly(actic acid) homo- and copolymers: 1. *Polymer*, 1979, **20**, 1459–1464.
5. Pitt, C.G., Poly-ϵ-caprolactone and its copolymers, in M. Chasin and R. Langer, *Biodegradable Polymers as Drug Delivery Systems*. New York: Marcel Dekker, 1990, pp. 71–120.
6. Barrera, D.A., *et al.*, Synthesis and RGD peptide modification of a new biodegradable copolymer: poly(lactic acid-*co*-lysine). *Journal of American Chemical Society*, 1993, **115**, 11010–11011.
7. Miller, R.A., J.M. Brady, and D.E. Cutright, Degradation rates of oral resorbable implants (polylactates and polyglycolates): rate modification with changes in PLA/PGA copolymer ratios. *Journal of Biomedical Materials Research*, 1977, **11**, 711–719.
8. Kulkarni, R.K., *et al.*, Biodegradable poly(lactic acid) polymers. *Journal of Biomedical Materials Research*, 1971, **5**, 169–181.
9. Schmitt, E.E. and R.A. Polistina, Surgical sutures. U.S. Patent, 1967; 3,297,033 (10 January, 1967).
10. Wasserman, D. and A.J. Levy, Nahtmaterials aus weichgemachten Lacttid-Glykolid-Copolymerisaten. German Patent, 1975; Offenlegungsschrift 24 06 539.
11. Schneider, M.A.K., Element de suture absorbable et son procede de fabrication. French Patent, 1967; 1,478,694.
12. Boswell, G.A. and R.M. Scribner, Polylactid-Arzneimittel-Mischungen. German Patent, 1971; Offenlegungsschrift 2 051 580.
13. Boswell, G.A. and R.M. Scribner, Polylactide–drug mixtures. U.S. Patent, 1973; 3,773,919.
14. Woodland, J.H., *et al.*, Long-acting delivery systems for narcotic antagonists. *Journal of Medicinal Chemistry*, 1973, **16**(8), 897–901.

15. Yolles, S., *et al.*, Long-acting delivery systems for narcotic antagonists. *Advances in Experimental Medicine and Biology*, 1973, **47**, 177–193.
16. Jacknicz, T.M., *et al.*, Polylactic acid as a biodegradable carrier for contraceptive steroids. *Contraception*, 1973, **8**, 227–234.
17. Wise, D.L., *et al.*, Sustained release of an antimalarial drug using a copolymer of glycolic/lactic acid. *Life Sciences*, 1976, **19**, 867–874.
18. Beck, L.R., *et al.*, New long-acting injectable microcapsule contraceptive system. *American Journal of Obstetrics and Gynecology*, 1979, **135**, 419–426.
19. Sinclair, R.G., Polymers of lactic and glycolic acids as ecologically beneficial, cost-effective encapsulating materials. *Environmental Science and Technology*, 1976, **7**(10), 955–956.
20. Heller, J., *et al.*, Controlled drug release by polymer dissolution. I. Partial esters of maleic anhydride copolymers. Properties and theory. *Journal of Applied Polymer Science*, 1978, **22**, 1991–2009.
21. Heller, J., Development of poly(ortho esters): a historical overview. *Biomaterials*, 1990, **11**, 659–665.
22. Heller, J., *et al.*, Controlled drug release from bioerodible hydrophobic ointments. *Biomaterials*, 1990, **11**, 235–237.
23. Tamada, J. and R. Langer, Review: The development of polyanhydrides for drug delivery. *Journal of Biomaterials Science Polymer Edition*, 1992, 3(4), 315–353.
24. Chasin, M., *et al.*, Polyanhydrides as drug delivery systems, in M. Chasin and R. Langer, *Biodegradable Polymers as Drug Delivery Systems*. New York: Marcel Dekker, 1990.
25. Sidman, K.R., *et al.*, Biodegradable, implantable sustained release systems based on glutamtic acid copolymers. *Journal of Membrane Science*, 1980, **7**, 277–291.
26. Kohn, J., Pseudopoly(amino acids), in M. Chasin and R. Langer, *Biodegradable Polymers as Drug Delivery Systems*. Editors. New York: Marcel Dekker, 1990, pp. 195–229.
27. Allcock, H.R., Polyphosphazenes as new biomedical and bioactive materials, in M. Chasin and R. Langer, *Biodegradable Polymers as Drug Delivery Systems*, New York: Marcel Dekker, 1990, pp. 163–193.
28. Kohn, J. and R. Langer, Poly(iminocarbonates) as potential biomaterials. *Biomaterials*, 1986, **7**, 176–181.
29. Pulapura, S., C. Li, and J. Kohn, Structure–property relationships for the design of polyiminocarbonates. *Biomaterials*, 1990, **11**, 666–678.
30. Ertel, S.I. and J. Kohn, Evaluation of a series of tyrosine-derived polycarbonates as degradable biomaterials. *Journal of Biomedical Materials Research*, 1994, **28**, 919–930.
31. Silver, F.H., *et al.*, Tissue compatibility of tyrosine-derived polcarbonates and poly-iminocarbonates: an initial evaluation. *Journal of Long-term Effects of Medical Implants*, 1992, **1**(4), 329–346.
32. Ertel, S.I. and J. Kohn, Evaluation of poly(DTH carbonate) a tyrosine-derived degradable polymer, for orthopedic applications. *Journal of Biomedical Materials Research*, 1995, **29**, 1337–1348.
33. Ray, J.A., *et al.*, Polydioxanone (PDS), a novel monofilament synthetic absorbable suture. *Surgery, Gynecology & Obstetrics*, 1981, **153**, 497–507.

Water-Soluble Polymers

I was born with a plastic spoon in my mouth.
Pete Townsend, *Substitute* (1966)

The previous two sections reviewed characteristics of polymers that, in general, are not soluble in water and, therefore, are typically used as solid materials: fibers, matrices, microspheres, or foams. Water-soluble polymers are also useful as biomaterials. Water-soluble polymers can be used in their molecular, water-soluble form as agents to modify other materials or as solid, dissolvable matrices (see the example of copolymers of methyl vinyl ether and maleic anhydride in Section A.2.2). Alternatively, water-soluble polymers may be cross-linked, by chemical or physical means, into solid materials (gels) that swell in water but do not dissolve.

A.3.1 NATURALLY OCCURRING POLYMERS

Saccharides

Because of their structural diversity, and the opportunity to create a variety of linkages between monomer units, polysaccharides are among the most diverse and important polymers in nature. Consider, for example, the structural similarities and differences in the disaccharides sucrose and lactose:

Sucrose (glucose-α(1,2)-fructose)   Lactose (galactose-β(1,4)-glucose)   (A.3-1)

Even these disaccharides, with similar molecular composition, have significant chemical differences: here, the fructose unit of sucrose is an example of a five-carbon sugar (pentose); the linkages of the monosaccharides occur between different carbons in sucrose and lactose; and the stereochemistry of the linkage between the sugar units is different. Branched polymers are also common among polysaccharides, because of the potential reactivity of all of the carbons within the saccharide ring. Because of this inherent flexibility in polysaccharide chemistry, it is possible to assemble a diverse group of macromolecules; these polymers have diverse functions in nature, serving as structural elements in cells, as energy storage, and as cell–cell recognition molecules. This range of structure and function also provides the opportunity to use polysaccharides as natural and versatile components of biomaterials.

Cellulose. Cellulose is the most abundant organic compound on earth; half of all of the organic carbon is in cellulose [1]. Cellulose is a polymer of glucose, with all of the glucose residues connected by β-(1,4) linkages.

$$
\left[\begin{array}{c} \text{CH}_2\text{OH} \quad \text{OH} \\ \text{HO} \\ \text{HO} \quad \text{OH} \quad \text{CH}_2\text{OH} \end{array} \right]_n
\qquad (A.3\text{-}2)
$$

Because this structural unit is stabilized by hydrogen bonds between the adjacent hexoses, cellulose forms a long straight polymer. Fibrils with high tensile strength can be formed by hydrogen binding between chains that are aligned together.

Chitin. Chitin, the second most abundant organic compound, is similar to cellulose, except it is composed of N-acetylglucosamine in a β-(1,4) linkage. (The chitin structure can be recovered from Equation (A.3-2) by replacing the –OH at carbon-2 with an –NH–CO–CH$_3$.) Because chitin is readily available and occurs naturally in many insects and marine organisms, it is a popular component of cosmetic and health-care products. The Japanese Ministry of Health and Welfare approved chitosan as an ingredient for hair-care products in 1986, carboxymethylchitosan as skin-care product in 1987, and chitin non-woven fabric as skin substitute in 1988 [2].

Dextran. Dextran is composed entirely of glucose residues; it is used as a storage polysaccharide in yeasts and bacteria. The linkages in dextran are almost exclusively α-(1,6), with an occasional α-(1,2), α-(1,3), and α-(1,4) for branching. The α-(1,6) arrangement leads to an open helix confirmation:

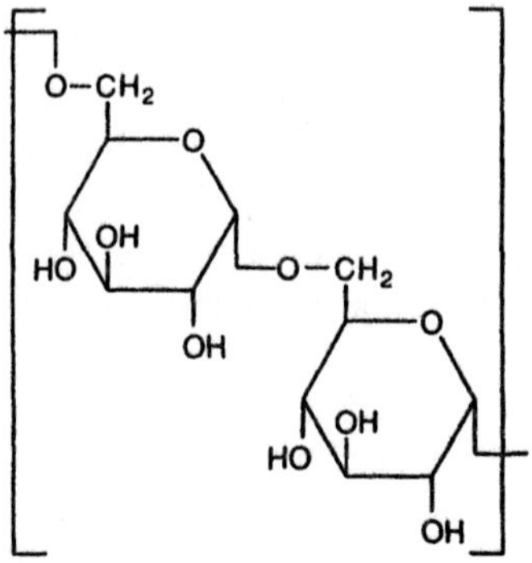

(A.3-3)

Ficoll is a synthetic polymer of sucrose. Alginate is a linear polysaccharide from marine organisms consisting of D-mannuronic and L-guluronic acid residues. Alginate has the important property that it forms a gel in the presence of bivalent cations, like Ca^{2+}.

Proteins

Collagen.   The extracellular matrix protein collagen is the most abundant protein in animals and, therefore, has been used in a variety of biomaterials.

Albumin.   Albumin is the most abundant serum protein and, because of its abundance, is frequently considered in the design of protein-based biomaterials.

A.3.2 ACRYLATES AND ACRYLAMIDES

Background and Nomenclature

Synthetic hydrogels have been frequently used in biomedical applications, because of the similarity of their physical properties to living tissues [3]. The most widely used synthetic hydrogels are polymers of acrylic acid, acrylamide, and 2-hydroxyethyl methacrylate (HEMA):

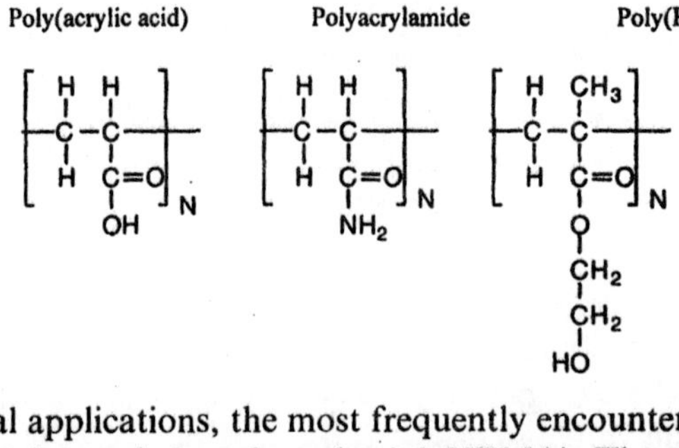

Poly(acrylic acid)   Polyacrylamide   Poly(HEMA)

(A.3-4)

For biomedical applications, the most frequently encountered synthetic hydrogel is poly(2-hydroxyethyl methacrylate) (pHEMA). The characterization and applications of pHEMA have been reviewed [4]. Both HEMA and pHEMA are easy and inexpensive to produce. Because of the primary alcohol, the monomer or polymer can be functionalized. In addition, HEMA can be copolymerized with other acrylic or methacrylic derivatives, as well as other vinyl monomers, resulting in hydrogels with a range of chemical and physical properties. HEMA has low toxicity in animals, but some preparations can produce local irritation, probably due to the presence of contaminants. pHEMA is among the most biocompatible of the synthetic polymers; implants of pHEMA produce minimal local reaction and the polymer resists degradation in tissues.

Biomedical Applications

PHEMA has been used extensively in ophthalmic applications, as a materials for contact or intraocular lenses. In general, cells cannot attach or spread on a purely pHEMA surface. This observation has been used to examine the role of attachment and spreading on cell growth [6, 7] and cell motility. This resistance to cell attachment may underlie pHEMA's excellent biocompatibility, and has led to many studies on the use of hydrogels, such as pHEMA, for the reduction of thrombosis at the surface of blood-contacting biomaterials. Since pHEMA has rather poor mechanical properties, however, it is not suitable for use as a vascular graft material. Therefore, hybrid materials have been produced by grafting pHEMA to the surface of polymeric materials with good mechanical properties, like polyurethanes, styrene–butadiene copolymers [8], or polysiloxanes [9].

Interesting properties can be engineered into these materials by copolymerization. Copolymerization of 2-ethylacrylic acid or MAA with N-[4-(phenylazo)phenyl]methacrylamide, a photosensitive monomer, results in polymers in which the interaction with lipid biolayers can be photoregulated [10, 11]. Thermally reversible polymers can be produced from poly(N-isopropylacrylamide), which is a water-soluble polymer at room temperature. The LCST (lower critical solution temperature) of the homopolymer is 32 °C; at this temperature a reversible phase separation occurs. This LCST can be adjusted by copolymerization with more or less hydrophilic monomers.

A.3.3 POLY(ETHYLENE GLYCOL)

Poly(ethylene glycol) (PEG) is a simple polymer:

$$HO\left[\begin{array}{c} H\ \ H \\ | \ \ \ | \\ C-C-O \\ | \ \ \ | \\ H\ \ H \end{array}\right]_n \begin{array}{c} H\ \ H \\ | \ \ \ | \\ C-C-OH \\ | \ \ \ | \\ H\ \ H \end{array} \qquad (A.3\text{-}5)$$

yet one of the most frequently used water-soluble polymers in biomedical applications. PEG is useful in biomedical applications because of its high solubility in water, where it behaves as a highly mobile molecule. In addition, it has a large exclusion volume, occupying a larger volume in aqueous solution than other polymers of comparable molecular weight. Because of these properties, PEG molecules in aqueous solution tend to exclude or reject other polymers. It is unusual among the group of water-soluble polymers in that it is also soluble in a variety of organic solvents, including methylene chloride, ethanol, and acetone (it is important to note that solubility characteristics depend on the molecular weight of the polymer). These properties lead to a number of useful applications: (i) addition of PEG to aqueous solutions of proteins and nucleic acids frequently induces crystallization; (ii) addition of high concentrations of PEG to cell suspensions induces cell fusion; (iii) immobilization of PEG to polymer surfaces greatly reduces protein adhesion; and (iv) covalent coupling of PEG to proteins decreases their immunogenicity and increases their half-life in plasma. PEG is non-toxic and biocompatible.

Low molecular weight PEGs (< 1000) are liquids at room temperature. Higher molecular weight PEGs are solids and, when the molecular weight is above 2×10^4, PEG is frequently referred to as poly(ethylene oxide) (PEO) or polyoxyethylene. In some cases this is a useful distinction, since PEG generally refers to molecules with terminal hydroxyl groups on each end of the molecule (as in Equation A.3-5), while PEO generally refers to units of sufficient molecular weight that the end groups can be neglected. PEG has been studied in great detail and its properties and applications have been reviewed [12]

References

1. Stryer, L., *Biochemistry*, 2nd ed. New York: W.H. Freeman, 1988.
2. Muzzarelli, R., *In vivo* biochemical significance of chitin-based medical items, in S,. Dumitriu, *Polymeric Biomaterials*. New York: Marcel Dekker, 1994, pp. 179–197.
3. Peppas, N.A., *Hydrogels in Medicine and Pharmacy*. Boca Raton, FL: CRC Press, 1987.
4. Montheard, J.-P., M. Chatzopoulos, and D. Chappard, 2-Hydroxethyl methacrylate (HEMA): Chemical properties and applications in biomedical fields. *J.M.S.— Rev. Macromol. Chem. Phys.*, 1992, C32(1), 1–34.
5. Pinchuk, L., E.C. Eckstein, and M.R. Van De Mark, The interaction of urea with the generic class of poly(2-hydroxyethyl methacrylate) hydrogels. *Journal of Biomedical Materials Research*, 1984, 18, 671–684.
6. Folkman, J. and A. Moscona, Role of cell shape in growth control. *Nature*, 1978, 273, 345–349.
7. Horbett, T., M. Schway, and B. Ratner, Hydrophilic-hydrophobic copolymers as cell substrates: Effect on 3T3 cell growth rates. *Journal of Colloid and Interface Science*, 1985, 104, 28–39.
8. Hsiue, G.-H., J.-M. Yang, and R.-L. Wu, Preparation and properties of a biomaterial: HEMA grafted SBS by gamma-ray irradiation. *Journal of Biomedical Materials Research*, 1988, 22, 405–415.
9. Seifert, L.M. and R.T. Greer, Evaluation of *in vivo* adsorption of blood elements onto hydrogel-coated silicone rubber by scanning electron microscopy and Fourier transform infrared spectroscopy. *Journal of Biomedical Materials Research*, 1985, 19, 1043–1071.
10. Ferritto, M.S. and D.A. Tirrell, Photoregulation of the binding of a synthetic polyelectrolyte to phosphatidylcholine bilayer membranes. *Macromolecules*, 1988, 21, 3117–3119.
11. Ferritto, M.S. and D.A. Tirrell, Photoregulation of the binding of an azobenzene-modified poly(methacrylic acid) to phosphatidylcholine bilayer membranes. *Biomaterials*, 1990, 11, 645–651.
12. Harris, J.M., ed., *Poly(Ethylene Glycol) Chemistry: Biotechnical and Biomedical Applications*. New York: Plenum Press, 1992.

APPENDIX B

USEFUL DATA AND NOMENCLATURE

Physiological Parameters

Table B.1 Standard data for man

Age	30 years
Height	1.73 m
Weight	68 kg
Surface area	1.80 m²
Normal body core temperature	37.0 °C
Normal mean skin temperature	34.2 °C
Heat capacity	0.86 Kcal/kg °C
Percentage body fat	12% (8.2 kg)
Subcutaneous fat layer	5 mm
Body fluids	41 L (60 wt% of body)
Intracellular	28 L
Interstitial	10.0 L
Transcellular	—
Plasma	3.0 L
Basal metabolism	40 kcal/m² · h, 72 kcal/h
O_2 consumption	250 mL/min[a]
CO_2 production	200 mL/min[a]
Respiratory quotient	0.80
Blood volume	5 L
Resting cardiac output	5 L/min
Systemic blood pressure	120/80 mmHg
Heart rate at rest	65/min
General cardiac output	3.0 + 8M L/min, where M = liters O_2 Consumed/min at STP
Total lung capacity	6 L[b]
Vital capacity	4.2 L[b]
Ventilation rate	6 L/min[b]
Alveolar ventilation rate	4 L/min[b]
Tidal volume	500 mL[b]
Dead space	150 mL[b]
Breathing frequency	12/min
Pulmonary capillary blood volume	75 mL
Arterial O_2 content	0.195 mL O_2/mL blood[a]
Arterial CO_2 content	0.492 mL CO_2/mL blood[a]
Venous O_2 content	0.145 mL O_2/mL blood[a]
Venous CO_2 content	0.532 mL CO_2/mL blood[a]

Adapted from [1], p. 66.

[a] At standard temperature and pressure (STP).

[b] At body temperature and pressure (37 °C, 1 atm).

Cardiovascular System

Table B.2 Estimates of blood distribution in vascular bed of a hypothetical man

Pulmonary	Volume (mL)	Systemic	Volume (mL)
Pulmonary arteries	400	Aorta	100
Pulmonary capillaries	60	Systemic arteries	450
Venules	140	Systemic capillaries	300
Pulmonary veins	700	Venules	200
		Systemic veins	2050
Total pulmonary system	1300	Total systemic vessels	3100
Heart	250	Unaccounted	550[a]

Age 30, weight 63 kg, height 178 cm, blood volume (assumed 5.2 L). From [2], p. 64.
[a] Probably represents extra blood in reservoirs of liver and spleen.

Table B.3 Systemic circulation of man

Structure	Diameter (cm)	Blood velocity (cm/s)	Tube Reynolds number[a]
Ascending aorta	2.0–3.2	63[b]	3,600–5,800
Descending aorta	1.6–2.0	27[b]	1,200–1,500
Large arteries	0.2–0.6	20–50[b]	110–850
Capillaries	0.0005–0.001	0.05–0.1[c]	0.0007–0.003
Large veins	0.5–1.0	15–20[c]	210–570
Venae cavae	2.0	11–16[c]	630–900

Data from [3].
[a] Assuming viscosity of blood is 0.035 P.
[b] Mean peak value.
[c] Mean velocity over indefinite period of time.

Table B.4 Systemic circulation of a dog

Structure	Diameter (cm)	Number	Total cross-sectional area (cm^2)	Length (cm)	Total volume (cm^3)	Blood velocity (cm/s)	Tube Reynolds number[a]
Left atrium					25		
Left ventricle					25		
Aorta	1.0	1	0.8	40	30	50	1,670
Large arteries	0.3	40	3.0	20	60	23	230
Main arterial branches	0.1	600	5.0	10	50	8	27
Terminal branches	0.06	1,800	5.0	1	5	6	12
Arterioles	0.002	40×10^6	125	0.2	25	0.3	0.02
Capillaries	0.0008	12×10^8	600	0.1	60	0.07	0.002
Venules	0.003	80×10^6	570	0.2	110	0.07	0.007
Terminal veins	0.15	1,800	30	1	30	1.3	6.5
Main venous branches	0.24	600	27	10	270	1.5	12
Large veins	0.6	40	11	20	220	3.6	72
Vena cavae	1.25	1	1.2	40	50	33.0	1,375
Right atrium					25		
Right ventricle					25		
Main pulmonary artery	1.2	1	1.1	2.4	24[b]	36.4	2,090
Lobar pulmonary artery branches	0.4	9	1.19	17.9		33.6	670
Smaller arteries and arterioles	—	—	—	—	18		
Pulmonary capillaries	0.0008	—	300	0.05	16	0.14	0.006
Pulmonary veins	—	600	—	—	52[c]		
Large pulmonary veins	—	4	—	—			

Adapted from [3], p. 92.

[a] Assuming viscosity of blood is 0.03 P.

[b] Includes both main pulmonary artery and lobar pulmonary artery branches.

[c] Includes both pulmonary veins and large pulmonary veins.

Table B.5 Volumes and blood supplies of different body regions for a standard man

Tissue	Volume (L)	Blood flow (mL/min)	Blood flow (mL/100 mL tissue/min)
Adrenals	0.02	100	500
Kidneys	0.3	1,240	410
Thyroid	0.02	80	400
Gray matter	0.75	600	80
Heart	0.3	240	80
Other small glands and organs	0.16	80	50
Liver plus portal system	3.9	1,580	41
White matter	0.75	160	21
Red marrow	1.4	120	9
Muscle	30	300/600/1,500	1/2/5
Skin			
Nutritive	3	30/60/150	1/2/5
Shunt		1,620/1,290/300	54/43/10
Non-fat subcutaneous	4.8	70	1.5
Fatty marrow	2.2	60	2.7
Fat	10	200	2
Bone cortex	6.4	0	0
Arterial blood	1.4		
Venous blood	4.0		
Lung parenchymal tissue	0.6		
Air in lungs	2.5 + half tidal volume		
Total	70.0[c]	6,480	

70 kg, 1.83 m^2, 30 to 39 years old, from [4].

[a] Arterial blood.

[b] Skin-shunt venous blood.

[c] Excluding the air in the lung.

Clinical Chemistry

Table B.6 Normal clinical values for blood and urine

	Blood	CSF	Urine
Water content	93%	99%	99%
Protein content	7,000 mg/dL	35 mg/dL	~ 0
Osmolarity	295 mOsm/L	295 mOsm/L	
Inorganic substances			
Ammonia	12–55 M		
Bicarbonate	22–26 mEq/L		
Calcium	4.8 mEq/L	2.1 mEq/L	0–300 mg/day
Carbon dioxide	24–30 mEq/L		
Chloride	100–106 mEq/L	119 mEq/L	
Copper	100–200 g/dL		0–60 g/day
Iron	50–150 g/dL		
Lead	< 10 g/dL		< 120 g/day
Magnesium	1.5–2.0 mEq/L	0.3 mEq/L	
P_{CO_2}	35–45 mmHg		
	4.7–6.0 kPa		
pH	7.35–7.45	7.33	
Phosphorus	3.0–4.5 mg/dL		
P_{O_2}	75–100 mmHg		
	10–13.3 kPa		
Potassium	3.5–5.0 mEq/L	2.8 mEq/L	
Sodium	135–145 mEq/L	135–145 mEq/L	
Organic molecules			
Acetoacetate	Negative		0
Ascorbic acid	0.4–15 mg/dL		
Bilirubin	Direct 0–0.4 mg/dL		
	Indirect 0.6 mg/dL		
Carotenoids	0.8–4.0 g/mL		
Creatinine	0.6–1.5 mg/dL		15–25 mg/kg
			body weight
Glucose	70–110 mg/dL	60 mg/dL	0
Lactic acid	0.5–2.2 mEq/L		
Lipids	Total 450–1000 mg/dL		
	Cholesterol 120–220 mg/dL		
	Phospholipids 9–16 mg/dL		
	(as lipid P)		
Fatty acids	190–420 mg/dL		
Triglycerides	40–150 mg/dL		
Phenylalanine	0–2 mg/dL		
Pyruvic acid	0–0.11 mEq/L		
Urea nitrogen (BUN)	8–25 mg/dL		
Uric acid	3–7 mg/dL		
Vitamin A	0.15–0.6 g/mL		

From Devlin, T. M., *Textbook of Biochemistry*, 4th ed, New York: Wiley-Liss, 1997, and *New England Journal of Medicine* 327:718, 1992. Deciliter (dL) is 100 mL.

Table B.7   pH range of body fluids

Blood	7.2–7.8
Colon	7.0–7.5
Conjuctival sac	7.8–8.0
Duodenum	4.8–8.2
Milk	8.5–8.7
Mouth	6.2–7.2
Stomach	1–3
Sweat	4.7–4.8
Urethra	5–7
Vagina	3.4–4.2

Permeation and Diffusion

Table B.8 Permeability of human red blood cells

Solute	M_w	Partition coefficient	Permeability (cm/s)
Carbon dioxide	34	—	0.2–0.6
Water	18	9×10^{-4}	0.005
Methanol	32	9×10^{-3}	0.003
Urea	60	2×10^{-4}	1.0–4.0×10^{-4}
			4×10^{-6}
Butanol	74	0.13	3.0×10^{-4}
Ethylene glycol	62	7×10^{-4}	2.0×10^{-4}
Ethanol	46	0.02	1.5×10^{-4}
Methylurea	74	7×10^{-4}	5.0×10^{-5}
Glycerol	92	9×10^{-5}	1.5×10^{-6}
			4×10^{-6}
Glucose	180	—	1.5×10^{-7}
			6×10^{-8}
Uric acid	158	—	1.3–3.6×10^{-8}
Fructose	180	—	1.5×10^{-9}
^{51}Cr-EDTA (ethylenediamine tetraacetate)	292		
Sucrose	342	—	0
Indole	117		4×10^{-4}
Tryptophan	204		1×10^{-7}
Na^+	23		1×10^{-12}
K^+	39		6×10^{-12}
Cl^-			7×10^{-11}

Permeabilities from [5], [6]; Goldstein and Solomon (1960); Jacobs *et al.* (1935)

Table B.9 Diffusion coefficients for solutes in water at 37 °C

Solute	Molecular weight	D (10^{-5} cm^2/s)
H_2	2.0	7.8
O_2	32.0	2.7
CO_2	44.0	2.7
NH_3	18.0	2.7
N_2	28.0	2.5
Acetylene	26.0	2.4
Cl_2	70.9	1.9
HCl	36.5	4.0
HNO_3	63.0	4.0
H_2SO_4	98.1	2.6
NaOH	40.0	2.3
NaCl	58.4	2.1
Ethyl alcohol	46.1	1.5
Acetic acid	60.1	1.3
Phenol	94.1	1.3
Glycerol	92.1	1.1
Glucose	180.0	0.9
Sucrose	342.3	0.7
Argon	39.9	2.7
Air	—	2.7
Bromine	159.8	1.6
Carbon dioxide	34.0	2.6
Carbon monoxide	28.0	2.7
Chlorine	71.0	1.7
Ethane	30.1	1.6
Ethylene	28.1	2.5
Helium	4.0	8.4
Hydrogen	2.0	6.0
Methane	16.0	2.0
Nitric oxide	30.0	3.5
Nitrogen	28.0	2.5
Oxygen	32.0	2.8
Propane	44.1	1.3
Ammonia	17.0	2.2
Benzene	78.1	1.4
Hydrogen sulfide	34.1	1.9
Sulfuric acid	98.1	2.3
Nitric acid	63.0	3.5
Acetylene	26.0	1.2
Methanol	32.0	1.1
Ethanol	46.1	1.1
1-Propanol	60.1	1.2
2-Propanol	60.1	1.2
n-Butanol	74.1	1.0
Benzyl alcohol	108.1	1.1
Formic acid	46.0	2.0

Table B.9 *(cont'd)*

Solute	Molecular weight	D (10^{-5} cm^2/s)
Acetic acid	60.0	1.6
Propionic acid	74.1	1.4
Benzoic acid	122.1	1.3
Glycine	75.1	1.4
Valine	117.2	1.1
Acetone	58.1	1.5
Urea	60.1	1.8
Fluorescein	331.0	0.7
[51]Cr-ethylenediamine tetraacetate (EDTA)	292	0.74

Protein Properties

Table B.10 Properties of some polypeptides and proteins

	M_w	N_{aa}	N_c	D_w $(10^{-7}\,cm^2/s)$	$t_{1/2}$ (min)
Polypeptide	1,200		1	21 ± 5	
GRGDS synthetic peptide		5	1		$\alpha = 1;\ \beta = 8$ [7]
YRGDS		5	1		$\alpha = 0.5;\ \beta = 160$ [8]
Insulin (bovine)	5,733	51	2		
Ribonuclease (bovine pancreas)	12,640	124	1		
Lysozyme (egg white)	13,930	129	1		
Lactalbumin	14,500			13 ± 3	
Interleukin 2 (recombinant, non-glycosylated, rIL-2)	$\sim 15,000$				~ 30 [9]
Myoglobin (horse heart)	16,890	153	1		
Chymotrypsin (bovine pancreas)	22,600	241	3		
Nerve growth factor (mouse)					
Fc fragment from IgG (human)	50,000		2	8.4 ± 0.4	
Fab fragment from IgG (human)	50,000		2	8.3 ± 1.0	
Hemoglobin (human)	64,500	574	4		
Serum albumin (human)	68,500	~ 550	1		
Serum albumin (cow)	68,000	~ 550	1	8.3 ± 1.7 [10]	
Hexokinase (yeast)	96,000	~ 800	4		
F(ab')$_2$ fragment from IgG (human)	100,000		4	6.7 ± 0.24	
Tryptophan synthetase (E. coli)	117,000	~ 975	4		
γ-Globulin (horse)	149,900	$\sim 1,250$	4		
IgA (human)	150,000	$\sim 1,250$	4	5.2 ± 0.3	
IgG (human)	150,000	$\sim 1,250$	4	4.4 ± 1.3	
Glycogen phosphorylase (rabbit muscle)	495,000	$\sim 4,100$	4		
IgM (human)	970,000		~ 20	3.2 ± 1.4	
Laminin (EHS sarcoma)	820,000	6,477	3		
Glutamate dehydrogenase (bovine liver)	10^6	$\sim 8,300$	~ 40		
Fatty acid synthetase (yeast)	2,300,000	$\sim 20,000$	~ 21		
Tobacco mosaic virus	$\sim 10^7$	$\sim 336,500$	2,130		

Molecular weight (Da), number of amino acids (N_{aa}), number of polypeptide chains (N_c), diffusion coefficients in water 25 °C (D_w). Protein diffusion coefficients from [11], unless otherwise indicated. Half-life in the plasma following i.v. injection, $t_{1/2}$.

Table B.11   Protein composition of human blood

	M_w	c_p (mg/100 mL)
Albumin	69,000	3,500–4,500
Prealbumin	61,000	28–35
Insulin	5,000	0–29 units/mL
Fibrinogen	341,000	200–600
Erythropoietin	34,000	0.1–0.5 ng/mL
α_1-*Globulins*		
α_1-Lipoprotein		
HDL$_2$	435,000	35–117
HDL$_3$	195,000	217–270
α_1-Acid glycoprotein (orosomucoid)	44,100	75–100
α_1-Antitrypsin (α_1-glycoprotein)	45,000	210–500
Transcortin		7
α_{1X}-Glycoprotein		14–25
Haptoglobin	100,000	30–190
α_2-*Globulins*		
Ceruloplasmin	160,000	27–39
α_2-Macroglobulin	820,000	220–380
α_2-Lipoprotein (low density)	5×10^6–2×10^7	150–230
α_{2HS}-Glycoprotein	49,000	8
Z_n-α_2-Glycoprotein	41,000	4
prothrombin	62,700	9
β-*Globulins*		
β-Lipoprotein	3–2×10^6	280–440
Transferrin	90,000	200–320
β_{1C}-Globulin		35
Hemoplexin (β_{1B}-globulin)	80,000	80–100
β_2-Glycoprotein		20–25
γ-*Globulins*		
γG-Immunoglobulin	160,000	1,200–1,800
γM-immunoglobulin	10^6	75
γA-immunoglobulin	350,000	100
Enzymes		
Aldolase		0–7 U/mL
Amylase		4–25 U/mL
Cholinesterase		0.5 pH U or mmole/h
Creatinine kinase		40–150 U/L
Lactate dehydrogenase		110–210 U/L
Lipase		< 2 U/mL
Nucleotidase		1–11 U/L
Phosphatase (acid)		0.1–0.63 Sigma U/mL
Phosphatase (alkaline)		13–39 U/L
Transaminase (SGOT)		9–40 U/mL
Total		6,726–9,819

Adapted from [5].

Table B.12 Diffusion coefficients in biological fluids at 37 °C

	M_w	D_{saline} or D_{water}	D_{plasma}
Urea		184[a]	146[a]
Creatinine		121.5[a]	87.1[a]
Glucose	180	90	
Uric acid		121[a]	74.5[a]
Sucrose		71.1[a]	59.0[a]
Fluorescein	332	74[b]	—
Inulin		24.0[a]	21.7[a]
Dextran	4,000	15	
	16,000	29.3[a]	24.1[a]
	70,000		
	150,000		
	9,300	11.5 at 23 °C [10]	
	73,000	5 at 23 °C [10]	
	526,000	4 at 23 °C [10]	
	2×10^6	3.7 at 23 °C [10]	

All diffusion coefficients are expressed as $10^{-7}\,cm^2/s$, and corrected to 37 °C by assuming $D\mu/T$ is constant. $D_1 = D_2(\mu_2/\mu_1)(T_1/T_2)$ (Correction factors from 25 °C: $D^{37} = (37 + 273)/(25 + 273) \times (892/695) = 1.335$; from 20 °C = $(37 + 273)/(20 + 273) \times (100/695) = 1.524$.
[a] From [5].
[b] From [12].

Table B.13 Plasma half-lives for proteins

Protein	Plasma half-life ($t_{1,2}$, min)	Source of material and species (ref.)
Basic fibroblast growth factor (bFGF)	1.5	Recombinant bFGF in rats [13]
γ-Interferon	11–32	Partially purified protein in humans [14]
Nerve growth factor (NGF)	2.4	Purified mouse NGF in rats [15]
Interleukin-2 (IL-2)	30	Recombinant human IL-2 in humans [16]
Erythropoietin (EPO)	300–360	Recombinant human EPO in humans [17]
Ciliary neurotrophic factor (CNTF)	3	Recombinant human CNTF in humans

Mathematical Tables and Functions

Table B.14 .Values of the error function, erf, and error function complement, erfc

x	erf(x)	erfc(x)
0	0	1
0.05	0.056372	0.943628
0.1	0.112463	0.887537
0.15	0.167996	0.832004
0.2	0.222703	0.777297
0.25	0.276326	0.723674
0.3	0.328627	0.671373
0.35	0.379382	0.620618
0.4	0.428392	0.571608
0.45	0.475482	0.524518
0.5	0.5205	0.4795
0.55	0.563323	0.436677
0.6	0.603856	0.396144
0.65	0.642029	0.357971
0.7	0.677801	0.322199
0.75	0.711155	0.288845
0.8	0.742101	0.257899
0.85	0.770668	0.229332
0.9	0.796908	0.203092
0.95	0.820891	0.179109
1	0.842701	0.157299
1.05	0.862436	0.137564
1.1	0.880205	0.119795
1.15	0.896124	0.103876
1.2	0.910314	0.089686
1.25	0.9229	0.0771
1.3	0.934008	0.065992
1.35	0.943762	0.056238
1.4	0.952285	0.047715
1.45	0.959695	0.040305
1.5	0.966105	0.033895
1.6	0.976348	0.023652
1.7	0.98379	0.01621
1.8	0.989091	0.010909
1.9	0.99279	0.00721
2	0.995322	0.004678
2.5	0.999593	0.000407
3	0.999978	2.21×10^{-5}
3.5	0.999999	7.43×10^{-7}
4	1	1.54×10^{-8}

Values obtained from [18], pp. 310–311.

Nomenclature

a	Solute radius or major axis length (cm)
A	Constant of integration
A_d	Virial concentration coefficient
b	Particle minor axis length (cm)
C, C_A, c, c_A	Concentration of solute A (g/mL)
C_0, c_0	Initial concentration (g/mL)
c_{ext}	Concentration in external reservoir (g/mL)
c_p	Concentration of solute in a polymer matrix or film (g/cm^3)
D, D_A	Binary diffusion coefficient of solute A in a system of interest (cm^2/s)
$D_{A,w}$	Binary diffusion coefficient of solute A in water (cm^2/s)
D_{ij}	Multi-component diffusion coefficient (cm^2/s)
$D_{i:p}$	Diffusion coefficient for solute i in a polymer film (cm^2/s)
E_{ion}	Equilibrium membrane potential for single ion (mV)
f	Frictional drag coefficient
F	Shape factor
F	Faraday's constant (2.3×10^4 cal/V $\cdot$ mol)
F_x	Force required to move particle through fluid (g $\cdot$ cm/s^2)
G	Lag coefficient for particle in a pore
J_0	Bessel function of the first kind of order 0
j_A	Mass flux of component A (mg/cm^2 s) in a binary mixture with respect to a coordinate system that is moving with the mass average velocity of the mixture
K	Enhanced friction for particle within a pore
	Equilibrium partition coefficient
k	First-order elimination rate constant (s^{-1})
k_a	Association constant (M^{-1})
K_b	Binding constant
k_B	Boltzmann's constant (1.3807×10^{-23} J/K)
K_d	Dissociation constant (M)

k_f	Rate constant for forward (association) reaction in binding (s^{-1})
K_m	Michaelis–Menten constant (M)
k_m	Hydraulic permeability (cm/s · mmHg)
k_n	Rate constant for reaction n
k_r	Rate constant for reverse (dissociation) reaction in binding (s^{-1})
k_s	Permeability (cm/s)
L	Length (cm) in Chapter 3
	Contour length of polymer chain (cm) in Chapter 4
L_{ij}, L_{ji}	Onsager coefficient
L_p	Hydraulic conductance (1/cm · s · mmHg)
m	Molecular mass (g/molecule)
M	Mass of drug in compartment (g)
M_t	Mass of drug released from a drug delivery system at time t (g)
M_w	Molecular weight (Da)
n_A	Mass flux of component A (mg/cm^2 s) in a binary mixture with respect to a stationary coordinate system
N_{Av}	Avagadro constant ($6.0220 \times 10^{23}\,\text{mol}^{-1}$)
p	Probability of an open site or bond in Chapter 4
	Hydrostatic pressure (mmHg) in Chapters 5 and 6
P	Persistence length on polymer chain (cm) in Chapter 4
	Membrane permeability (cm/s) in Chapter 5
$p(\cdot)$	Probability density function
p_C	Critical probability
Pe	Péclet number
q_{elim_i}	Rate of elimination in organ i (mol/s · cm^3)
Q_i	Plasma flow rate through organ i (cm^3/s)
r	Displacement in two or three dimensions (cm)
R	Radius of sphere or cylinder (cm)
R	Gas constant (8.314 J/mol · K or 2 cal/mol-K)
$\Re$	Geometric description of sphere
r_f	Fiber radius (cm)
R_G	Radius of gyration (cm)
r_p	Pore radius (cm)
S	Surface area (cm^2)
T	Temperature (K)
t	Time (s)
$t_{1/2}$	Half-life (s)
u	Transform variable ($= rc_A$)
$\tilde{V}$	Molar volume (cm^3/mol)
V_d	Volume of distribution (L)
V_h	Partial molar volume (cm^3/mol)
V_m	Net membrane potential (mV)
V_{max}	Maximum velocity of enzyme reaction (cm/s)
v_{rms}	Root-mean-square velocity (cm/s)
V_w	Transmembrane volumetric flux (cm/s)
v_x, v_z, v_0	Velocity of particle movement (cm/s)

w	Width (cm)
x	Displacement (cm)
Y_0	Bessel function of the second kind of order 0
z	Coordination number for a lattice
z_{ion}	Valence of ion

Greek symbols

α	Constant
β	Dimensionless radial position within pore
γ_i	Activity coefficient of component i
δ	Distance moved between changes in direction for a particle executing a random walk (cm)
δ_{ij}	Dirac delta function
ϵ	Exponent
ϵ	Volume fraction of pore space
ζ	Coordination number fora Bethe lattice
η	Effectiveness of drug delivery
θ	Pore trajectory
κ	Hydraulic permeability (cm^2)
λ	Ratio of solute to pore radius (cm/cm)
λ_f	Length density of fibers (cm/cm^2)
μ	Fluid viscosity (g/cm · s)
μ_i	Chemical potential of component i
ν	Constant
π	Osmotic pressure
ρ	Mass density (mg/cm^3)
ρ	Total density (g/cm^3)
ρ_A	Mass concentration of A (g/cm^3)
σ_i	Reflection coefficient for solute i
τ	Time interval between changes in direction for a particle executing a random walk (s)
τ	Tortuosity
ϕ	Diffusion–elimination modulus
ϕ, ϕ_v	Volume fraction
Φ	Equilibrium partition coefficient for particles in pores
ψ_A	Rate of generation of A per volume (g/cm^3 · s)
ψ_A^m	Molar rate of generation of A per volume (mol/cm^3 · s)
$\psi_{solvent}$	Association constant
ω_A	Mass fraction of A (mg/mg)

References

1. Seagrave, R.C., *Biomedical Applications of Heat and Mass Transfer*. Ames, IA: The Iowa State University Press, 1971.

2. Burton, A.C., *Physiology and Biophysics of the Circulation*. Chicago: Yearbook Medical Publications, 1965.

3. Whitmore, R.L., *Rheology of the Circulation*. Oxford: Pergamon, 1968.

4. Dedrick, R.L. and K.B. Bischoff, *Chemical Engineering Progress Symposium Series*, No. 84, 1968, **64**, 32.

5. Colton, C., *et al.*, Diffusion of organic solutes in stagnant plasma and red cell suspensions. *Chemical Engineering Progress Symposium Series*, 1970, **66**, 85–99.

6. Stryer, L., *Biochemistry*, 2nd ed. New York: W.H. Freeman, 1988.

7. Humphries, M.J., K. Olden, and K.M. Yamada, A synthetic peptide from fibronectin inhibits experimental metastasis of murine melanoma cells. *Science*, 1986, **233**, 467–470.

8. Braatz, J.A., *et al.*, Functional peptide–polyurethane conjugates with extended circulatory half-lives. *Bioconjugate Chemistry*, 1993, **4**, 262–267.

9. Katre, N.V., M.J. Knauf, and W.J. Laird, Chemical modification of recombinant interleukin 2 by polyethylene glycol increases its potency in the murine Meth A sarcoma model. *Proceedings of the National Academy of Sciences USA*, 1987, **84**, 1487–1491.

10. Kosar, T.F. and R.J. Phillips, Measurement of protein diffusion in dextran solutions by holographic interferometry. *AIChE Journal*, 1995, **41**(3), 707–711.

11. Saltzman, W.M., *et al.*, Antibody diffusion in human cervical mucus. *Biophysical Journal*, 1994, **66**, 508–515.

12. Radomsky, M.L., *et al.*, Macromolecules released from polymers: diffusion into unstirred fluids. *Biomaterials*, 1990, **11**, 619–624.

13. Whalen, G.F., Y. Shing, and J. Folkman, The rate of intravenously administered bFGF and the effect of heparin. *Growth Factors*, 1989, **1**, 157–164.

14. Gutterman, J.U., *et al.*, Pharmacokinetic study of partially pure gamma-interferon in cancer patients. *Cancer Research*, 1984, **44**, 4164–4171.

15. Poduslo, J.F., G.L. Curran, and C.T. Berg, Macromolecular permeability across the blood–nerve and blood–brain barriers. *Proceedings of the National Academy of Sciences USA*, 1994, **91**, 5705–5709.

16. Konrad, M.W., *et al.*, *Cancer Research*, 1990, **50**, 2009–2017.

17. Cohen, A.M., Erythropoietin and G-CSF, in A.H.C. Kung, R.A. Baughman, and J.W. Larrick, *Therapeutic Proteins: Pharmacokinetics and Pharmacodynamics*. New York: W.H. Freeman, 1993, pp. 165–186.

18. Abramowitz, M. and I.A. Stegun, *Handbook of Mathematical Functions with Formulas, Graphs, and Mathematical Tables*. Washington, DC: National Bureau of Standards, 1964.

Index